Fungicide Resistance in Crop Protection

Risk and Management

Fungicide Resistance in Crop Protection

Risk and Management

Edited by

Tarlochan S. Thind

Punjab Agricultural University, India

CABI is a trading name of CAB International

CABI	CABI
Nosworthy Way	875 Massachusetts Avenue
Wallingford	7th Floor
Oxfordshire OX10 8DE	Cambridge, MA 02139
UK	USA
Tel: +44 (0)1491 832111	Tel: +1 617 395 4056
Fax: +44 (0)1491 833508	Fax: +1 617 354 6875
E-mail: cabi@cabi.org	E-mail: cabi-nao@cabi.org
Website: www.cabi.org	

A catalogue record for this book is available from the British Library, London, UK.

Library of Congress Cataloging-in-Publication Data

Fungicide resistance in crop protection : risk and management / edited by Tarlochan S. Thind.
    p. cm.
  Includes bibliographical references and index.
  ISBN 978-1-84593-905-2 (alk. paper)
1. Fungicide resistance.   2. Fungicides.   3. Fungal diseases of plants. I. Thind, T. S. II. Title.

  SB951.3.F8445 2011
  632'.4--dc23

                                        2011026517

ISBN-13: 978 1 84593 905 2

Commissioning editor: Nigel Farrar
Editorial assistant: Alexandra Lainsbury
Production editor: Fiona Chippendale

Typeset by SPi, Pondicherry, India
Printed and bound in the UK by the MPG Books Group, Bodmin

Contents

Contributors

Editor: Tarlochan S. Thind, Department of Plant Pathology, Punjab Agricultural University, Ludhiana - 141004, India. E-mail: thind_ts@yahoo.co.in

Achour Amiri, Plant Pathology, Gulf Coast REC, 14625 CR 672, Wimauma, FL 33598, USA. E-mail: aamiri@ufl.edu

Domenico Bertetti, Agroinnova, University of Torino, Via Leonardo da Vinci 44, 10095 Grugliasco, Italy. E-mail: domenico.bertetti@unito.it

Phillip M. Brannen, Plant Pathology Department, 2106 Miller Plant Sciences, University of Georgia, Athens, GA 30602, USA. E-mail: pbrannen@uga.edu

Keith Brent, International Consultant (Plant Protection), Ex Director of Long Ashton Research Station, University of Bristol, UK. E-mail: brent@st-raphael.wanadoo.co.uk

Yigal R. Cohen, Faculty of Life Sciences, Bar Ilan University, 52100 Ramat Gan, Israel. E-mail: ycohen@mail.biu.ac.il

Hans J. Cools, Fungicide Resistance Group, Department of Plant Pathology and Microbiology, Rothamsted Research, Harpenden, Hertfordshire AL5 2JQ, UK. E-mail: hans.cools@bbsrc.ac.uk

Marie-France Corio-Costet, Research Manager, INRA, UMR SAVE (Health and Agroecology of Vineyard, 1065), ISVV, Rue E. Bourlaud, BP 81, 33883 Villenave d'Ornon, France. E-mail: coriocos@bordeaux.inra.fr

Rita Milvia De Miccolis Angelini, Department of Environmental and Agro-Forestry Biology and Chemistry, University of Bari, Via Amendola 165/a, I-70126 Bari, Italy. E-mail: milvia.demiccolis@agr.uniba.it

Eva Edin, Department of Forest Mycology and Plant Pathology, Swedish University of Agricultural Sciences (SLU), Box 7026, 750 07 Uppsala, Sweden. E-mail: eva.edin@slu.se

Franco Faretra, Department of Environmental and Agro-Forestry Biology and Chemistry, University of Bari, Via Amendola 165/a, I-70126 Bari, Italy. E-mail: faretra@agr.uniba.it

Bart A. Fraaije, Fungicide Resistance Group, Department of Plant Pathology and Microbiology, Rothamsted Research, Harpenden, Hertfordshire AL5 2JQ, UK. E-mail: bart.fraaije@bbsrc.ac.uk

Angelo Garibaldi, Agroinnova, University of Torino, Via Leonardo da Vinci 44, 10095 Grugliasco, Italy. E-mail: angelo.garibaldi@unito.it

Ulrich Gisi, Syngenta Crop Protection Münchwilen AG, Research Biology, WST 540.1.71, CH 4332 Stein, Switzerland. E-mail: ulrich.gisi@syngenta.com

Cláudia Vieira Godoy, Empresa Brasileira de Pesquisa Agropecuária (Embrapa), National Center of Soybean Research, Caixa Postal 231, CEP 86001-970, Londrina, PR – Brazil. E-mail: godoy@cnpso.embrapa.br

Maria Lodovica Gullino, Director, Agroinnova, University of Torino, Via Leonardo da Vinci 44, 10095 Grugliasco, Italy. E-mail: marialodovica.gullino@unito.it

Dietrich Hermann, Syngenta Crop Protection Münchwilen AG, Research Biology, WST 540.1.71, CH 4332 Stein, Switzerland. E-mail : dietrich.hermann@syngenta.com

Derek W. Hollomon, Technical Editor, Pest Management Science, Orchard House, Bristol Road, Chew Stoke, Bristol BS40 8UB, UK. E-mail: agdr17@ukgateway.net

Hideo Ishii, Senior Scientist (Pesticides), National Institute for Agro-Environmental Sciences, Kannondai 3-1-3, Tsukuba, Ibaraki 305-8604, Japan. E-mail: hideo@niaes.affrc.go.jp

Matthias Kretschmer, Michael Smith Laboratories, University of British Columbia, 2185 East Mall, Vancouver, British Columbia, Canada, V6T 1Z4. E-mail: kretschm@interchange.ubc.ca

Andy Leadbeater, Chairman, FRAC, Syngenta Crop Protection AG, WRO-1004.4.31, Schwarzwaldallee 215, CH-4058 Basel, Switzerland. E-mail: andy.leadbeater@syngenta.com

Aleš Lebeda, Department of Botany, Faculty of Science, Palacký University, Slechtitelu 11, 783 71 Olomouc-Holice, Czech Republic. E-mail: ales.lebeda@upol.cz

Margaret T. McGrath, Department of Plant Pathology and Plant-Microbe Biology, Cornell University, Long Island Horticultural Research and Extension Center, 3059 Sound Avenue, Riverhead, NY 11901-1098, USA. E-mail: mtm3@cornell.edu

David Ouimette, Crop Protection Research and Development, Dow AgroSciences, 9330 Zionsville Road, Indianapolis, IN 46268, USA. E-mail: DGOuimette@dow.com

Stefania Pollastro, Department of Environmental and Agro-Forestry Biology and Chemistry, University of Bari, Via Amendola 165/a, I-70126 Bari, Italy. E-mail: stefania.pollastro@agr.uniba.it

G. Schnabel, Associate Professor, Fruit Pathologist, Department of Entomology, Soils and Plant Sciences, B02 Long Hall, Clemson University, Clemson SC 29634, USA. E-mail: SCHNABE@clemson.edu

Silvia Laura Toffolatti, Department of Plant Production, Plant Pathology Section, University of Milan, Via Celoria 2, 20133 Milano, Italy. E-mail: silvia.toffolatti@unimi.it

Stefano Torriani, Plant Pathology Group, Institute of Integrative Biology, ETH Zurich, Universitaetstrasse 2, 8092 Zurich, Switzerland. E-mail: stefano.torriani@agrl.ethz.ch

Annamaria Vercesi, Department of Plant Production, Plant Pathology Section, University of Milan, Via Celoria 2, 20133 Milano, Italy. E-mail: annamaria.vercesi@unimi.it

About the Editor

———————————

Dr Tarlochan S. Thind is presently working as Senior Plant Pathologist and Head in the Department of Plant Pathology, Punjab Agricultural University, Ludhiana, India. He received his MSc (1974) and PhD (1977) degrees in botany from the University of Saugar, India, and did his postdoctoral research (1985–1987) at the National Institute of Agricultural Research (INRA) at Bordeaux, France.

Dr Thind has 33 years' experience of working on disease problems in vegetables, fruit and cereals at the Punjab Agricultural University, Ludhiana, India. His areas of research include the use of epidemiological parameters in the need-based application of fungicides in managing plant diseases, variability in plant pathogens and the problem of fungicide resistance in plant pathogens. He has handled several research projects on different aspects of fungicide use in agriculture.

He has provided very useful recommendations to farmers for the effective and economical management of disease problems in different crops. He has played a significant role in disseminating disease management technology to farmers and has also developed a web-based decision support system on timing fungicide application for potato late blight. He is actively involved in teaching plant pathology courses to graduate students, as well as in various extension education programmes for state farmers.

Dr Thind worked at INRA, Bordeaux, France, in 1985–1987 and 1992–1993 on the problem of fungicide resistance to SBI fungicides in viticultural pathogens. He worked as a Fulbright Fellow at the University of California, Davis, in 2001 and 2005 on the problem of young grapevine decline. He has also served as an FAO consultant for disease diagnosis in vegetable crops in Cape Varde, West Africa. He has participated in many national and international conferences in different countries.

Dr Thind has published 126 research articles in national and international journals, 30 extension articles, 12 book chapters and 2 bulletins. He has edited five books and co-authored one on *Disease Problems in Vegetable Production*. He is a life member of seven plant pathological societies and is a member of the Chemical Control Committee of the International Society for Plant Pathology. He is a member of the editorial board of the international journal, *Pest Management Science*. He was elected President of the Indian Society of Mycology and Plant Pathology (2007) and the Indian Phytopathological Society (2010) and is a fellow of three societies. Dr Thind has been given several awards and honours by professional societies for his contributions in plant pathology. Croplife India has nominated Dr Thind as Advisor for FRAC (India Chapter) and he has written one technical bulletin on the problem of fungicide resistance in India.

Preface

Fungicides serve as an important means for effective management of many serious plant diseases, but their indiscriminate use has resulted in the development of resistance in several pathogens. Development of resistance to fungicides has become a serious concern in the management of crop diseases and has threatened the potential of some highly effective commercial fungicides. This has led to poor disease control in many instances. Unlike insecticides, where resistance problems are known to occur much earlier, the practical problem of fungicide resistance emerged much later, in the 1970s and thereafter. The problem is more common with site-specific fungicides, and the performance of many of the systemic fungicides developed in the past four decades has been affected adversely.

In the last decade, more such compounds with novel modes of action have been developed and several of the modern selective fungicides have become vulnerable to the risk of resistance development in the target pathogens in different countries. This calls for the implementation of suitable resistance management strategies to achieve the expected disease control levels and to prolong the active life of the potential fungicides. Much research has been carried out in recent years on the detection, characterization and mechanism of resistance build-up to different fungicides, including the recently developed novel action compounds, and its management strategies in several laboratories across the world, as evident from the vast information available on these aspects in different research journals. There has been a strong need to compile this scattered information in a consolidated form.

This book contains 21 chapters contributed by active researchers from several countries including Brazil, Canada, the Czech Republic, France, Greece, India, Israel, Italy, Japan, Sweden, Switzerland, the UK and the USA. The book chapters have been divided into five sections. The first section on 'History of Resistance Development' has one chapter dealing with the historical perspectives of the development of fungicide resistance. The second section on 'Status, Detection and Management of Resistance' includes eleven chapters on resistance development and its management in important pathogens to various groups of fungicides such as benzimidazoles, SBIs, phenylamides, carboxylic acid amides, QoIs and other novel action compounds. One chapter in this section also gives information on field kits and Internet-supported fungicide resistance monitoring. In the section on 'Resistance Cases in Different Countries', there are six chapters dealing with resistance development in several countries such as Italy, France, Japan, India and the USA. The section on 'Genetics and Multi-Drug Resistance' includes two chapters on the genetics of fungicide resistance and the emergence of multi-drug resistance in fungal pathogens. In the final section on 'Role of FRAC' there is one chapter that highlights the role of the Fungicide Resistance Action Committee (FRAC) in managing resistance to different groups of risky fungicides. The book therefore covers different aspects of the problem of fungicide resistance. Each chapter has a concluding paragraph at the end highlighting the present status and future projections of the problem and its management.

This book will be useful to research scientists, teachers, students, extension workers and pesticide industry personnel. The author is thankful to all the contributors for their cooperation and efforts in compiling the chapters and submitting the manuscripts in time. I am also grateful to CABI for entrusting me with the task of editing this book. I am thankful to my wife Sanjeev for her moral support and my children, Guramrinder, Navleen and Gagandeep, for their encouragement during the compilation of this book.

Part I

History of Resistance Development

1 Historical Perspectives of Fungicide Resistance

Keith Brent
St Raphael, Chew Magna, Bristol, UK

1.1 Introduction

More than 40 years have now passed since the development of the resistance of plant pathogens to fungicide treatments emerged as a major threat to crop protection. This opening chapter describes the appearance of resistant pathogen populations during this time, indicates their impact on crop disease management and reviews progress in detecting resistance and preventing or delaying its onset and spread. These are big topics to try to cover in a single chapter. A huge amount of research has been done in many countries, so the author has had to be highly selective in producing this short account. However, many of his inevitable omissions will be rectified and his condensations and generalizations discussed in more detail in the more specific chapters that follow.

Historical perspectives of fungicide resistance can be considered to start with the work of Charles Darwin, who pointed out in *On the Origin of Species* (Darwin, 1859) the capacity for variation that exists in living organisms and the role of the selection of variants as a driving force in evolution. The change of a pathogen population from sensitive to resistant, through the selection of resistant mutants in response to external change (i.e. fungicide application), is clearly an evolutionary process, a very rapid one that greatly affects the survival of the fungus (the 'survival of the fittest'). The Darwinian connection with fungicide resistance and other problems affecting plant disease control has been discussed recently by Hollomon and Brent (2009).

1.2 The Use of Fungicides in Crop Protection

For over 200 years, growers have applied chemicals in order to prevent damage to crops being caused by fungi. The application of brine (aqueous solution of sodium chloride) to wheat seed to protect against infection by bunt fungus (*Tilletia caries*) dates back to the mid-18th century (Tillet, 1755). Later, copper sulfate solution was found to give better results (Prevost, 1807) and copper preparations became widely used until they were superseded by organomercurial seed treatments (Riehm, 1913). To combat diseases of foliage and fruit, sprays based on sulfur, first recommended by Forsyth (1802), were applied increasingly throughout the 19th century, mainly against powdery mildews. Then the discovery of Bordeaux mixture (Millardet, 1885) led to widespread use of copper compounds to control grape downy mildew (*Plasmopara viticola*), potato blight (caused by *Phytophthora infestans*) and other diseases not controlled by sulfur.

Subsequently, a range of organic compounds was introduced, the most important probably being dithiocarbamates (e.g. thiram, maneb, mancozeb) which appeared in the 1940s, phthalimides (e.g. captan, captafol) from the 1950s and chlorothalonil from the 1960s. These fungicides proved more effective against many diseases and were less injurious to crop plants but, like the older materials, they acted only at the plant surface, forming a protective layer. They penetrated little into plants and so could not affect established infections, and they had to be applied frequently to replace losses through weathering and to protect new plant growth.

Systemic fungicides which would penetrate into plants and hence could eradicate existing infections and resist weathering, and might also move into new areas of growth, were sought for many years in industrial and academic research centres (Wain and Carter, 1977). Griseofulvin, an antifungal antibiotic derived from *Penicillium griseofulvum*, showed considerable promise in glasshouse tests, being readily translocated in plants and active against a range of pathogens (Brian *et al.*, 1951). It was used commercially to a small extent, for example in Japan to control diseases in high-value apple and melon crops (see Misato *et al.*, 1977), but it was too costly and showed no great advantage over established fungicides. Nevertheless, the work on griseofulvin showed that systemic fungicidal action was an achievable objective and spurred on further research to find more effective systemic fungicides. Griseofulvin did find widespread medical use as an oral treatment against fungal infections of the skin and nails.

Eventually, in the late 1960s, six different classes of systemic fungicide, all highly effective against certain diseases, appeared in very rapid succession: benzimidazoles, oxathiins, amines (morpholines), 2-aminopyrimidines, organophosphorus and antibiotic rice blast fungicides. The 1970s saw the introduction of four more highly active fungicide classes, with different degrees of systemicity: dicarboximides, phenylamides, triazoles and fosetyl-aluminium. Over the past 20 years, quinone outside inhibitor (QoI) fungicides, anilinopyrimidines, phenylpyrroles, benzamides, carboxylic acid amides and new succinic dehydrogenase inhibitors have emerged.

This is a remarkable record of invention, involving very wide ranges of chemistry, mechanisms of action and practical uses, and originating from the laboratories of many companies. Moreover, the achievement of fungicide treatments of higher potency has been accompanied by marked decreases in application rates and increased specificities of action against target organisms, conferring greater margins of safety to crops, human health and the environment. About 150 different fungicidal compounds are now used in crop protection worldwide, with a total sales value of the order of US$10 billion/year and accounting for 23% of the total crop protection chemicals market.

A study of eight major crops concluded that globally crop disease causes a potential yield loss of 17.5% and that this is reduced to 13.5% through the application of control measures (Oerke, 1996). While the use of hygienic cultural practices, especially crop rotation, and of disease-resistant varieties have decreased or prevented disease-induced losses in many situations, fungicide application is still the predominant means of disease control in most crops worldwide, and is likely to remain an important component for decades to come.

1.3 The Occurrence of Fungicide Resistance Worldwide

This section considers the evolution of resistance in target organisms in the context of the history of fungicide development and use outlined above. More information on the development of resistance against particular fungicide classes, and on the actions taken to detect and to combat such resistance, is given in subsequent sections.

In the early 1960s, the development of resistance was not widely regarded as a serious potential problem. Fungicides had been used over many years, intensively in some situations, for example orchards, vineyards and potato fields, with no signs of trouble. At that time, publications appeared reporting the lack of efficacy of diphenyl-impregnated paper wraps in preventing the decay of lemons during shipment from the USA to Europe, caused by *Penicillium digitatum* (Harding, 1962). This problem was associated with earlier treatment with sodium *o*-phenylphenate and with the detection of strains of *P. digitatum* that were resistant to both diphenyl and sodium *o*-phenylphenate. Failure of control of a wheat bunt pathogen (*Tilletia foetida*) in Australia, linked to resistance to another aromatic hydrocarbon fungicide, hexachlorbenzene, was reported by Kuiper (1965). Then there was a report of the detection of resistance to organo-mercurial treatment in isolates of *Pyrenophora avenae* obtained from oat seeds in Scotland (Noble *et al.*, 1966). These reports were widely noted, but raised little alarm since they were isolated cases involving fungicides that had already been in use for many years.

Much more concern was raised by a report by Schroeder and Provvidenti (1969) of resistance to benomyl in powdery mildew of cucurbits (*Podosphaera fusca*, formerly *Sphaerotheca fuliginea*) in field plots in New York State, USA. At the time, this benzimidazole fungicide had been in commercial use for less than 2 years. Other reports of resistance to benomyl and other benzimidazole fungicides in various pathogens soon followed.

In the same year, Szkolnik and Gilpatrick (1969) reported the widespread failure of dodine in the control of *Venturia inaequalis*, the apple scab pathogen, in orchards in New York State, USA. Failures were associated with the finding of isolates which required two to four times higher concentrations of dodine for 50% inhibition (LD_{50}) in spore germination tests. Dodine, a long-chain guanidine fungicide with uncertain mode of action, had been in widespread, exclusive and repetitive use for about 10 years because it had shown excellent protectant and curative properties. Following these failures, use of dodine declined rapidly, causing substantial loss to the manufacturer. However, in areas where dodine had only limited use, or was used in mixed programmes with other fungicides, resistance was not found after more than 20 years of use (Gilpatrick, 1982).

In 1970, resistance of cucumber powdery mildew to dimethirimol, a 2-aminopyrimidine

fungicide, with consequent loss of performance, arose in a number of commercial glasshouses in Holland during the second year of its use (Bent *et al.*, 1971). The manufacturer immediately withdrew dimethirimol from use in affected regions.

Thus, in the very early 1970s, the rapid appearance of fungicide resistance in target pathogens quickly became recognized as a major potential threat to the use of new fungicides, particularly those with penetrative or systemic properties associated with selective fungitoxicity. Subsequently, problems of resistance affected the use of most new classes of fungicides, to greater or lesser degrees, within a few years of their commercial introduction. Table 1.1 gives a much condensed history of the occurrence of resistance worldwide, listing those fungicide classes for which resistance is well documented, together with examples of the more important diseases involved.

Usually, the first indications of a resistance problem came from reports from growers of failure of disease control following fungicide treatment. Many possible reasons other than resistance could underlie a perceived loss of control, such as incorrect application, use of a deteriorated or mistaken product, wrongly identified pathogen, or exceptionally heavy disease pressure, and at times growers and advisers attributed difficulties of disease control to fungicide resistance in the absence of evidence that resistance was the main cause. However, in many cases there was no other obvious explanation and loss of control was soon shown to be associated with greatly decreased sensitivity of the pathogen, as revealed by laboratory or glasshouse tests on samples taken from the problem sites.

It was soon learned that the evolution of resistance to one particular fungicide generally conferred resistance to some degree against all other fungicides that were within the same chemical class or which had the same biochemical mode of action. As an early example, strains resistant to benomyl were found to be cross-resistant to other benzimidazole fungicides such as carbendazim, thiabendazole or thiophanate-methyl.

Resistance affecting a particular fungicide–pathogen combination generally appeared

Table 1.1. Occurrence of practical fungicide resistance in crops.

Date first observed (approx.)	Fungicide or fungicide class[a]	Years of commercial use before resistance observed (approx.)	Main crop diseases and pathogens affected	References[b]
1960	Aromatic hydrocarbons	20	Citrus storage rots, *Penicillium* spp.	1
1964	Organomercurials	40	Cereal leaf spot and stripe, *Pyrenophora* spp.	2
1969	Dodine	10	Apple scab, *Venturia inaequalis*	3
1970	Benzimidazoles	2	Many target pathogens	4
1971	2-Aminopyrimidines	2	Cucumber and barley powdery mildews *Podosphaera fusca* and *Blumeria graminis*	5
1971	Kasugamycin	6	Rice blast, *Magnaporthe grisea*	6
1976	Phosphorothiolates	9	Rice blast, *Magnaporthe grisea*	6
1977	Triphenyltins	13	Sugarbeet leaf spot, *Cercospora beticola*	7
1980	Phenylamides	2	Potato late blight and grape downy mildew, *Phytophthora infestans* and *Plasmopara viticola*	8
1982	Dicarboximides	5	Grape grey mould, *Botrytis cinerea*	9
1982	Sterol demethylation inhibitors (DMIs)	7	Many target pathogens	10
1985	Oxathiins	15	Barley loose smut, *Ustilago nuda*	11
1998	Quinone outside inhibitors (QoIs)	2	Many target pathogens	12
2002	Melanin biosynthesis inhibitors (dehydratase)	2	Rice blast, *Magnaporthe grisea*	13

Notes: [a]More recent reports of field resistance to anilinopyrimidines, carboxylic acid amides and newer succinic dehydrogenase inhibitors (SDHIs) are not included. [b]References: (1) Eckert (1982); (2) Noble *et al.* (1966); (3) Gilpatrick (1982); (4) Smith (1988); (5) Brent (1982); (6) Kato (1988); (7) Giannopolitis (1978); (8) Staub (1994); (9) Lorenz (1988); (10) De Waard (1994); (11) Locke (1986); (12) Heaney *et al.* (2000), (13) Kaku *et al.* (2003).
Source: FRAC Monograph No. 1, 2nd edn, Brent and Hollomon (2007), with modification.

first in regions where disease intensity, and hence fungicide use, were relatively heavy. Resistance tended to be less prevalent or absent in regions where disease incidence was lower and fungicide use less intense. Many fungicides were used against a range of target pathogens which could differ widely in their taxonomic position. The occurrence of resistance in one of these pathogens was often followed by its appearance in others, after varying periods of time, but seldom in all of them. Possible reasons for this variation in the capacity of different target pathogens to acquire resistance are discussed later.

In many cases, the occurrence of fungicide resistance was signalled by a sudden and obvious loss of disease control in treated crops. Pathogen isolates obtained from such crops were found generally to differ greatly from sensitive ones in their response to fungicide exposure, being unaffected by concentrations 100-fold or greater than those that inhibited the growth of sensitive isolates completely. This clear-cut type of resistance has commonly been called qualitative, single-step, bimodal, discrete or major-gene resistance and has been shown for several fungicides to result from a single-locus mutation in the gene coding for the target site.

In contrast, resistance to the 2-aminopyrimidine fungicide ethirimol, to sterol demethylation inhibitors (DMI fungicides) and to dicarboximides was found to build up more gradually and to be partial, leading to a slow and sometimes fluctuating decline in disease control over several years. Clear detection of this type of resistance required sustained monitoring for sensitivity differences between and within populations in treated and untreated crops. Many isolates showed intermediate responses in bioassays and resistance involved a shift in the whole spectrum of responses, with the average sensitivity decreasing gradually. Such resistance was referred to as quantitative, multi-step, unimodal, continuous, or polygenic resistance. The multiplicity of isolate sensitivities and evidence from genetic crossing experiments were taken to indicate that the cumulative effects of mutation of several genes were involved, and the term 'polygenic' was widely used for many years. However, recent research on DMI resistance mechanisms indicates that a number of single-base changes at different loci within a single target-related gene give rise to alleles that confer different degrees of resistance. So, this type of resistance may not necessarily be polygenic in origin and the use of other terms, for example quantitative resistance, is now preferable.

Once it has evolved, qualitative fungicide resistance has usually proved to be persistent. Some pathogen populations have been shown to retain their resistance for many years, even after treatment with the affected fungicide was stopped or much restricted. In other cases, resistance has declined gradually after fungicide withdrawal, but has returned quickly if treatment is resumed. On the other hand, quantitative resistance appears to revert relatively quickly to a more sensitive condition when the fungicide concerned is replaced or used less intensively.

While most of the fungicides introduced since the late 1960s have encountered resistance problems, some have not done so after many years of extensive use. Exceptions include anilinopyrimidines, phenylpyrroles, fosetyl-aluminium and certain rice blast fungicides, e.g. probenazole, isoprothiolane and tricyclazole. Resistance to amine fungicides ('morpholines') has developed very slowly, with relatively little practical impact. It is remarkable that the older materials, such as sulfur, copper fungicides, dithiocarbamates, phthalimides and chlorothalonil, have never been affected by resistance and remain fully effective despite their widespread and often exclusive use for many years. It has long been assumed that this durability results from their multi-site action, which cannot be nullified by single target-site mutation in the pathogen, and from an absence of mutations that could confer other feasible mechanisms of resistance such as fungicide exclusion or metabolism.

1.4 The Evolution and Management of Resistance in Different Classes of Fungicides

It is impossible within the confines of this chapter to document the histories of resistance development and management for all fungicide classes. This section endeavours to outline events relevant to certain classes, selected in view of their historical, scientific and/or widespread practical significance, and to include information that exemplifies or extends points made in the general account given above.

Benzimidazoles

As mentioned above, resistance to benomyl was first detected in cucurbit powdery mildew (*P. fusca*) in 1969, within 2 years of its introduction. This was revealed through applying spores to cucumber leaves detached from plants grown on soil drenched with standard amounts of benomyl solution. The early onset of resistance, and its early detection, in this particular plant pathogen has also happened with other fungicides (e.g. DMIs, 2-aminopyrimidines). *P. fusca* has very obvious and rapidly developing symptoms, so failure of disease control is seen quickly and easily. It produces many spores and has a short generation time, which favours selection of mutants. It often occurs, and is exposed to fungicide treatment, in the confines of glasshouses or tunnels, so there is little or no ingress of sensitive wild-type spores.

Further developments have been described by Smith (1988). High levels of resistance to benomyl and other benzimidazoles were soon observed in isolates of many other target pathogens, for example *Botrytis cinerea* on grapes in Europe and *Cercospora* spp. on peanuts in the USA and sugarbeet in Greece. Some pathogens took much longer to develop detectable benzimidazole resistance: about 10 years for *Oculimacula* spp., causing cereal eyespot disease (Locke, 1986), and 15 years for *Rhynchosporium secalis*, causing barley leaf scald (Kendall *et al.*, 1993). The less abundant spore production, longer generation times and less frequent exposure to fungicide application associated with these pathogens might have accounted for their slower development of resistance.

The earlier reports of benzimidazole resistance were all published by public sector or academic organizations. Faced with this unforeseen and worrying development, the companies involved in manufacture and sales took a low-key approach, dealing with complaints on an individual basis and putting general warnings of the existence of resistant strains and disclaimer notices on product labels. The results of monitoring or other studies and information regarding resistance management were not published by these companies until 10 years after resistance was first detected.

In some situations and regions, the use of mixtures or alternations with non-benzimidazole fungicides was encouraged by the companies concerned and by advisory services, although often this was done too late. Once established, benzimidazole resistance generally persisted. A major exception was in the control of black Sigatoka disease of bananas (Delp, 1988; Smith, 1988). In this case, benzimidazole resistance declined after benzimidazole application was stopped, presumably because the resistant strains were less fit than the sensitive ones and effective re-entry in mixture or alternation with non-benzimidazole fungicides became possible.

An early example of the successful use of a mixture strategy to avoid resistance development was shown in the control of *Cercospora* leaf spots of peanuts in the USA. Resistance problems soon developed in the south-eastern states, where there was sole use of benomyl. In Texas, where benzimidazole–mancozeb mixtures were used from the start, there was no resistance over many years, except in some trial plots where benzimidazole alone was applied repeatedly (Smith, 1988).

Benzimidazoles are still in use worldwide, often in mixture or rotation with other fungicides. With a lack of recent trials or monitoring data, it is hard to judge the degree of current effectiveness in the various uses. Benzimidazole-resistant strains of *Oculimacula* spp. (eyespot) and *Mycosphaerella graminicola* (leaf blotch) were still found to be common in wheat in France in 1997–2003 (Leroux *et al.*, 2003, 2005a).

2-Aminopyrimidines

Dimethirimol was first used in 1968, as a soil drench for the control of cucumber powdery mildew (*P. fusca*) in glasshouses. One application provided excellent protection for several weeks, and this striking example of systemic action was seen as a major breakthrough in plant disease control. In 1969, dimethirimol treatment was used more widely, especially in Holland, again with excellent results. However, in the spring of 1970, many reports of inadequate control were received from Dutch growers. Leaf-disc tests indicating sensitivity of the fungus to a range of dimethirimol concentrations, based on a method used by Dekker (1963) to test for the systemic action of chemicals, were established quickly by the manufacturing company and applied to mildew samples taken from many sites of use. A close correlation between lower fungicide sensitivity and lack of disease control was revealed (Bent *et al.*, 1971). Isolates varied much in sensitivity, the most resistant tolerating at least 100-fold higher fungicide concentrations than the most sensitive. Resistance was also detectable in glasshouses in the UK and Germany, although not in protected or open-air cucurbit crops in Spain and Israel over many years. The product was withdrawn from use in affected countries and limited evidence indicated a gradual decline in resistance. Dimethirimol was reintroduced in Holland in the late 1970s, with restriction to one application per year, unrelated fungicides being used at other times. Initially, this was successful but, after 2 years, resistance problems returned.

Another 2-aminopyrimidine, ethirimol, became widely used in the UK from 1971 as a barley seed treatment to control powdery mildew (*Blumeria graminis*). Much variation in sensitivity between mildew isolates was revealed in leaf segment tests (Wolfe and Dinoor, 1973; Shephard *et al.*, 1975). Ethirimol was withdrawn from use on winter barley crops in 1973 in order to avoid selection for resistance in the periods between successive spring barley crops (at the time, spring crops were the more important economically). Good control was maintained for many years. Sustained monitoring showed that the sensitivity range of isolates from treated fields was always lower than that of isolates from untreated fields, indicating that a limited degree of selection for resistance occurred within treated crops. However, mean sensitivity values changed little from 1973 to 1975 (Shephard *et al.*, 1975) and had increased slightly by 1984 (Heaney *et al.*, 1984). By this time, ethirimol was again being used on both winter and spring barley, although less intensively and along with a range of other mildew fungicides. Evidence that some loss of fitness was associated with the less sensitive forms was obtained by Hollomon (1978), and crossing experiments indicated that several mutations contributed to decrease of sensitivity (Hollomon, 1981).

The difference between the rapid loss of control of cucumber mildew to dimethirimol in Holland and the continued good control of barley mildew to ethirimol in the UK was striking. With hindsight, it was realized that the year-round, almost universal use of dimethirimol, coupled with near perfect levels of control, in a protected environment, was a recipe for disaster, whereas the risk of control

failure with single, annual applications of ethirimol in a proportion of open barley crops was clearly lower (Brent, 1982). The diversity of responses of isolates, and the genetic data, indicated that 2-aminopyrimidine resistance was quantitative. However, the rapid and severe loss of performance of dimethirimol suggested that in situations where selection pressure was exceptionally high, then development of quantitative resistance might not be slow and gradual.

These studies on the 2-aminopyrimidines have some historical interest, having yielded the first systematic surveys of fungicide resistance in relation to fungicide performance, the first publication of fungicide resistance data by a manufacturer and the first evidence for quantitative or unimodal resistance.

Phenylamides

These fungicides, which act specifically against Oomycetes, were first used in 1977. In 1980, serious loss of control by metalaxyl occurred in downy mildew of cucumbers (*Pseudoperonospora cubensis*) in Israel and Greece. Isolates of the fungus were shown to be highly resistant to metalaxyl in sprayed plant and detached leaf assays, in reports from five different research centres (Malathrakis, 1980; Pappas, 1980; Reuveni *et al.*, 1980; Georgopoulos and Grigoriu, 1981; Katan and Bashi, 1981). Also in 1980, failures of control of potato late blight (*P. infestans*) by metalaxyl occurred in Holland and Ireland, and resistance was confirmed in leaf-disc assays (Davidse, 1981). Resistance appeared in the following year in grape downy mildew (*P. viticola*) in France and South Africa and in tobacco blue mould (*Peronospora tabacina*) in Central America, and subsequently in other crop pathogens.

This dramatic onset of widespread resistance was unexpected because previous experiments involving repeated exposure of *P. infestans* to phenylamides, carried out by the manufacturer of metalaxyl, had failed to select for resistance and had been taken as a sign of low risk (Staub *et al.*, 1979). However, exposure of *Phytophthora megasperma* to a mutagenic chemical (a nitrosoguanidine) produced many highly phenylamide-resistant mutant strains, suggesting that mutagenesis could be more revealing than 'training' experiments in resistance risk studies (Davidse *et al.*, 1981).

Because the development of resistance in *P. infestans* was associated with the solo use of metalaxyl, as in Holland, and had not occurred in countries where only formulated mixtures with mancozeb were applied, as in the UK, the manufacturer immediately withdrew the single product from use against foliar diseases and recommended the application of mixtures of metalaxyl with multi-site fungicides.

The intercompany Fungicide Resistance Action Committee (FRAC) was established soon after the onset of phenylamide resistance. Its Phenylamides Working Group issued a set of guidelines for phenylamide resistance management, which included using mixtures for foliar application, avoiding curative use and limiting the number of sprays per season. These guidelines were implemented by all the companies involved, and this fungicide class continued in use against all target diseases (Staub, 1994). The appearance of several newer oomycete-active fungicides over the years has much increased the options for diversified application programmes.

The FRAC recommendations did not delay for long the appearance and spread of resistant strains of *P. infestans* and other target pathogens in many regions. Nevertheless, data from field experiments indicated that phenylamide–mancozeb mixtures continued to perform better than mancozeb alone against *P. infestans* (Staub, 1994) and against *Bremia lactucae*, lettuce downy mildew (Wicks *et al.*, 1994). Reasons for this are not well understood. Mutant frequency in crops may have been overestimated in leaf-disc assays with multiple spore inocula. Also, the proportion of nuclei in hyphal cells and sporangia with a resistant gene could have been a critical factor (Cooke *et al.*, 2006).

Sterol demethylation inhibitors (DMI fungicides)

This large class of fungicides, which includes triazoles and imidazoles, was first introduced in the mid-1970s and since then over 30 different DMIs have been used in crop protection. DMI-resistant mutants were obtained readily in the laboratory by mutagenic treatment of *Cladosporium cucumerinum* (Fuchs and Drandarevski, 1976), and soon this was shown in studies on several other fungi. However, the mutants generally showed a decrease in pathogenicity and other fitness parameters, and the development of practical resistance was deemed unlikely.

Declines in the sensitivity of field populations of a number of pathogens were reported in the 1980s; for example, in barley powdery mildew, *B. graminis* (Fletcher and Wolfe, 1981; Heaney *et al.*, 1984), cucurbit powdery mildew, *P. fusca* (Huggenberger *et al.*, 1984), barley leaf scald, *R. secalis* (Hollomon, 1984) and apple scab, *V. inaequalis* (Stanis and Jones, 1985; Thind *et al.*, 1986). However, resistance development tended to be relatively slow and gradual and to fluctuate in severity according to the intensity and exclusivity of DMI treatment, as was typical of quantitative (unimodal) resistance. Erosion of performance against certain pathogens, to varying degrees, was experienced. Azole resistance in several pathogens is now known to be associated with one or a combination of several point mutations in the *CYP51* gene that codes for the target sterol demethylase, and possibly also with overexpression of efflux transporters (e.g. Leroux and Walker, 2011). The introduction in recent years of azole fungicides (e.g. prothioconazole), which exhibit much lower resistance factors, has largely overcome the decrease in effectiveness of older products (see Brent and Hollomon, 2007a).

In 1987, FRAC issued general guidelines regarding the use of DMIs and other sterol biosynthesis inhibitors, which included avoidance of repeated solo applications in one season against a high-risk pathogen in areas of high disease pressure, use of mixtures or rotation with effective non-cross-resistant fungicides against diseases requiring multiple spray applications or, where performance was declining and less sensitive forms were detected, avoidance of application of reduced doses. These recommendations have been widely implemented and DMIs have continued to give good control of most target pathogens after some 35 years of widespread use.

Dicarboximides

Since the mid-1970s, fungicides of this class (e.g. iprodione, vinclozolin and procymidone) have been used mainly to control *Botrytis* spp. and closely related pathogens, largely replacing the benzimidazoles to which resistance had become widespread. Spontaneous dicarboximide-resistant variants were frequently observed in laboratory cultures and, after about 3 years of intensive commercial use, were detected in vineyards in Germany (Holz, 1979; Lorenz and Eichhorn, 1980), and later in a range of other crops. The proportions of resistant strains fluctuated, tending to decline after dicarboximide treatment ceased and to increase again when it was resumed. Isolates showed various levels of resistance, and loss of performance was found to be associated with moderately resistant pathogen strains which were fitter than the most highly resistant ones. Resistance gradually increased in frequency and severity during the 1980s, especially in grapevines in those regions of France and Germany where *Botrytis* was most prevalent and the dicarboximides used most intensively (Lorenz, 1988).

FRAC recommendations included restriction of dicarboximide application to two or three times per crop per season and use of mixtures with non-carboximide fungicides in areas where resistance was established. Mixture or rotation of dicarboximides with earlier companion compounds, such as captan, dichlofluanid and chlorothalonil, did not give fully adequate control, but use with more recently introduced fungicides, such as fluazinam, fludioxonil, fenhexamid and anilinopyrimidines, has proved effective (Leroux *et al.*, 2005b).

Quinone outside inhibitors (QoI fungicides)

Fungicides in this class have a common mode of action: inhibition of electron transfer at the QoI site in mitochondrial complex III. They were introduced in the late 1990s for the control of a wide range of crop pathogens. Within 2 years, failures of control of wheat powdery mildew (*B. graminis* f. sp. *tritici*), associated with the presence of highly resistant populations, were experienced in Germany and soon after in other countries (Chin *et al.*, 2001). Serious problems of resistance in many major pathogens, for example *Mycosphaerella fijiensis* var. *difformis* (cause of black Sigatoka disease of bananas), *M. graminicola* (cause of wheat leaf spot) and *P. viticola* (grape downy mildew) soon appeared. Resistance did not develop in *P. infestans* or in rust fungi. The latter have been found in general to carry a low resistance risk (although triazole resistance in soybean rust (*Phakopsora pachyrhizi*) has caused concern recently). As with other fungicide classes, the incidence and severity of problems caused by QoI resistance has varied greatly between regions, being greater where the target disease is more prevalent and use of QoIs more intense.

QoI resistance in many pathogens was soon shown to be qualitative (bimodal), with resistance factors generally greater than 100, and to originate mainly from a single mutation (G143A) in the cytochrome bc_1 gene that codes for the target site (Gisi *et al.*, 2002). Mutations at other loci on the cytochrome bc_1 gene were found in isolates from a few pathogens; these were associated with lower resistance factors.

General recommendations were made by FRAC to limit the number of QoI applications, whether applied solo or in mixture with non-QoI fungicides, and to rotate applications with those of effective fungicides from other non-cross-resistant classes. Specific recommendations were made for use on certain crops; for example, on cereals and bananas QoIs should always be used mixed with non-cross-resistant fungicides.

1.5 Progress in Resistance Research

Under this heading some comments are made on achievements over the years, and lessons learned, in the main sectors of fungicide resistance research. Important efforts that have been made to promote interaction and collaboration between workers from the different sectors and organizations concerned with fungicide performance and resistance are also mentioned.

Monitoring

Our knowledge of the occurrence and practical impact of fungicide resistance has been gained almost entirely through monitoring, i.e. testing samples of field populations of pathogens for their sensitivity to particular fungicides. Monitoring dates back to the 1960s and at that time was done locally to investigate the cause of reported failures of disease control. This type of monitoring is still done as problems arise. Since the early 1970s, larger-scale surveys of sensitivity, sometimes continued over several years, have been made in treated and untreated crops, mainly to obtain early warning of impending resistance problems for high-risk fungicides and to check whether management strategies are working. For qualitative resistance, useful early warning has rarely been attained through such bioassay tests, because many tests need to be done even to detect mutants at 1% frequency, which would be dangerously high. However, monitoring has given useful early indications of gradual sensitivity shifts in whole populations that are involved in the development of quantitative resistance.

The selection of sampling method can affect the cost and effectiveness of monitoring operations greatly. Methods have ranged from driving a car through crop-growing areas with test plants mounted on top, to measure the response of whole spore populations, to taking single-pustule or even single-spore isolates to indicate the detailed distribution of resistant mutants.

A wide range of assay methods has been used, including spore germination tests, mycelial growth tests on liquid or solid media and floating leaf-disc and leaf-segment tests (for obligate pathogens), involving exposure either to a single discriminating fungicide concentration or, more accurately, to a range of concentrations. Results of the tests have been scored in various ways, the ED_{50} value

being the commonest and in many situations the most reliable and comparable form. Resistance factors (ED_{50} for resistant isolate divided by baseline ED_{50}) have proved a useful way to express the magnitude of acquired resistance. The importance of examining resistant strains for fitness parameters and for their cross-resistance to other fungicides has long been recognized and acted upon. FRAC has published detailed monitoring procedures for a number of fungicide classes from 1991 onwards.

A major step in monitoring research was the establishment of PCR (polymerase chain reaction) detection tests for specific fungicide-resistant mutants in the late 1990s. They have been used mainly in monitoring for QoI resistance caused by G143A mutation (much the commonest cause of QoI resistance). These tests can detect resistant mutants at frequencies as low as 1 in 10,000 and have been used to test for the risk of QoI resistance arising in a range of pathogens before the possible appearance of disease control failures. PCR tests have been done successfully on DNA extracts from diseased plant material so that pathogen isolation is not required.

The interpretation and presentation of the results of monitoring tests has required much care. It has proved a huge help to have access to 'baseline data', information on the range of sensitivity existing in field samples of target pathogens obtained before commercial use of a fungicide starts. Absence of such data hindered interpretation of early monitoring operations, but for many years now baseline testing of important target pathogens has been done routinely prior to the introduction of new fungicides, and baseline data are required by some registration authorities. In the early years, some undue concern and confusion was caused by authors attaching practical significance too readily to their detection of some variation in the sensitivity of field isolates. However, there is now general acceptance of the need to determine whether the less sensitive forms are sufficiently resistant and pathogenic, and present in sufficient quantity to affect field performance of the fungicide, and whether the incidence of such populations correlates with the loss or reduction of disease control in treated crops.

More information on the methodology and results of fungicide resistance monitoring is provided by Brent (1992) and Brent and Hollomon (2007a,b), and a review of baseline research is given by Russell (2003).

Resistance management strategies

On the basis of theoretical argument and practical experience regarding the development of resistance to fungicides, and also to antibiotics and antimicrobial drugs, it became generally accepted in the 1970s that sustained and sole use of fungicides with site-specific mechanisms of action conferred a high risk of resistance. Conversely, occasional use, complemented by the use of other unrelated fungicides, was unlikely to lead to resistance problems.

In order to avoid or lessen the sole and/or sustained application of at-risk fungicides, various use strategies were adopted. These included application only in preformulated or tank mixes or in rotation with effective non-cross-resistant fungicides, preferably multi-site fungicides with low risk, limitation of number of treatments per season, avoidance of eradicant use and integrated use with available non-chemical control measures, such as disease-resistant crop varieties, biological control agents and crop rotation. Some examples are given in previous sections. Strategy design and implementation were (and still are) not easy matters for the companies concerned, especially because the degree of risk associated with a particular use of a novel fungicide was often hard to judge and because some limitation on sales was generally involved. Requirements for long-term conservation of product effectiveness had to be balanced with the need to establish an amount and pattern of use that would satisfy the needs of the grower and provide a reasonable payback for the company.

Use strategies to minimize resistance risk needed to be applied uniformly over large areas in order to exert their full effect. Where cross-resistant fungicides were marketed by different companies, it was necessary to try to formulate and then publish through labels and literature an agreed common-use strategy and to promote its implementation through growers and advisers. In 1980, a time when serious resistance problems with phenylamides and signs of resistance to dicarboximides were causing much renewed concern to a number of manufacturers, a group of industrial scientists attending a fungicide resistance course at Wageningen proposed formation of an intercompany group that would cooperate in investigating possible resistance problems and establishing countermeasures. As a result, FRAC was set up in 1981. It has remained active and influential ever since, through its guidelines for fungicide use, reports from working groups, updated list of fungicides according to mode of action and resistance risk, approved methods for sensitivity testing and monographs on resistance management, assessment of risk and baseline research. All this large amount of information has been made freely available on its website. Fungicide resistance action groups organized by public sector research and advisory workers also have been formed in several countries and have played a useful role in providing information and in promoting interaction with and between the agricultural and chemical industries in formulating strategies and identifying priorities for further research.

Answering the obvious question 'have the management strategies worked?' is not easy. Being applied over whole regions, comparison of commercial-scale 'managed' and 'unmanaged' disease control with respect to resistance build-up has seldom been possible. Field experimentation also has proved difficult, requiring large replicated plots, often extending over several years, and beset by problems such as variation in disease pressure from year to year, variation in the sensitivity of initial inoculum and ingress of external inoculum. Overall, experimentation, mathematical modelling and also some commercial experiences (as mentioned above for benzimidazoles and phenylamides) have shown that mixture and rotation strategies do delay but not totally prevent resistance arising against at-risk fungicides. Maintenance of the recommended dose has often been given as a guideline for avoiding resistance risk, but the concept that lowering dose hastens the onset of fungicide resistance has been questioned and there is some evidence that it can delay the development of qualitative resistance (see Brent and Hollomon, 2007a,b). Further work on the effects of dose is still needed.

Aspects of resistance management have been reviewed fairly recently by Kuck (2005) and Brent and Hollomon (2007b).

Assessment of risk

This is another subject with a rich history, but only a very brief outline can be offered here. A FRAC monograph (Brent and Hollomon, 2007a), a guideline of the European and Mediterranean Plant Protection Organization (EPPO) (Anon., 2002) and earlier reviews by Gisi and Staehle-Csech (1988) and Brent *et al.* (1990) provide fuller accounts of the subject in general.

Ever since widespread fungicide resistance was first experienced in 1970, the assessment of risk has been recognized as important to the fungicide manufacturer. It has influenced decisions on whether a candidate product is worth developing and marketing (with the huge costs involved), on what use strategies are adopted and on what resistance monitoring should be done. It has guided advisers and growers in selecting and scheduling treatments, and on the need for vigilance. Since the 1990s, registration officials increasingly have required information on risk as a part of efficacy assessment.

Experience of the durability of field performance and the incidence of resistance soon made it clear that overall resistance risk comprised three main components: fungicide-associated risks, target pathogen-associated risks and risks attached to regional conditions of use. Thus, a combination of a high-risk fungicide plus a high-risk disease plus high-risk use conditions would indicate a very high likelihood of resistance developing. The following factors have been found repeatedly to act as positive indicators of each of these three risk components.

- Fungicide-associated: chemical class has a resistance record; single site of action; cross-resistance with other resistance-prone fungicides; mutagenic treatment elicits resistant-fit mutants of target pathogen; sexual crossing causes target pathogen to produce resistant-fit recombinants; repetitive treatment of target pathogen in laboratory or field elicits resistant-fit strains.
- Pathogen-associated: short generation time; abundant sporulation; rapid, long-distance spore dispersal; genetic adaptability (haploid pathogen; gene structure allows expression of resistant mutation; frequent sexual reproduction).
- Conditions of use: application will be repetitive or continuous and/or intensive in the region; other types of fungicide will not be used in mixture or rotation; non-chemical control measures will not be used; disease pressure is generally severe in the region; pathogen is confined (e.g. in glasshouses), preventing entry of sensitive forms.

For many years now, resistance risk assessment, and associated biochemical and genetic studies, have been an essential part of industrial fungicide research and development. Each main usage of a new fungicide, against a particular target pathogen in a key region of use, has required a separate assessment, integrating all the risk indications as a basis for overall judgement. This has been only very approximate, indicating high, medium or low risk. After new fungicides have entered commercial use, risk assessments have been adjusted in the light of experience and have become more reliable. Up-to-date assessments of fungicide-associated risk are presented on the FRAC website for all fungicides in current use.

A number of mathematical models have been proposed for predicting the rate of build-up of resistance in relation to different fungicide application regimes. Early examples were those of Kable and Jeffery (1980) and Josepovits and Dobrovolszky (1985). While modelling has provided a valuable theoretical background to resistance studies, it has not been used in the practical assessment of resistance risk because of a lack of verification and the difficulty of getting all the data required to work the models.

With hindsight, most fungicide–pathogen combinations can be seen to have yielded on this basis a reasonably realistic initial risk assessment. There are a few exceptions. For example, *P. infestans* would have scored as a high-risk pathogen, whereas with the notable exception of phenylamide resistance, there has been no sign of its resistance to most fungicides, including QoIs, carboxylic acid amides and cymoxanil. Surprisingly, rust fungi have proved to be low-risk pathogens, although they are well known to carry a high risk of overcoming host plant resistance. The melanin biosynthesis inhibitor (reductase) fungicides might not have been expected to remain so clear of problems over their long periods of use.

1.6 Overview

Over the past 40 years, fungicide application has maintained its key importance in agriculture, despite the problems of resistance described above. The linked processes of risk assessment, baseline research, monitoring and design and implementation of resistance management strategies, at first applied sporadically but now operating almost universally, undoubtedly have prevented more serious losses of disease control than those which have actually occurred. They are being applied very actively to maintain the performance of newer fungicides, including succinic dehydrogenase inhibitors and carboxylic acid amides, against pathogens known to produce resistant mutants in the field. Our knowledge of the biochemical and genetic mechanisms that underlie the evolution of resistance in initially sensitive pathogens has grown greatly, and continues to do so. Understanding of the factors that determine the emergence, build-up and persistence of resistant pathogen populations in crops, and how they relate to use strategies, has not grown to the same extent. Field-related research has proved difficult, but more is needed and the availability of rapid and sensitive molecular tests to detect mutants will help.

The integrated use of disease-resistant plant varieties together with chemical treatment so far has played only a limited role in fungicide resistance management and in plant resistance management, although it may well increase in importance as recent developments in plant breeding technology lead to new forms of plant resistance. Chemical and biochemical diversity has proved essential to conserving disease control. It is vital that industry continues to invent new fungicides, especially ones with novel mechanisms of action and including multi-site fungicides, despite the huge costs and long development periods involved. Current legal action in the European Union aiming to remove many agrochemicals, including some major fungicides, on the grounds of alleged toxic and environmental threats is a cause for serious concern.

The value of good communication and interaction between all concerned with fungicide resistance has been shown many times and in many ways. The part played by FRAC in promoting intercompany cooperation and in publishing information has been impressive. Very useful contact and interaction between industrial and public sector researchers has arisen from the formation of joint action groups and through conferences and symposia. The series of international courses led by Johann Dekker and held at several centres worldwide in the 1980s and the ongoing series of Rheinhardsbrunn Symposia, initiated by Horst Lyr, have been especially influential. Fungicide resistance remains both a major scientific topic, from which much can be learned, and a serious challenge to reliable, long-term crop protection, and it is vital that research continues and that the knowledge gained is well disseminated and used.

Acknowledgements

Thanks are due to the Fungicide Resistance Action Committee (FRAC) for permission to reproduce Table 1.1, taken from FRAC Monograph No. 1 (2nd edn), and to Dr Derek Hollomon for reviewing the manuscript and for helpful advice.

References

Anon. (2002) Standard PP 1/213(2): *Efficacy Evaluation of Plant Protection Products: Resistance Risk Analysis*. European Plant Protection Organization, Paris.

Bent, K.J., Cole, A.M., Turner, J.A.W. and Woolner, M. (1971) Resistance of cucumber powdery mildew to dimethirimol. In: *Proceedings 6th British Insecticides and Fungicides Conference*. British Crop Protection Council, London, pp. 274–282.

Brent, K.J. (1982) Case study 4: powdery mildews of barley and cucumber. In: Dekker, J. and Georgopoulos, S.G. (eds) *Fungicide Resistance in Crop Protection*. Pudoc, Wageningen, The Netherlands, pp. 219–230.

Brent, K.J. (1992) Monitoring fungicide resistance: purpose, procedures and progress. In: Denholm, I., Devonshire, A.L. and Hollomon, D.W. (eds) *Resistance '91: Achievements and Developments in Combating Pesticide Resistance*. Elsevier, London and New York, pp. 1–18.

Brent, K.J. and Hollomon, D.W. (2007a) *Fungicide Resistance: The Assessment of Risk*. FRAC Monograph No. 2 (2nd edn). CropLife International, Brussels, 52 pp.

Brent, K.J. and Hollomon, D.W. (2007b) *Fungicide Resistance in Crop Pathogens: How Can It Be Managed?* FRAC Monograph No. 1 (2nd edn). CropLife International, Brussels, 55 pp.

Brent, K.J., Hollomon, D.W. and Shaw, M.W. (1990) Predicting the evolution of fungicide resistance. In: Green, M.B., Le Baron, H.M. and Moberg, W.K. (eds) *Managing Resistance to Agrochemicals*. American Chemical Society, Washington, DC, pp. 303–319.

Brian, P.W., Wright, J.M., Stubbs, J. and Way, A.M. (1951) Uptake of antibiotic metabolites of soil micro-organisms by plants. *Nature, London* 167, 347–348.

Chin, K.M., Chavaillaz, D., Kasbohrer, M., Staub, T. and Felsenstein, F.G. (2001) Characterising resistance risk of *Erysiphe graminis* f. sp. *tritici* to strobilurins. *Crop Protection* 20, 87–96.

Cooke, L.R., Carlisle, D.J., Donaghy, D.J., Quinn, C., Perez, F.M. and Deahl, K.L. (2006) The Northern Ireland *Phytophthora infestans* population 1998–2002 characterised by phenotypic and genotypic markers. *Plant Pathology* 55, 320–330.

Darwin, C. (1859) *On the Origin of Species*. Murray, London.

Davidse, L.C. (1981) Resistance to acylalanine fungicides in *Phytophthora megasperma* f. sp. *medicaginis*. *Netherlands Journal of Plant Pathology* 87, 11–24.

Davidse, L.C., Looyen, D., Turkensteen, L.J. and van der Wal, D. (1981) Occurrence of metalaxyl-resistant strains of *Phytophthora infestans* in Dutch potato fields. *Netherlands Journal of Plant Pathology* 87, 65–68.

Dekker, J. (1963) Effect of kinetin on powdery mildew. *Nature, London* 197, 1027–1028.

Delp, C.J. (1988) Resistance management strategies for benzimidazoles. In: Delp, C.J. (ed.) *Fungicide Resistance in North America*. American Phytopathological Society, St Paul, Minnesota, pp. 41–43.

De Waard, M.A. (1994) Resistance to fungicides which inhibit sterol 14α-demethylation, an historical perspective. In: Heaney, S., Slawson, D., Hollomon, D.W., Smith, M., Russell, P.E. and Parry, D.W. (eds) *Fungicide Resistance*. British Crop Protection Council, Farnham, Surrey, UK, pp. 3–10.

Eckert, J.W. (1982) Case study 5: Penicillium decay of citrus fruits. In: Dekker, J. and Georgopoulos, S.G. (eds) *Fungicide Resistance in Crop Protection*. Pudoc, Wageningen, The Netherlands, pp. 231–250.

Fletcher, J.T. and Wolfe, M.S. (1981) Insensitivity of *Erysiphe graminis* f. sp. *hordei* to triadimefon, triadimenol and other fungicides. In: *Proceedings of Brighton Crop Protection Conference – Pests and Diseases, 1981*. British Crop Protection Council, Croydon, UK, pp. 633–640.

Forsyth, W. (1802) *A Treatise on the Culture and Management of Fruit Trees*. Nichols, London.

Fuchs, A. and Drandarevski, C.A. (1976) The likelihood of development of resistance to systemic fungicides which inhibit ergosterol biosynthesis. *Netherlands Journal of Plant Pathology* 82, 85–87.

Georgopoulos, S.G. and Grigoriu, A.C. (1981) Metalaxyl-resistant strains of *Pseudoperonospora cubensis* in cucumber greenhouses of southern Greece. *Plant Disease* 65, 729–731.

Giannopolitis, C.N. (1978) Occurrence of strains of *Cercospora beticola* resistant to triphenyltin fungicides in Greece. *Plant Disease Reporter* 62, 205–208.

Gilpatrick, J.D. (1982) Case study 2: *Venturia* of pome fruits and *Monilinia* of stone fruits. In: Dekker, J. and Georgopoulos, S.G. (eds) *Fungicide Resistance in Crop Protection*. Pudoc, Wageningen, The Netherlands, pp. 195–206.

Gisi, U. and Staehle-Csech, U. (1988) Resistance risk evaluation of new candidates for disease control. In: Delp, C.J. (ed.) *Fungicide Resistance in North America*. American Phytopathological Society, St Paul, Minnesota, pp. 101–106.

Gisi, U., Sierotzki, H., Cook, A. and McCaffery, A. (2002) Mechanisms influencing the evolution of resistance to QoI inhibitor fungicides. *Pest Management Science* 58, 859–867.

Harding, P.R. (1962) Differential sensitivity to sodium orthophenylphenate by biphenyl-sensitive and biphenyl-resistant strains of *Penicillium digitatum*. *Plant Disease Reporter* 46, 100–104.

Heaney, S.P., Humphreys, G.J., Hutt, P., Montiel, P. and Jegerings, P.M.F.E. (1984) Sensitivity of barley powdery mildew to systemic fungicides in the UK. In: *1984 British Crop Protection Conference – Pests and Diseases*. British Crop Protection Council, Croydon, UK, pp. 459–464.

Heaney, S.P., Hall, A.A., Davis, S.A. and Olaya, G. (2000) Resistance to fungicides in the QoI-STAR cross-resistance group: current perspectives. In: *Proceedings British Crop Protection Conference – Pests and Diseases, 2000*. British Crop Protection Council, Farnham, Surrey, UK, pp. 755–762.

Hollomon, D.W. (1978) Competitive ability and ethirimol sensitivity in strains of barley powdery mildew. *Annals of Applied Biology* 90, 195–204.

Hollomon, D.W. (1981) Genetic control of ethirimol resistance in a natural population of *Erysiphe graminis* f. sp. *hordei*. *Phytopathology* 71, 536–540.

Hollomon, D.W. (1984) A laboratory assay to determine the sensitivity of *Rhynchosporium secalis* to the fungicide triadimenol. *Plant Pathology* 33, 65–70.

Hollomon, D.W. and Brent, K.J. (2009) Combating plant diseases – the Darwin connection. *Pest Management Science* 65, 1156–1163.

Holz, B. (1979) Uber eine Resistenzerscheinung von *Botrytis cinerea* an Reben gegen die neuen Kontaktbotrytizide im Gebiet der Mittelmosel. *Weinberg Keller* 26, 18–25.

Huggenberger, F., Collins, M.A. and Skylakakis, G. (1984) Decreased sensitivity of *Sphaerotheca fuliginea* to fenarimol and other ergosterol-biosynthesis inhibitors. *Crop Protection* 3, 137–149.

Josepovits, G. and Dobrovolszky, A. (1985) A novel mathematical approach to the prevention of fungicide resistance. *Pesticide Science* 16, 17–22.

Kable, P.F. and Jeffery, H. (1980) Selection for tolerance in organisms exposed to sprays of biocide mixtures: a theoretical model. *Phytopathology* 70, 8–12.

Kaku, K., Takagaki, M., Shimizu, M. and Nagayama, K. (2003) Diagnosis of dehydratase inhibitors in melanin biosynthesis inhibitor (MBI-D) resistance by primer-introduced restriction enzyme analysis in scytalone dehydratase gene of *Magnaporthe grisea*. *Pest Management Science* 59, 843–846.

Katan, T. and Bashi, E. (1981) Resistance to metalaxyl in isolates of *Pseudoperonospora cubensis*, the downy mildew pathogen of cucurbits. *Plant Disease* 65, 798–800.

Kato, T. (1988) Resistance experiences in Japan. In: Delp, C.J. (ed.) *Fungicide Resistance in North America*. American Phytopathological Society, St Paul, Minnesota, pp. 16–18.

Kendall, S., Hollomon, D.W., Ishii, H. and Heaney, S.P. (1993) Characterisation of benzimidazole-resistant strains of *Rhynchosporium secalis*. *Pesticide Science* 40, 175–181.

Kuck, K.-H. (2005) Fungicide resistance management in a new regulatory environment. In: Dehne, H.-W., Gisi, U., Kuck, K.-H., Russell, P.E. and Lyr, H. (eds) *Modern Fungicides and Antifungal Compounds IV*. British Crop Protection Council, Alton, UK, pp. 35–43.

Kuiper, J. (1965) Failure of hexachlorobenzene to control common bunt of wheat. *Nature* 206, 1219–1220.

Leroux, P. and Walker, A.S. (2011) Multiple mechanisms account for resistance to sterol demethylation inhibitors in field isolates of *Mycosphaerella graminicola*. *Pest Management Science* 67, 44–59.

Leroux, P., Albertini, C., Arnold, A. and Gredt, M. (2003) Le pietin verse: caracteristiques et distribution des souches de *Tapesia acuformis* et *Tapesia yalundae* resistantes aux fongicides. *Phytoma* 557, 8–13.

Leroux, P., Gredt, M., Walker, A.S., Monard, J.M. and Caron, D. (2005a) Resistance of the wheat leaf blotch pathogen *Septoria tritici* to fungicides in France. In: Dehne, H.-W., Gisi, U., Kuck, K.-H., Russell, P.E. and Lyr, H. (eds) *Modern Fungicides and Antifungal Compounds IV*. BCPC, Alton, UK, pp. 115–124.

Leroux, P., Gredt, M., Walker, A.S. and Panon, A.L. (2005b) Evolution of resistance of *Botrytis cinerea* to fungicides. In: Dehne, H.-W., Gisi, U., Kuck, K.-H., Russell, P.E. and Lyr, H. (eds) *Modern Fungicides and Antifungal Compounds IV*. BCPC, Alton, UK, pp. 133–143.

Locke, T. (1986) Current incidence in the UK of fungicide resistance in pathogens of cereals. In: *Proceedings 1986 British Crop Protection Conference, Pests and Diseases*. British Crop Protection Council, Thornton Heath, Surrey, UK, pp. 781–786.

Lorenz, G. (1988) Dicarboximide fungicides: history of resistance development and monitoring methods. In: Delp, C.J. (ed.) *Fungicide Resistance in North America*. American Phytopathological Society, St Paul, Minnesota, pp. 45–51.

Lorenz, G. and Eichhorn, K.W. (1980) Vorkommen und Verbreitung der Resistenz von *Botrytis cinerea* gegen Dicarboximid-Fungizide im Anbaugebiet der Rheinpfalz. *Die Weinwissenschaft* 35, 199–210.

Malathrakis, N.E. (1980) Control of downy mildew of cucumber by systemic and non-systemic fungicides. In: *Proceedings of the 5th Congress of the Mediterranean Phytopathological Union, Patras, Greece*. Hellenic Phytopathological Society, Athens, pp. 145–146.

Millardet, P.M.A. (1885) Traitement du mildiou par le melange de sulfate de cuivre et de chaux. *Journal d'Agriculture Pratique* 49, 513–516.

Misato, T., Ko, K. and Yamaguchi, I. (1977) Use of antibiotics in agriculture. *Advances in Applied Microbiology* 21, 53–86.

Noble, M., Maggarvie, Q.D., Hams, A.F. and Leaf, L.L. (1966) Resistance to mercury of *Pyrenophora avenae* in Scottish seed oats. *Plant Pathology* 15, 23–28.

Oerke, E.-C. (1996) The impact of diseases and of disease control on crop production. In: Lyr, H., Russell, P.E. and Sisler, H.D. (eds) *Modern Fungicides and Antifungal Compounds*. Intercept, Andover, UK, pp. 17–24.

Pappas, A.C. (1980) Effectiveness of metalaxyl and phosetyl-Al against *Pseudoperonospora cubensis* (Berk. & Curt.) Rostow isolates from cucumbers. *Proceedings of the 5th Congress of the Mediterranean Phytopathological Union, Patras, Greece*, pp. 146–148.

Prevost, B. (1807) *Memoire sur la Cause Immediate de la Carie ou Charbon des Bles, et de Plusieurs Autres Maladies des Plantes, et sur les Preservatifs de la Carie*. Bernard, Paris.

Reuveni, M., Eyal, H. and Cohen, Y. (1980). Development of resistance to metalaxyl in *Pseudoperonospora cubensis*. *Plant Disease* 64, 1108–1109.

Riehm, E. (1913) Prufung einiger mittel zur bekampfung des steinbrandes. *Mitteiling der Kaiserlich Biologischen Anstalt fur Land- u Forstwirtschaft* 14, 8–9.

Russell, P.E. (2003) *Sensitivity Baselines in Fungicide Resistance Research and Management*. FRAC Monograph No. 3. CropLife International, Brussels.

Schroeder, W.T. and Provvidenti, R. (1969) Resistance to benomyl in powdery mildew of cucurbits. *Plant Disease Reporter* 53, 271–275.

Shephard, M.C., Bent, K.J., Woolner, M. and Cole, A.M. (1975) Sensitivity to ethirimol of powdery mildew from UK barley crops. In: *Proceedings of 8th British Crop Protection Conference*. British Crop Protection Council, London, pp. 59–60.

Smith, C.M. (1988) History of benzimidazole use and resistance. In: Delp, C.J. (ed.) *Fungicide Resistance in North America*. American Phytopathological Society, St Paul, Minnesota.

Stanis, V.F. and Jones, A.L. (1985). Reduced sensitivity to sterol-inhibiting fungicides in field isolates of *Venturia inaequalis*. *Phytopathology* 75, 1098–1101.

Staub, T. (1994) Early experiences with phenylamide resistance and lessons for continued successful use. In: Heaney, S., Slawson, D., Hollomon, D.W., Smith, M., Russell, P.E. and Parry, D.W. (eds) *Fungicide Resistance*. British Crop Protection Council, Farnham, Surrey, UK, pp. 131–138.

Staub, T., Dahmen, H., Urech, P.A. and Schwinn, F. (1979) Failure to select for *in vivo* resistance in *Phytophthora infestans* to acylalanine fungicides. *Plant Disease Reporter* 64, 385–389.

Szkolnik, M. and Gilpatrick, J.D. (1969) Apparent resistance of *Venturia inaequalis* in New York apple orchards. *Plant Disease Reporter* 53, 861–84.

Thind, T.S., Clerjeau, M. and Olivier, J.M. 1986) First observations on the resistance in *Venturia inaequalis* and *Guignaria bidwellii* to ergosterol-biosynthesis inhibitors in France. In: *Proceedings of 1986 British Crop Protection Conference – Pests and Diseases*. British Crop Protection Council, Thornton Heath, Surrey, UK, pp. 491–498.

Tillet, M. (1755) Dissertation sur la cause qui corrompt et noircit les grains de ble dans les epis; et sur les moyens de prevenir ces accidents. P. Bruin, Bordeaux, France.

Wain, R.L. and Carter, G.A. (1977) Historical aspects. In: Marsh, R.W. (ed.) *Systemic Fungicides*. Longman, London, pp. 6–31.

Wicks, T.G., Hall, B. and Pezzaniti, P. (1994) Fungicidal control of metalaxyl-insensitive strains of *Bremia lactucae* on lettuce. *Crop Protection* 13, 617–624.

Wolfe, M.S. and Dinoor, A. (1973) Pathogen response to fungicide use. In: *Proceedings 7th British Insecticide and Fungicide Conference*. British Crop Protection Council, London, pp. 813–822.

Part II

Status, Detection and Management of Resistance

2 Resistance in *Venturia nashicola* to Benzimidazoles and Sterol Demethylation Inhibitors

Hideo Ishii

National Institute for Agro-Environmental Sciences, Tsukuba, Ibaraki, Japan

2.1 Introduction

Scab caused by *Venturia nashicola* is the most important disease on Asian pears such as Japanese pear (*Pyrus pylifolia* var. *culta*) and Chinese pear (*Pyrus ussuriensis* and others). Control of this disease is conducted mainly by spray applications of chemical fungicides, about 15 times a year in Japan, as scab-resistant commercial cultivars of pear are very few. Such intensive fungicide applications have resulted in the loss or decrease of control efficacy by the two major groups of fungicides, benzimidazoles and sterol demethylation inhibitors (DMIs).

2.2 Resistance to Benzimidazole Fungicides

In the early 1970s, the two benzimidazole fungicides, thiophanate-methyl and benomyl, were sprayed intensively, around ten times a year, and were highly effective for the control of pear scab in Japan. In 1975, however, growers claimed heavy occurrence of the disease in several prefectures. Shortly after, isolates of the pear scab fungus *V. nashicola*, cross-resistant to both fungicides, were frequently detected from pear orchards and their role in

control failure by fungicides was elucidated (Ishii and Yamaguchi, 1977).

Changes in the proportion of benzimidazole-resistant isolates in populations of *V. nashicola* were monitored in pear orchards. Successive applications of thiophanate-methyl rapidly increased the level of resistance in the populations. Similar results were obtained when the application of thiophanate-methyl was combined with an unrelated dithiocarbamate fungicide, indicating that strategies such as alternating or mixing benzimidazoles with an unrelated fungicide would only delay, not avoid, the build-up of resistance (Ishii *et al.*, 1985).

Methods for testing benzimidazole sensitivity

Mycelial growth of single-spore or mass-spore isolates was examined on potato dextrose agar (PDA) plates amended with benzimidazole fungicides. Although not registered in Japan, carbendazim (formerly referred to as MBC) was employed for the test, as this fungicide was known to be the active degradation component of thiophanate-methyl and benomyl (Ishii, 2003). When thiophanate-methyl was used as an alternative, PDA generally was amended with this fungicide before autoclaving to accelerate the conversion to carbendazim. Using this method, benzimidazole-resistant isolates of *V. nashicola* were divided into three groups, i.e. highly resistant (HR, MIC of carbendazim > 100 mg/l), intermediately resistant (IR, 100 mg/l ≥ MIC > 10 mg/l) and weakly resistant (WR, 10 mg/l ≥ MIC > 1 mg/l), according to the difference in resistance level and separate from sensitive (S, 1 mg/l ≥ MIC) isolates (Ishii *et al.*, 1984).

However, mycelial growth of *V. nashicola* is very slow on culture medium. To identify benzimidazole resistance rapidly, a simple method, named the 'germ tube septum method', was proposed (Umemoto and Nagai, 1979), and this was suggested to be useful for distinguishing between resistant and sensitive conidia, whether or not the conidial germ tubes septate on the media supplemented with fungicide. Germ tube elongation of sensitive conidia was also remarkably suppressed by thiophanate-methyl or benomyl, and high correlation was observed between germ tube length and the number of septa formed in the tube (Ishii and Yamaguchi, 1981). On potted pear trees pretreated with fungicide, lowered control efficacy of thiophanate-methyl was recorded when not only highly benzimidazole-resistant and intermediately resistant isolates but also weakly resistant ones were inoculated, indicating that weakly resistant isolates were not to be disregarded (Table 2.1).

Table 2.1. Control efficacy of thiophanate-methyl against pear scab fungus in inoculation tests.

Ratio of isolates	Protection (%) of thiophanate-methyl[a] at:	
	IR:S	WR:S
1:0	−62.0	−11.0
3:1	−20.0	39.0
1:1	21.0	35.0
1:3	20.0	45.0
0:1	100	100

Notes: [a]Applied at 467 mg/l. IR, intermediately benzimidazole-resistant; WR, weakly resistant; S, sensitive.

Stability of benzimidazole resistance

It is likely that fungicide-resistant strains of *V. nashicola* are distributed irregularly in pear orchards, as ascospores as well as conidia of this fungus are not disseminated over long distances. It is thus desirable to sample isolates from several designated trees in orchards when distribution and/or fluctuation of resistant strains are/is examined. Based on this, changes in the proportion of benzimidazole-resistant isolates in fungal populations were monitored.

When the application of benzimidazoles was stopped and other fungicides alone were applied in orchards where highly resistant isolates predominated, the proportion of highly resistant isolates gradually decreased and that of intermediately resistant, weakly resistant and sensitive isolates increased (Fig. 2.1; Ishii *et al.*, 1985). This phenomenon was thought to be an example of genetic homeostasis in microbial populations; however, resistant isolates comprised 80% of the total, 5 years after the last application of benzimidazoles, demonstrating well that benzimidazole resistance tended to persist for a long time. Ishizaki *et al.* (1983) also reported persistence of high benzimidazole resistance in *V. nashicola*.

However, differences in the fitness of resistant and sensitive fungal populations in individual regions will probably influence resistance frequencies in the absence of selection pressure exerted by fungicides. At present, benzimidazole fungicides are used occasionally for the control of other pear diseases such as fruit core rot caused by *Phomopsis fukusii*. This is probably the major reason why highly benzimidazole-resistant isolates of pear scab fungus are still widely distributed in the field. Resistant isolates have also been found recently in several provinces in China.

Genetics of benzimidazole resistance

Because benzimidazole resistance of *V. nashicola* was stable even when resistant isolates were spray-inoculated to pear seedlings

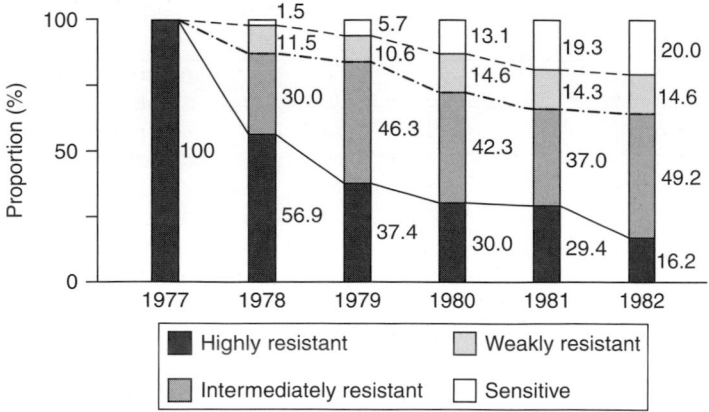

Fig. 2.1. Decline of fungicide resistance in an orchard where benzimidazole fungicides were replaced by other fungicides in 1977. Reprinted from Ishii *et al.* (1985) *Plant Pathology* 34, 363–368.

repeatedly or subcultured on a culture medium for many years, resistance was suspected to be inheritable. At first, eight monoascosporic isolates were obtained from individual mature asci on overwintered diseased leaves in the spring, and unordered tetrad analysis from these natural crosses was carried out. Of eight single-ascospore isolates, four were benzimidazole resistant and four were benzimidazole sensitive, suggesting that naturally occurring resistance was controlled by a major gene (Ishii and Yanase, 1983).

To confirm this, methods for artificially crossing heterothallic *V. nashicola* were developed. Mature ascospores were produced when mycelial suspensions of two compatible isolates were mixed with the culture medium and incubated at 5°C in the dark for about 6 months. The inheritance of benzimidazole resistance was demonstrated experimentally using this technique. Progeny of the cross between a resistant isolate and a sensitive one segregated at a 1:1 ratio, indicating that resistance was under the control of a major gene (Table 2.2; Ishii *et al.*, 1984). Additional ascospore analysis suggests that manifestation of three levels of benzimidazole resistance (i.e. high, intermediate and weak) is due to three allelic mutations in a single gene on a chromosome, and that each level is controlled by one of the multiple alleles.

Biochemical mechanism of benzimidazole resistance

The primary mode of action of benzimidazole fungicides is through specific binding to the β-tubulin subunit of fungal tubulin, which consequently interferes with microtubule assembly, which in turn is essential for numerous cellular processes such as mitosis and cytoskeleton formation (Davidse, 1986). In the presence of benzimidazoles, conidial germ tubes of benzimidazole-resistant isolates of *V. nashicola* elongated normally because the fungicide did not affect mitosis. With fluorescent 4,6-diaminophenylindole (DAPI) staining, each nucleus was arranged regularly in individual cells. In contrast, in benzimidazole-sensitive isolates, a nucleus-like structure that was not readily visible was scattered randomly in cells.

Decreased binding of the fungicide to tubulin-like proteins was generally involved in benzimidazole resistance (Davidse, 1986; Davides and Ishii, 1995). In *V. nashicola*, *Botrytis cinerea* (grey mould) and *Gibberella fujikuroi* (Bakanae disease on rice), the binding of ^{14}C-carbendazim to tubulin-like proteins was much lower in benzimidazole-resistant isolates than in sensitive isolates (Fig. 2.2; Ishii and Davidse, 1986; Ishii and Takeda, 1989), suggesting that decreased affinity of the fungicide with the target

Table 2.2. Random ascospore analysis of crosses between carbendazim-resistant and -sensitive isolates of *Venturia nashicola*.

Cross (isolate)	Number of progeny isolates[a]				Chi-square values (1:1)[b]
	HR	IR	WR	S	
HR × S (JS-135) (JS-115)	58	0	0	42	2.56
IR × S (JS- 41) (JS-4)	0	59	0	41	3.24
WR × S (JS- 27) (CS-1)	0	0	55	45	1.00
HR × HR (JS-111) (JS-140)	100	0	0	0	
HR × IR (JS-111) (CS-22)	57	43	0	0	1.96
HR × WR (JS-11) (JS-77)	48	1	51	0	0.09
IR × WR (JS-41) (JS-77)	1	57	42	0	2.27
WR × WR (JS-132) (JS-27)	0	0	100	0	
S × S (CS-1) (CS-11)	0	0	0	100	

Notes: [a]HR, highly resistant; IR, intermediately resistant; WR, weakly resistant; S, sensitive; [b]expected value at $P = 0.05$ is 3.84. Reprinted from Ishii *et al.* (1984) *Mededelingen Faculteit van de Landbouwwetenschappen, Rijksuniversiteit, Gent* 49/2a, 163–172.

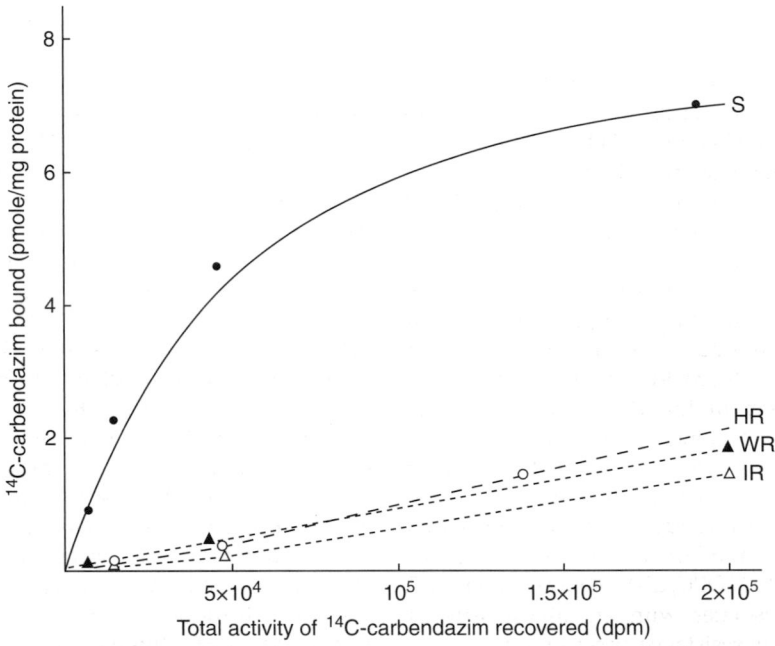

Fig. 2.2. Binding of ^{14}C-carbendazim to cell-free mycelial extracts from isolates of *Venturia nashicola* showing different levels of resistance to this fungicide. HR, highly resistant; IR, intermediately resistant; WR, weakly resistant; S, sensitive. Reprinted from Ishii and Davidse (1986) *Proceedings of 1986 British Crop Protection Conference – Pests and Diseases*, Volume 2, pp. 567–573.

protein (i.e. modification in the target site) was a major factor in the resistance. Furthermore, the binding activity of the fungicide with extracts of isolates with high, intermediate or weak resistance did not differ. As mentioned above, the levels of benzimidazole resistance were shown to be controlled by one of the multiple alleles. It was likely that allelic mutations changed the structure of the fungicide binding site on the β-tubulin molecule. This report was the first on the mode of inheritance of benzimidazole resistance related to the biochemical characterization of resistance using field isolates of phytopathogenic fungi.

Negative cross-resistance to N-phenylcarbamate and other groups of compounds

Based on the important discovery of Leroux and Gredt (1979) that benzimidazole-resistant isolates of *B. cinerea* and *Penicillium expansum* showed negative cross-resistance to the herbicidal N-phenylcarbamate compounds barban, chlorpropham and chlorbufam, the Sumitomo Chemical Co Ltd, Japan, found the N-phenylcarbamate compounds, methyl N-(3,5-dichlorophenyl)carbamate (MDPC) and diethofencarb (Suzuki *et al.*, 1984; Kato, 1988; Takahashi *et al.*, 1988). Diethofencarb was developed commercially and its mixture with thiophanate-methyl has been used subsequently for the control of grey mould and other fungal isolates with high resistance to benzimidazoles.

Meanwhile, the Nippon Soda Co Ltd, Japan, was also considering the development of N-phenylformamidoxime compounds such as N-(3,5-dichloro-4-propynyloxyphenyl)-N-methoxyformamidine (DCPF) (Nakata *et al.*, 1987, 1992), although eventually these compounds were not commercialized. Most isolates of *V. nashicola*, highly resistant to benzimidazole fungicides, exhibited increased sensitivity, in other words negative cross-resistance, to MDPC, diethofencarb and DCPF. However, isolates with an intermediate or weak level of resistance or sensitivity to benzimidazoles were inherently insensitive to these compounds (Ishii and van Raak, 1988). As intermediately or weakly benzimidazole-resistant isolates were widely distributed in the field populations of *V. nashicola*, diethofencarb was not registered for pear scab control (Ishii *et al.*, 1992).

Benzimidazole-resistant isolates of *V. nashicola* also showed negative cross-resistance to the ring-substituted N-phenylanilines, N-(3-chlorophenyl)aniline (MC-1) and N-(3,5-

dichlorophenyl)aniline (MC-2). The increased sensitivity to MC-1 and MC-2 was observed specifically in highly carbendazim-resistant isolates, but not in intermediately resistant and sensitive isolates (Ishii *et al.*, 1995). In contrast, the β-tubulin inhibitor, rhizoxin, inhibited mycelial growth of *V. nashicola* and other fungi, regardless of the presence and the level of benzimidazole resistance. Interestingly, this macrolide antibiotic is also active against fungal species which are naturally insensitive to benzimidazole fungicides, such as *Phytophthora, Pythium* and *Alternaria* spp., indicating that the binding site of rhizoxin differs from that of benzimidazole fungicides (Ishii *et al.*, 1990a; Ishii, 1992).

Mechanism of negative cross-resistance

Also important to determine was whether the increased sensitivity to N-phenylcarbamate, N-phenylformamidoxime and N-phenylaniline compounds was determined by pleiotropism of the same allele that coded for high benzimidazole resistance. Inheritance of sensitivity to these compounds was then examined by random ascospore analysis. A segregation ratio of 1:1 was obtained from the cross between MDPC-sensitive and -resistant (insensitive) isolates. In an allelism test, no MDPC-resistant progenies appeared from the cross between two MDPC-sensitive isolates, and no MDPC-sensitive progenies resulted from the cross between two MDPC-resistant isolates. The increased sensitivity of *V. nashicola* isolates to these compounds was controlled by a single major gene (Ishii and van Raak, 1988; Ishii *et al.*, 1992). In random ascospore progeny from *V. nashicola* crosses, those progeny with a high level of carbendazim resistance and MC-2 sensitivity always segregated together, indicating that the sensitivity to MC-2 was controlled by a single gene which was either identical or linked very closely to one conferring high-level resistance to carbendazim (Ishii *et al.*, 1995).

The binding of ^{14}C-DCPF to cell-free protein derived from highly benzimidazole-resistant field isolates of *V. nashicola* was higher than to protein from benzimidazole-sensitive isolates or from isolates with an intermediate or low level of benzimidazole resistance, all of which were insensitive to DCPF (Fig. 2.3; Ishii and Takeda, 1989). The mechanism of negative cross-resistance between benzimidazoles and N-phenylformamidoximes was thus clarified, and the molecular structure of N-phenylformamidoximes was assumed to fit well to the altered binding site for β-tubulin molecules in the highly benzimidazole-resistant isolates. Using a benzimidazole-resistant mutant of the model fungus *Neurospora crassa*, Fujimura *et al.* (1992) strongly suggested that diethofencarb was selectively toxic to the mutant by binding to the tubulin.

Molecular mechanisms of benzimidazole resistance and gene-level detection of resistance

Most benzimidazole-resistant isolates collected from the field had codon changes at position 198 or 200 in β-tubulin genes (Table 2.3; Ishii, 2002). For *V. nashicola*, substitution of glutamic acid (GAG) at codon 198 by alanine (GCG) resulted in high resistance to benzimidazoles (Ishii *et al.*, 1997). This mutation seemed to be further involved in negative cross-resistance to diethofencarb and DCPF, although biochemical evidence could not be obtained for diethofencarb. In isolates with intermediate resistance to benzimidazoles, the amino acid at codon 200 was altered from phenylalanine (TTC) to tyrosine (TAC). These single-base substitutions in the targeted β-tubulin gene possibly change the physico-chemical nature of the binding site for benzimidazoles. Actually, decreased binding of ^{14}C-carbendazim to cell-free mycelial extracts from resistant isolates has been demonstrated as already described. In contrast, mutations were not found at codon 198 nor 200 in weakly benzimidazole-resistant isolates.

Using the nucleotide sequence data for the β-tubulin gene of the slow-growing fungus *V. nashicola*, enzyme restriction sites were

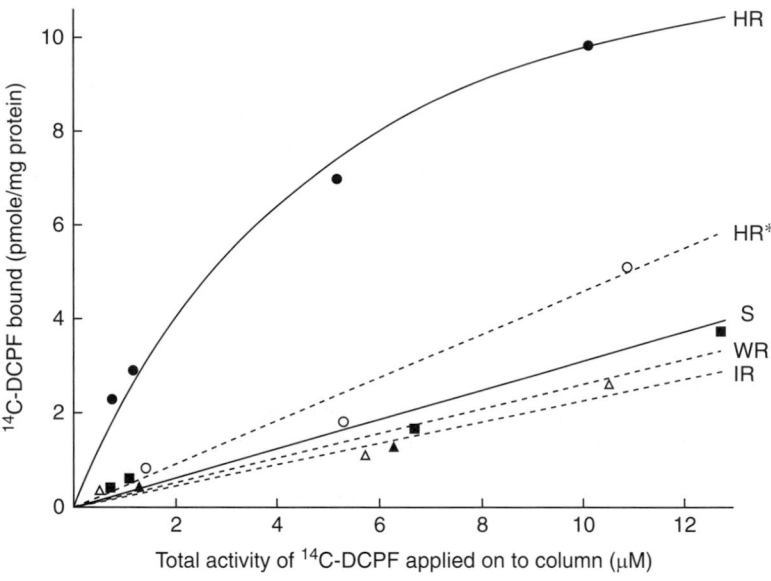

Fig. 2.3. Binding of ^{14}C-DCPF in cell-free mycelial extracts of *Venturia nashicola* isolates with different levels of resistance to carbendazim and DCPF. HR, highly resistant to carbendazim and sensitive to DCPF; HR*, highly resistant to carbendazim and resistant to DCPF; IR, intermediately resistant to carbendazim and insensitive to DCPF; WR, weakly resistant to carbendazim and insensitive to DCPF; S, sensitive to carbendazim and insensitive to DCPF. Reprinted from Ishii and Takeda (1989) *Netherlands Journal of Plant Pathology* 95 (Supplement 1), 99–108.

identified (Fig. 2.4; Ishii *et al.*, 1997). The base change (GAG to GCG) in codon 198 created a *Tha*I restriction site together with a *Hga*I site. Two restriction fragments were generated from the PCR (polymerase chain reaction) amplified DNA samples bearing such a mutation after treatment with each enzyme. Highly benzimidazole-resistant isolates, which showed negative cross-resistance to diethofencarb, were clearly identified by this PCR-RFLP (restriction fragment length polymorphism) method.

Allele-specific PCR (ASPCR) primers were also designed based on the sequence differences (Table 2.4; Ishii *et al.*, 1997) and were used to detect benzimidazole resistance. Mutations in codon 198 and 200 were both identified by ASPCR. Furthermore, with *V. nashicola*, we first introduced single-strand

DNA conformation polymorphism (SSCP) analysis for the diagnosis of fungicide resistance. When denatured, single-strand DNA carrying one base difference can be separated on a gel according to the distinct patterns of electrophoresis. PCR products of the β-tubulin gene fragment were separated using a capillary gel electrophoresis and fluorescence detection system. A mutation at either codon 198 or 200 was identified by the difference in migration time of the DNA fragment (Ishii *et al.*, 1997). When the mixture of DNA samples from both highly benzimidazole-resistant and -sensitive isolates was applied on to a gel, each sample was separated as a result of the different conformation of each DNA strand (Fig. 2.5).

Table 2.3. Point mutations and deduced amino acid substitutions in the β-tubulin gene of *Venturia nashicola* with resistance to benzimidazole fungicides.

Sensitivity to: benzimidazoles/ diethofencarb	Nucleotides and amino acids in position:	
	198	200
S/LS[a]	GAG(Glu)	TTC(Phe)
WR/LS	GAG(Glu)	TTC(Phe)
IR/LS	GAG(Glu)	TAC(Tyr)
HR/S	GCG(Ala)	TTC(Phe)
HR/HR	AAG(Lys)	TTC(Phe)

Notes: [a]S, sensitive; WR, weakly resistant; IR, intermediately resistant; HR, highly resistant; LS, less sensitive.

2.3 Resistance to DMI Fungicides

DMI use and outbreak of resistance

Two sterol demethylation inhibiting (DMI) fungicides, triflumizole and bitertanol, were first introduced in 1986 to control Japanese pear scab after benzimidazole fungicides lost their efficacy against this disease. Since then, many other DMI fungicides have been registered and these fungicides have become essential agents to control the disease. Their performance has been maintained in the field and they have also played an important role in diminishing the application of other conventional fungicides.

Benzimidazole-sensitive and weakly resistant isolates:
5′···GAC GAG ACCTTCTGCATT···GGTCTCCGCTGT···3′
Intermeadiately benzimidazole-resistant isolates:
5′···GACGAGACCT A CTGCATT···GGTCTCCGCTGT···3′
Highly benzimidazole-resistant and diethofencarb-sensitive isolates:
5′···GAC GCG ACCTTCTGCATT···GGTCTCCGCTGT···3′

↑

*Tha*I (or *Hga*I)
Highly benzimidazole-resistant and diethofencarb-resistant isolates:
5′···GAC AAG ACCTTCTGCATT···GGTCTCCGCTGT···3′

Fig. 2.4. Enzyme restriction site created by a single point mutation in β-tubulin gene of highly benzimidazole-resistant and diethofencarb-sensitive isolates of *Venturia nashicola*. Nucleotide sequences corresponding to position 198 of β-tubulin protein are highlighted by shaded boxes. The site of a point mutation is underlined. Arrow indicates the restriction site of the enzyme *Tha*I or *Hga*I.

Table 2.4. Nucleotide sequence of primers used for allele-specific PCR to identify *Venturia nashicola* isolates differing sensitivity to benzimidazoles and diethofencarb.[a]

Forward primer:
S/LS (WR/LS)	5'-ACCAGCTTGTCGAGAATTCGGACGA-3'
(IR/LS)	5'-TTGTCGAGAATTCGGACGAGACCTA-3'
(HR/S)	5'-ACCAGCTTGTCGAGAATTCGGACGC-3'
(HR/HR)	5'-CACCAGCTTGTCGAGAATTCGGACA-3'

Reverse primer:
B	5'-AGAGGAGCAAAACCGACC-3'

Notes: [a]S, sensitive; WR, weakly resistant; IR, intermediately resistant; HR, highly resistant; LS, less sensitive. Refer to Table 2.3. Site of point mutations is underlined.

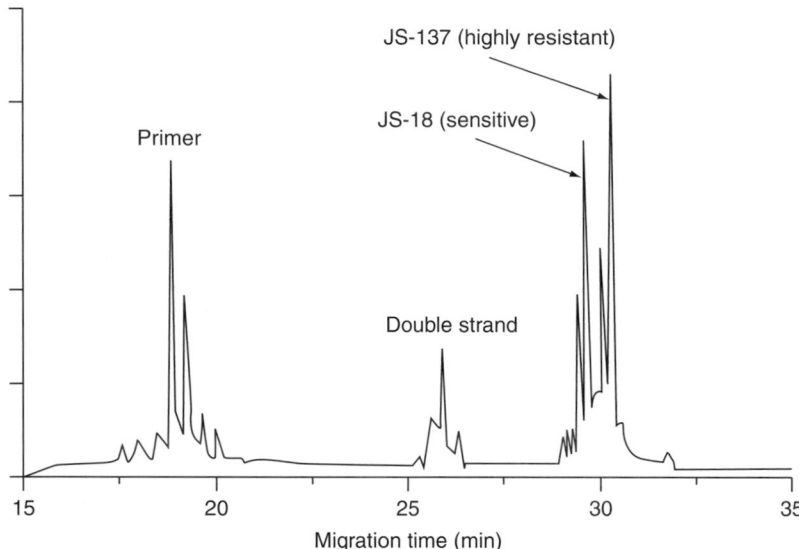

Fig. 2.5. Detection of one base change in the β-tubulin gene of *Venturia nashicola* by SSCP analysis. Reprinted from Kendall and Hollomon (1998) in *Fungicidal Activity – Chemical and Biological Approaches to Plant Protection*, pp. 87–108.

Because resistance problems with *V. inaequalis*, the causal fungus of apple scab disease, had already occurred in Europe (Thind *et al.*, 1986), strategies for resistance management, such as limiting DMI fungicide applications to a maximum of three times per year in a mixture or alternative use with other effective fungicides having a distinct mode of action, were advised and the field performance of DMI fungicides has therefore been maintained.

Results from field trials also indicated that any shifts to lower DMI sensitivity would not proceed very quickly, and it was thought the potential problems of DMI resistance could perhaps be avoided and the long-term effectiveness of DMI fungicides maintained by applying them sparingly (Ishii *et al.*, 1990b). However, in the spring of 2005, heavy attack of young pear leaves or fruit with scab was found in Fukuoka Prefecture, Kyushu, Japan. Application of DMI fungicides reached four to six times per year and these fungicides were often sprayed successively in the areas where the control failure had occurred. In Korea, DMI fungicides were used more extensively, resulting in the serious loss of fungicide efficacy against pear scab control (Yeo *et al.*, 2006).

Demonstration of DMI resistance

Pear leaves bearing abundant conidia of *V. nashicola* were collected from orchards where field performance of DMI fungicides decreased and scab inoculation tests were conducted using potted pear trees sprayed previously with fungicides such as fenarimol, hexaconazole and difenoconazole. The results showed the first occurrence of DMI-resistant isolates of *V. nashicola* in Japan (Ishii and Kikuhara, 2007). Resistant isolates were widely distributed in Fukuoka Prefecture and were also detected in some other prefectures. Difenoconazole still exhibited a high level of efficacy and complete disease control was maintained, as inherent activity of this fungicide seemed to be superior to other DMI fungicides. Subsequently, however, fungal isolates less sensitive to difenoconazole have also been found in fungus inoculation tests, although the performance of this fungicide does not seem to have decreased yet in the field.

Methods for testing DMI fungicide sensitivity of *V. nashicola*

In vitro methods for testing DMI sensitivity of *V. nashicola* were proposed (Ishii *et al.*, 1990b). For triflumizole or bitertanol, sensitivity was determined in mycelial growth tests and comparisons were based on EC_{50} rather than MIC values. Measurements of germ tube length rather than observation of conidial germination were needed to assess sensitivity to these fungicides as a simple method. Baseline sensitivity data on fenarimol were obtained using fungal isolates which were collected from non-DMI-treated trees in Japan, and even in China. In these two cases, mean EC_{50} values of fenarimol for mycelial growth on PDA medium were 0.199 mg/l and 0.12 mg/l, respectively. Based on these values, fenarimol sensitivity was monitored in the 1990s and a shift to lower sensitivity was observed in many isolates derived from DMI-treated pear orchards (Tomita and Ishii, 1998). To examine whether the efficacy of fenarimol had also declined, spore samples were collected from commercial orchards from which less fenarimol-sensitive isolates had been found in tests *in vitro*. However, in inoculation tests on pear seedlings, fenarimol still showed adequate control, indicating that the performance of fenarimol was still maintained in the field at that time.

Meanwhile, we found difficulties in monitoring the DMI sensitivity of this fungus. Recent studies have indicated that methods such as mycelial growth tests on DMI-amended PDA medium are not appropriate to determine DMI resistance development precisely in *V. nashicola*. Fenarimol sensitivity of single-spore isolates tended to return to sensitivity when stored and/or subcultured on fungicide-free PDA medium. Therefore, it was doubtful whether our monitoring studies reflected the real situation of the DMI sensitivity of the fungal populations in the field. And this phenomenon, recovery of DMI sensitivity, probably is related to the mechanism of resistance in this slow-growing fungus. Elucidation of the resistance mechanism and development of molecular methods will be useful, if available, to identify resistant isolates rapidly and precisely in the future.

Analysis of sterol demethylase gene

Nucleotide sequences of the gene *CYP51*, encoding sterol 14α-demethylase, the target enzyme of DMIs, were analysed using *V. nashicola* (Cools *et al.*, 2002). Subsequently, point mutations were found in this gene amplified from some but not all fenarimol-resistant isolates; however, it is not yet known whether these mutations are the cause of resistance. Mechanisms other than point mutations in *CYP51* probably contribute to reducing DMI sensitivity in *V. nashicola*.

Strategies for combating DMI resistance of *V. nashicola*

As alternatives to DMI fungicides, QoI fungicides (inhibitors of mitochondrial respiration at Qo site of cytochrome *bc1* enzyme complex) such as kresoxim-methyl and azoxystrobin are commercially available in Japan. A mixture of pyraclostrobin with the SDHI (succinate dehydrogenase inhibitors) fungicide, boscalid, has also been registered for the control of pear scab. Although QoI-resistant strains of *V. nashicola* have not yet been detected in Japan, they have been reported in the close fungus *Venturia pirina*, the cause of scab on European pear overseas (Pitman *et al.*, 2007). Furthermore, strains of *Alternaria alternata* doubly resistant to QoI and SDHI fungicides are widely distributed on pistachio in California, USA (Avenot

et al., 2008), clearly indicating that these fungicides possess high risk for resistance development in fungi. Their application needs to be reduced to as little as possible.

DMI fungicides, difenoconazole in particular, are still effective for the control of pear scab disease. Therefore, difenoconazole, as well as other representative DMIs, for example hexaconazole, imibenzonazole and fenbuconazole, are used as a solo or tank mix with other fungicides. Iminoctadine-albesilate or cyprodinil has been recommended as a partner fungicide in the mixture.

References

Avenot, H., Morgan, D.P. and Michailides, T.J. (2008) Resistance to pyraclostrobin, boscalid and multiple resistance to Pristine® (pyraclostrobin + boscalid) fungicide in *Alternaria alternata* causing alternaria late blight of pistachios in California. *Plant Pathology* 57, 135–140.

Cools, H.J., Ishii, H., Butters, J.A. and Hollomon, D.W. (2002) Cloning and sequence analysis of the eburicol 14α-demethylase encoding gene (*CYP51*) from the Japanese pear scab fungus *Venturia nashicola*. *Journal of Phytopathology* 150, 444–450.

Davidse, L.C. (1986) Benzimidazole fungicides: mechanism of action and biological impact. *Annual Review of Phytopathology* 24, 43–65.

Davidse, L.C. and Ishii, H. (1995) Biochemical and molecular aspects of the mechanisms of action of benzimidazoles, *N*-phenylcarbamates and *N*-phenylformamidoximes and the mechanisms of resistance to these compounds in fungi. In: Lyr, H. (ed.) *Modern Selective Fungicides*. Gustav Fischer Verlag, Jena, Germany, pp. 305–322.

Fujimura, M., Kamakura, T. and Yamaguchi, I. (1992) Action mechanism of diethofencarb to a benzimidazole-resistant mutant in *Neurospora crassa. Journal of Pesticide Science* 17, 237–242.

Ishii, H. (1992) Target sites of tubulin-binding fungicides. In: Köller, W. (ed.) *Target Sites of Fungicide Action*. CRC Press, Boca Raton, Florida, pp. 43–52.

Ishii, H. (2002) DNA-based approaches for diagnosis of fungicide resistance. In: Clark, J.M. and Yamaguchi, I. (eds) *Agrochemical Resistance – Extent, Mechanism, and Detection*. American Chemical Society, Washington, DC, pp. 242–259.

Ishii, H. (2003) Fungicides, tubulin binding compounds. In: Plimmer, J.R., Gammon, D.W. and Ragsdale, N.N. (eds) *Encyclopedia of Agrochemicals* 2. John Wiley & Sons, Hoboken, New Jersey, pp. 640–648.

Ishii, H. and Davidse, L.C. (1986) Decreased binding of carbendazim to cellular protein from *Venturia nashicola* and its involvement in benzimidazole resistance. In: *Proceedings of 1986 British Crop Protection Conference – Pests and Diseases*, Volume 2. BCPC Publications, Surrey, UK, pp. 567–573.

Ishii, H. and Kikuhara, K. (2007) Occurrence of DMI resistance in *Venturia nashicola*, the scab fungus of Asian pears. In: *Abstracts of the 17th Symposium of Research Committee on Fungicide Resistance*. The Phytopathological Society of Japan, Utsunomiya, Japan, pp. 49–60.

Ishii, H. and Takeda, H. (1989) Differential binding of a *N*-phenylformamidoxime compound in cell-free extracts of benzimidazole-resistant and -sensitive isolates of *Venturia nashicola, Botrytis cinerea* and *Gibberella fujikuroi. Netherlands Journal of Plant Pathology* 95 (Supplement 1), 99–108.

Ishii, H. and van Raak, M. (1988) Inheritance of increased sensitivity to *N*-phenylcarbamates in benzimidazole-resistant *Venturia nashicola. Phytopathology* 78, 695–698.

Ishii, H. and Yamaguchi, A. (1977) Tolerance of *Venturia nashicola* to thiophanate-methyl and benomyl in Japan. *Annals of the Phytopathological Society of Japan* 43, 557–561.

Ishii, H. and Yamaguchi, A. (1981) Resistance of *Venturia nashicola* to thiophanate-methyl and benomyl: existence of weakly resistant isolates and its practical significance. *Annals of the Phytopathological Society of Japan* 47, 528–533.

Ishii, H. and Yanase, H. (1983) Resistance of *Venturia nashicola* to thiophanate-methyl and benomyl: forma-tion of the perfect state in culture and its application to genetic analysis of the resistance. *Annals of the Phytopathological Society of Japan* 49, 153–159.

Ishii, H., Yanase, H. and Dekker, J. (1984) Resistance of *Venturia nashicola* to benzimidazole fungicides. *Mededelingen Faculteit van de Landbouwwetenschappen, Rijksuniversiteit, Gent* 49/2a, 163–172.

Ishii, H., Udagawa, H., Yanase, H. and Yamaguchi, A. (1985) Resistance of *Venturia nashicola* to thiophanate-methyl and benomyl: build-up and decline of resistance in the field. *Plant Pathology* 34, 363–368.

Ishii, H., Takeda, H., Iwasaki, S. and Sato, Z. (1990a) Antifungal activity of the microtubule inhibitor rhizoxin against the benzimidazole-resistant plant pathogenic fungi. *Tagungsbericht Berlin Akademie der Landwirtschaftswissenschaften* 291, 247–252.

Ishii, H., Takeda, H., Nagamatsu, Y. and Nakashima, H. (1990b) Sensitivity of the pear scab fungus (*Venturia nashicola*) to three ergosterol biosynthesis-inhibiting fungicides. *Pesticide Science* 30, 405–413.

Ishii, H., van Raak, M., Inoue, I. and Tomikawa, A. (1992) Limitations in the exploitation of *N*-phenylcarbamates and *N*-phenylformamidoximes to control benzimidazole-resistant *Venturia nashicola* on Japanese pear. *Plant Pathology* 41, 543–553.

Ishii, H., Josepovits, G., Gasztonyi, M. and Miura, T. (1995) Further studies in increased sensitivity to *N*-phenylanilines in benzimidazole-resistant strains of *Botrytis cinerea* and *Venturia nashicola*. *Pesticide Science* 43, 189–193.

Ishii, H., Kamahori, M., Hollomon, D.W. and Narusaka, Y. (1997) DNA-based approaches for diagnosis of benzimidazole resistance in *Venturia nashicola*, the scab fungus of Japanese pear. *Abstracts of the International Conference Resistance '97*. IACR-Rothamsted, Harpenden, UK.

Ishizaki, H., Kohno, M., Tsuchida, M., Umino, M., Hazumi, Y., Katoh, S., *et al.* (1983) Yearly fluctuation in the occurrence of thiophanate-methyl-resistant strains of the pear scab fungus in pear orchards at Karasu Town, Mie Prefecture. *Annals of the Phytopathological Society of Japan* 49, 347–351.

Kato, T. (1988) Negative cross-resistance activity of MDPC and diethofencarb against benzimidazole-resistant fungi. In: Delp, C.J. (ed.) *Fungicide Resistance in North America*. APS Press, St Paul, Minnesota, 40 pp.

Kendalll, S.J. and Hollomon, D.W. (1998) Fungicide resistance. In: Hutson, D. and Miyamoto, J. (eds) *Fungicidal Activity – Chemical and Biological Approaches to Plant Protection*. John Wiley & Sons, Chichester, UK, pp. 87–108.

Leroux, P. and Gredt, M. (1979) Effets du barbane, du chlorbufame, du chlorprophame et du prophame sur diverses souches de *Botrytis cinerea* Pers et de *Penicillium expansum* Link sensibles ou resistantes au carbendazime et au thiabendazole. *Comptes Rendus de l'Académie des Sciences Paris* 289, 691–693.

Nakata, A., Sano, S., Hashimoto, S., Hayakawa, K., Nishikawa, H. and Yasuda, Y. (1987) Negatively cor-related cross-resistance to *N*-phenylformamidoximes in benzimidazole-resistant phytopathogenic fungi. *Annals of the Phytopathological Society of Japan* 53, 659–662.

Nakata, A., Hashimoto, S., Sano, S. and Hayakawa, K. (1992) Fungitoxic properties of *N*- phenylformami-doximes in control of benzimidazole-resistant isolates of fungi. *Journal of Pesticide Science* 17, 243–249.

Pitman, T.L., Elkins, R. and Gulber, W.D. (2007) Detection of strobilurin resistant *Venturia pirina* popula-tions in northern California. *Phytopathology* 97, S93.

Suzuki, K., Kato, T., Takahashi, J. and Kamoshita, K. (1984) Mode of action of *N*-(3,5-dichlorophenyl) carbamate in the benzimidazole-resistant isolate of *Botrytis cinerea*. *Journal of Pesticide Science* 9, 497–501.

Takahashi, J., Nakamura, S., Noguchi, H., Kato, T. and Kamoshita, K. (1988) Fungicidal activity of *N*-phenylcarbamates against benzimidazole resistant fungi. *Journal of Pesticide Science* 13, 63–69.

Thind, T.S., Clerjau, M. and Olivier, J.M. (1986) First observations on resistance in *Venturia inaequalis* and *Guignardia bidwellii* to ergosterol-biosynthesis inhibitors in France. In: *Proceedings of 1986 British Crop Protection Conference – Pests and Diseases*, Volume 2. BCPC Publications, Surrey, UK, pp. 491–498.

Tomita, Y. and Ishii, H. (1998) Reduced sensitivity to fenarimol in Japanese field strains of *Venturia nash-icola*. *Pesticide Science* 54, 150–156.

Umemoto, S. and Nagai, Y. (1979) Simple method for determining benomyl tolerance of *Venturia nashicola*. *Annals of the Phytopathological Society of Japan* 45, 430–435.

Yeo, M.I., Kwon, S.M., Lee, I.K., Choi, Y.M., Park, E.K. and Uhm, J.Y. (2006) Assessment of sensitivity to sterol biosynthesis inhibitors in *Venturia nashicola*, the causal fungus of pear scab collected in Naju and Ulsan. In: *Program and Abstracts of the 2006 Annual Meeting of the Korean Society of Plant Pathology*, pp. 151–152.

3 Fungicide Resistance in *Erysiphe necator* – Monitoring, Detection and Management Strategies

David Ouimette

Crop Disease Management R&D, Dow AgroSciences, Indianapolis, USA

3.1 Introduction

Grape powdery mildew (GPM), caused by the pathogen *Erysiphe necator* (previously named *Uncinula necator* (Schwein) Burrill) is one of the most damaging diseases of grapes throughout the world. Infection of grapevines with powdery mildew results in a reduction of vine growth, yield, fruit quality and winter hardiness (Pearson and Goheen, 1988). For the production of table grapes and for wine-making, only extremely low levels of fruit infection are tolerated. The fungus can infect all green tissues of the host, sending haustoria into the epidermal cells to absorb nutrients. Depending on geographical location, there are two types of overwintering methods which may overlap: (i) as chasmothecia (previously known as cleistothecia), which are produced in late summer or early autumn and result in the release of ascospores beginning the following spring; or (ii) as dormant mycelium in infected buds that leads to the production of 'flag shoots', which are newly emerged shoots highly infected with GPM. Both types of overwintering are responsible for initiating early season disease epidemics when weather conditions are favourable. Abundant spore production and a relatively short latent period (approximately 7–10 days depending on temperature) can result in many disease cycles per season, and hence

high potential for severe disease epidemics. Disease development is favoured by temperatures between 20 and 27°C, but fungal growth can occur between 6 and 32°C. In contrast to most foliar fungal diseases, free water on the leaf surface is not required for infection, but high humidity levels favour disease infection and development. Grape berries are susceptible to fungal infection until the sugar content reaches approximately 8%, but powdery mildew mycelia on infected berries can continue to sporulate until the internal concentration reaches 15%.

3.2 Grape Powdery Mildew Management

Agronomic cultural practices

There are various agronomic cultural practices that can reduce GPM severity within a vineyard, and some of the more important include the following: (i) canopy management that allows for good air circulation (reduced humidity), fungicide spray penetration and sunlight exposure; (ii) avoiding high nitrogen fertilizer applications to deter excess vegetative growth, which can increase humidity within the canopy and favour powdery mildew development; (iii) orienting vineyard rows in the direction of prevailing winds to increase air movement within the canopy and reduce humidity; and (iv) irrigation early in the morning to facilitate the drying of foliage and soil, and hence reducing the relative humidity within the vineyard (Pearson and Goheen, 1988). However, even when utilizing cultural practices to help manage GPM, the high level of control required in commercial vineyards is achieved primarily by the use of chemical solutions such as fungicides.

Chemical control

An overarching disease control strategy should be focused on reducing initial inoculum and protection of foliage from subsequent infection. Preventive fungicide control of GPM is critical, starting early in the season. Fungicide applications prior to flowering and for approximately 4–6 weeks afterwards are the most critical, because during this period berries are most susceptible to GPM infection. When vines are growing rapidly and environmental conditions are favourable for GPM development, a spray interval of 7–10 days is common to protect new growth from infection. If temperatures rise above 35°C, GPM development slows markedly and spray intervals can be increased to 14–21 days. For wine grape production, fungicide applications are not usually necessary after verasion because increasing sugar content in the berries inhibits further GPM development. However, for table grape production it is often necessary to continue fungicide applications after verasion to protect the rachis and pedicel from infection, which can reduce significantly the storage life and quality of the grape cluster.

There are a number of fungicide groups (using nomenclature from the Fungicide Resistance Action Committee [FRAC], www.frac.info/) that are used for GPM management and which will be discussed in more detail. With the exception of sulfur (multi-site/inorganics) and meptyldinocap (uncouplers), all other prominent FRAC groupings used for GPM are prone to resistance development to varying degrees. These include the

methyl benzimidazole carbamates (MBCs); demethylation inhibitors (DMIs); amines; aza-naphthalenes; benzophenones; and quinone outside inhibitors (QoIs).

3.3 Chemical Groupings for Grape Powdery Mildew Control

Sulfur

Sulfur was the first fungicide used for GPM control and is still widely used due to attributes such as lack of resistance development, relatively low cost and preventive and curative action. It acts primarily as a protectant fungicide by inhibiting spore germination. Sulfur is a multi-site inhibitor and via its breakdown to hydrogen sulfide inhibits respiration, disrupts protein function and inhibits biochemical pathways by chelation of metals important for normal enzyme function within the fungal cells (Hewitt, 1998). Much of the fungicidal activity of sulfur is associated with the vapour action of its breakdown product, hydrogen sulfide. As a result, its effectiveness depends on the formulation used as well as environmental factors such as temperature. The optimal temperature range for hydrogen sulfide production is 25–30°C. At temperatures greater than 30°C, the broad and indiscriminate mode of action (MOA) of sulfur may pose a risk of phytotoxicity to the crop. Sulfur is commonly applied as a dust or flowable or wettable powder formulation, depending on the climate and rainfall. In dry climates sulfur dust is preferred, while in wetter climates wettable powder or flowable formulations are used to aid retention on the foliage (Pearson and Goheen, 1988). Using sulfur early in the spray programme is not only an economical way to control powdery mildew but also has the advantage of controlling phytophagous mites.

Meptyldinocap

Meptyldinocap is a contact fungicide which acts as an uncoupler of oxidative phosphorylation, thus preventing the formation of adenosine triphosphate (ATP). Meptyldinocap is made up of the most potent single isomer of the well-known but now discontinued molecule dinocap. Meptyldinocap is active against all stages of the *E. necator* life cycle and as a result provides protectant, curative and eradicant activity against powdery mildew. In spite of many years of commercial use, no instances of *E. necator* resistance to meptyldinocap have been reported, making it a key fungicide in a resistance management strategy. In contrast to sulfur, disease control is equally effective at low and high temperatures. Use recommendations in Europe include four protectant applications at 10-day intervals and/or two eradicant applications at 5-day intervals (Hufnagl, 2007).

Methyl benzamidazole carbamates (MBCs)

The MBC fungicides were introduced in the early 1970s for control of a wide variety of diseases and were the first target site-specific fungicides with systemic properties. The group includes benomyl, thiabendazole, carbendazim, thiophanate-methyl and fuberidazole. MBCs have a mode of action which has been well characterized and involves the disruption of cell division at metaphase. This mitotic disruption is the result of MBC binding to the β-tubulin protein, which inhibits tubulin biosynthesis. Modification of a single amino acid in the β-tubulin protein renders the fungus insensitive to MBCs (Hewitt, 1998).

Benomyl was introduced into the GPM market in the early 1970s and initially provided excellent disease control. However, widespread resistance developed within a short period of time after introduction. For example, benomyl provided 97–100% control of GPM in western New York when initially introduced (from 1973 to 1976), but reduced effectiveness was observed in some vineyards in 1977 and 1978. Isolates of *E. necator* that were collected in 1978 and 1979 and assayed for benomyl sensitivity showed a complete loss of sensitivity to the fungicide. Subsequent field trials in 1979 demonstrated that benomyl provided only 5 and 8% disease control on foliage and fruit clusters, respectively

(Pearson and Taschenberg, 1980). In Turkey, the sensitivity of 32 isolates of *E. necator* to benomyl and carbendazim was determined in pot trials, and reduced sensitivity was observed as a result of continuous MBC usage, with cross-resistance occuring between benomyl and carbendazim (Ari and Delan, 1988). In India, the sensitivity to carbendazim of 22 *E. necator* isolates which were collected from different grape growing areas in Andhra Pradesh was determined. Of the 22 isolates, the ED_{50} values ranged from sensitive (12 μg/ml) to highly resistant (387 μg/ml) (Kumar *et al.*, 2003).

Due to the relatively sudden development of resistance to MBCs by *E. necator* and the introduction of fungicides with more desirable attributes, in particular the DMIs, the use of MBCs for GPM control has dropped considerably since their initial introduction into the marketplace.

Demethylation inhibitors (DMIs)

DMI fungicides are used globally on a wide variety of crops, and reduced sensitivity has been documented with a number of key diseases. The mode of action of DMIs is the inhibition of the C-14 demethylase enzyme in the ergosterol biosynthesis pathway (Delye and Corio-Costet, 1994; Delye *et al.*, 1997; Gisi *et al.*, 2000). Resistance development is quantitative (gradual) and involves multiple genes (De Waard, 1994).

DMI fungicides have been a crucial tool for GPM control since they were introduced in the early 1980s, starting with triadimefon. Additional DMI introductions followed, including myclobutanil, fenarimol, penconazole and pyrifenox. To varying degrees, DMI fungicides possess attributes such as curative and protectant efficacy and redistribution within the plant through xylem mobility. Because most of the biomass of *E. necator* resides on the leaf surface, some DMIs can also control powdery mildew through vapour action (Pearson *et al.*, 1994).

In California, triadimefon was introduced in 1982 and was followed by other DMIs such as myclobutanil and fenarimol. By 1986, reduced sensitivity of *E. necator* to triadime-fon was documented. Gubler *et al.* collected *E. necator* isolates from 19 vineyards in four regions in California and analysed them for sensitivity to triadimefon, myclobutanil and fenarimol. A time-course study performed in one vineyard, where resistant strains were reported, demonstrated a steady and signifi-cant increase in EC_{50} values for all three fungi-cides during the growing season after multiple applications of triadimefon. Increased resist-ance to triadimefon, but not to myclobutanil and fenarimol, was maintained in ascospores released from chasmothecia collected after the growing season (Gubler *et al.*, 1994, 1996). Ypema *et al.* conducted additional DMI resist-ance monitoring with *E. necator* isolates col-lected from vineyards throughout California to triadimefon, myclobutanil and fenarimol. The highest means and ranges of EC_{50} values found were those for triadimefon. Means and ranges were lower for myclobutanil and lowest for fenarimol, reflecting differences in the inherent activities of the fungicides and the potential for development of resistance (Ypema *et al.*, 1997). Aloi *et al.* (1990) also determined the susceptibility of various California vineyard isolates of *E. necator* to fenarimol, myclobutanil and triadimefon and found that *E. necator* populations within vineyards where DMI fungicides were used frequently, or where a loss of triadimefon efficacy had been reported, had reduced sen-sitivity to triadimefon, myclobutanil and fenarimol. Resistance reversion of California *E. necator* populations previously documented as resistant to triadimefon was measured after 14 years of non-use. The research indi-cated that, due to higher EC_{50} values within the populations than previously reported, resistance within populations had increased (Miller and Gubler, 2003).

A discriminatory germ tube length assay was used to distinguish between sensitive and tolerant isolates to triadimenol, penco-nazole and flusilazole in South African vineyards which previously had been exposed to triadimefon or triadimenol. The level of resist-ance in the subpopulations was compared with those from a distant vineyard with little previous exposure to triadimefon. All of the populations evaluated demonstrated reduced

sensitivity to triadimenol, indicating an earlier shift in triadimenol sensitivity within the subpopulations that then became well estab-lished in the treated vineyards (Halleen *et al.*, 2000). Erickson and Wilcox (1997) determined the sensitivity of *E. necator* populations col-lected from New York vineyards which had no previous exposure to DMIs or had received prolonged DMI use with triadimenol, myclob-utanil and fenarimol. They found differential levels of cross-resistance between the three fungicides, especially between fenarimol, and the two triazole fungicides, triadimenol and myclobutanil. They concluded that this phe-nomenon might have significant implications with respect to fungicide spray programmes for managing GPM and DMI resistance. Steva *et al.* (1989) found that after repeated applica-tions of triadimefon and triadimenol, the fungicides became less effective in south and central Portuguese vineyards and con-firmed that reduced efficacy was a result of decreased sensitivity to triadimenol.

Fungicide sensitivity of *E. necator* isolates from vineyards in Vienna and lower Austria was determined with triadimefon, triadime-nol, myclobutanil, penconazole and pyrifenox, and all populations assessed demonstrated reduced sensitivity to the fungicides tested (Steinkellner and Redl, 2002). The penconа-zole sensitivity of *E. necator* isolates from France, Portugal, Germany, Switzerland and Italy was evaluated and it was found that the baseline sensitivity showed only a very small variation within and between different coun-tries and that isolates from penconazole-treated vineyards were less sensitive compared to the baseline population (Steden *et al.*, 1994). Myclobutanil-resistant *E. necator* pop-ulations were confirmed in a research trial in 2000 in Ontario, Canada. When comparing isolates using a discriminatory dose of 0.18 µg/ml of myclobutanil, the proportion of resistant isolates increased from 53% in con-trol plots in June to 97% in myclobutanil-treated plots in September (Northover and Homeyer, 2001).

Steva and Cazenave report that there are two types of resistance development depend-ing on the primary mode of overwintering by *E. necato*r: mycelium within dormant infected buds or as ascospores within chasmothecia.

When *E. necator* overwinters primarily as mycelium within infected buds, populations collected in the beginning of the season from flag shoots are more sensitive to DMIs than those sampled the previous year at the end of the season. In contrast, when chasmothecia are the primary overwintering method, populations maintain their level of resistance characterized the previous year (Steva and Cazenave, 1996).

Using a random amplified polymorphic DNA (RAPD) assay, the genetic polymorphism existing among 62 *E. necator* isolates collected from a vineyard was studied. Isolates overwintering as mycelium in buds were genetically distinct from isolates overwintering as ascospores within chasmothecia, suggesting the existence of two genetically isolated *E. necator* populations and, consequently, of two independent sources of inoculum in the vineyard. Isolates resistant to DMI fungicides were found in both populations, suggesting that resistance may have arisen independently in the two powdery mildew populations (Delye *et al.*, 1997).

Despite the prolonged use of DMI fungicides for GPM control in a wide variety of geographies, and numerous reports of reduced sensitivity within populations of *E. necator*, the DMIs still retain relatively high levels of efficacy and are an integral part of a disease control programme. In many cases, resistance development has stabilized, and when used at full label rates with the appropriate spray interval and in alternation with other fungicide groups, DMI fungicides are expected to remain as valuable tools for GPM control.

Amines

Spiroxamine is the sole representative of the spiroketal-amine chemistry class in the amine grouping of sterol biosynthesis inhibitors. (It is currently the only amine in the group that is used for GPM control.) Other amines such as the morpholines, fenpropimorph, tridemorph and fenpropiden, are used primarily in Europe on wheat and barley for powdery mildew and rust disease control. The amines all share the same mode of action, which is inhibition of the enzymes Δ^{14}-reductase and Δ^8, Δ^7-isomerase, in the sterol biosynthesis pathway (Kuck, 1997). Spiroxamine is a very useful tool for resistance management since, as with the DMIs, resistance development is gradual and of a quantitative nature. Spiroxamine also possesses such attributes as protectant, curative, systemic and eradicant efficacy.

Spiroxamine was registered and introduced into the South African GPM market in 1999 for use on wine and table grapes and was recommended to be used as a block of 2–3 consecutive applications (Fourie and Zahn, 2001). In California, the sensitivity of 36 single-spore isolates of *E. necator* to spiroxamine had a mean EC_{50} value of 0.365 μg/ml, with values distributed in a log-normal manner (Miller and Gubler, 2004).

Resistance management recommendations for both the DMIs and amines are as follows: they should be used at full label rates in a preventive manner and curative uses should be avoided; there should be no more than four applications per season, either alone or in combination or alternation with non-cross-resistant fungicides; adhere to the recommended timing and application volume as specified on the product label. No instances of reduced sensitivity have been reported thus far and sensitivity monitoring results indicate no disease control issues due to resistance to spiroxamine (www.frac.info/frac/index.htm, FRAC SBI Working Group 2010 update).

Aza-naphthalenes

The aza-naphthalene group includes the quinoline quinoxyfen and the quinolinone proquinazid, whose proposed mode of action is interference with signal transduction via an unknown mechanism. The exact site of action of these two fungicides is not known. Quinoxyfen effects have been linked to early cell signalling events in wheat powdery mildew (*Blumeria graminis*) during germling differentiation (Wheeler *et al.*, 2003). Recent studies have shown that quinoxyfen appears to target serine esterase activity, with a downstream perturbation in signal transduction (Lee *et al.*, 2008). Studies conducted on wheat powdery mildew have shown that proquinazid stimulates the expression of host genes classically associated with resistance responses, including genes in ethylene-mediated response pathways, phytoalexin biosynthesis, cell-wall strengthening and active oxygen production (Crane *et al.*, 2008).

Powdery mildew spores that are deposited on aza-naphthalene-treated foliage germinate but fail to form functional appressoria and haustoria, which is essential for infection (Gilbert *et al.*, 2009). Due to this mode of action, the aza-naphthalenes provide only protectant efficacy, with no curative or eradicant attributes.

Prior to the launch of quinoxyfen in European vines in 1998, Green and Gustafson (2006) used a leaf-disc sporulation assay to evaluate the sensitivity of *E. necator* to quinoxyfen and found that the distribution of EC_{50} values from 56 isolates collected from six countries ranged from <0.03 to 2.6 µg/ml. A discriminatory dose screen was later used to evaluate large populations of isolates with decreased sensitivity to quinoxyfen, and the individual isolates selected by this method were then further characterized. A further examination of a subset of these isolates using a quantitative spore germination and germ tube elongation inhibition assay suggested that, for some of these isolates, resistance was much less than predicted by the sporulation assay, indicating that for quinoxyfen a leaf-disc sporulation assay might overestimate the frequency of isolates with significantly reduced sensitivity and the threat of decreased performance due to resistance. Attempts to select for isolates or populations of *E. necator* with measurable declines in sensitivity to quinoxyfen were unsuccessful, even with repetitive field applications over a 3-year period (Gustafson *et al.*, 2000). Small increases in the frequency of *E. necator* isolates with reduced sensitivity to quinoxyfen were first detected during routine monitoring in 2003 (Green and Duriatti, 2005). Isolates able to grow actively at a fungicide discriminatory dose controlling baseline isolates can be found across Europe. The frequency of these isolates is seen to vary significantly between monitored regions within countries and also from year to year.

Genet and Jaworska (2009) developed baseline sensitivity data to proquinazid from 174 *E. necator* isolates collected from Italy, France, Germany, Austria, Portugal and Spain and demonstrated that the isolates had EC_{50} values ranging from 0.001 to 0.3 µg/ml. The EC_{50} values for proquinazid were significantly lower than those for quinoxyfen, indicating a higher intrinsic activity of proquinazid. A strong sensitivity relationship (cross-resistance) between proquinazid and quinoxyfen was observed among 65 *E. necator* isolates tested, indicating both fungicides should be managed together for resistance management. The aza-naphthalenes are considered as medium-risk fungicides for resistance development.

Benzophenones

Metrafenone is the first fungicide from the benzophenone chemical class to be developed for GPM control and has been registered in key grape production areas only (Europe, South America, Australia and the USA) in the last several years. Laboratory research indicates that the mode of action is interference with hyphal morphogenesis, polarized cell growth and the establishment and maintenance of cell polarity, likely by disturbing a pathway-regulating organization of the actin cytoskeleton. In the presence of metrafenone, powdery mildew spores germinated and formed primary appressoria, but further fungal development was prevented. When metrafenone was applied in a curative fashion (after the fungus had penetrated the leaf and formed haustoria), it caused rapid collapse of the mycelium and strongly inhibited sporulation (Opalski *et al.*, 2006). Additional mode of action research conducted with an experimental benzophenone fungicide against several powdery mildew species, including *E. necator*, demonstrated that protectant treatments prevented fungal development after formation of the primary appressoria, with no subsequent development of surface hyphae (Schmitt *et al.*, 2006). This same research also demonstrated that curative treatments of the benzophenone molecule produced bifurcated hyphal tips, collapsed hyphae and release of cytoplasmic globules. The mechanism of action of the benzophenones provides protectant, curative and antisporulant activity against *E. necator* and other powdery mildews.

There have been no reports of *E. necator* resistance to metrafenone and it is unclear whether this is due to the recent introduction of the fungicide into the marketplace or as a result of the mode of action, or a combination of both.

For resistance management, no more than three applications per year are recommended, with no more than two consecutive applications before switching to a fungicide with an alternate mode of action. Although metrafenone has curative activity, it is recommended that it is not used in this manner, to avoid resistance development (http://www.agproducts. basf.us/products/vivando-fungicide).

Quinone outside inhibitors (QoIs)

The QoIs bind at the ubiquinol oxidase (Qo) site of cytochrome *b*, thereby stopping electron transfer between cytochrome *b* and cytochrome *c*, which halts nicotinamide adenine dinucleotide (NADH) oxidation and adenosine triphosphate (ATP) synthesis. The group is dominated by the strobilurin chemical class, but also includes the structurally unrelated molecules fenamidone and famoxadone (Bartlett *et al.*, 2002). The strobilurins have emerged as arguably the most important class of fungicides globally since their introduction in 1996. A critical attribute contributing to their success is that they are the first site-specific fungicide class which effectively controls plant pathogens within the Ascomycetes, Basidiomycetes, Deuteromycetes and Oomycetes – a true broad-spectrum, site-specific fungicide. To date, there have been more than 12 different strobilurins registered globally for use against a wide variety of diseases. The strobilurins can vary considerably in regards to systemicity, disease spectrum and primary crop usage. They are most effective when used in a protectant manner due to their primary strength as potent spore germination inhibitors (Bartlett *et al.*, 2002). For control of GPM, the main strobilurins used are azoxystrobin, pyraclostrobin, kresoxym-methyl and trifloxystrobin. Although fenamidone and famoxadone belong to the QoI group, they differ from the strobilurins in that they are used primarily against oomycete diseases and are not used for GPM control.

Since the introduction of the strobilurins beginning in the mid-1990s, resistance development has been documented in a number of important plant diseases. Research has shown that the major mechanism of resistance is the substitution of the amino acid glycine with alanine at position 143 of the cytochrome b protein, referred to as the G143A mutation (Gisi *et al.*, 2002; Kuck and Mehl, 2003; Grasso *et al.*, 2006). A second target-site mutation has been described for *Alternaria solani* and involves the substitution of phenylalanine with leucine at position 129 of the cytochrome b protein, referred to as the F129L mutation (Kuck and Mehl, 2003; Pasche *et al.*, 2004).

Although the overall performance of the strobilurin fungicides for GPM control is still very high, instances of resistance in *E. necator* populations have been documented in various locations in the USA and Europe. Widespread *E. necator* resistance to azoxystrobin was found in the eastern USA region, in particular in the state of Virginia. Resistance monitoring in Virginia and nearby states demonstrated that of 154 *E. necator* isolates evaluated, 28 were considered sensitive to azoxystrobin, 47 had intermediate resistance and 79 isolates were considered highly resistant. The G143A mutation was detected in some, but not all, of the isolates considered as resistant (Colcol, 2008).

In New York, USA, 285 isolates of *E. necator* collected from vineyards with no history of fungicide use or with a known DMI resistance were assayed for their sensitivity to azoxystrobin and myclobutanil. Interestingly, mean ED_{50} values for azoxystrobin were higher in the DMI-resistant population compared to the non-fungicide-treated isolates, with the greatest ED_{50} increases observed for isolates treated exclusively with myclobutanil, leading to the term 'cross-sensitivity' instead of 'cross-resistance'. Although there are various possible mechanisms proposed to explain the cross-sensitivity to the two fungicides, the actual mechanism has not yet been characterized (Wong and Wilcox, 2002).

Field trials were carried out in southern Italy to compare the effectiveness of spray schedules based on trifloxystrobin and azoxystrobin, which were used alone and/or in alternation with the DMI fungicides penconazole or tebuconazole. As an indication of negligible fungicide resistance in the field trial vineyards, application schedules based on the exclusive use of each of the two strobilurins or in alternations with penconazole or tebuconazole showed similar levels of effectiveness, even under high disease pressure (Santomauro *et al.*, 2003).

Intensive monitoring in Europe in 2009 detected no strobilurin resistance in Portugal and most regions of France, Italy and Germany. Resistant populations were detected in Hungary and commercial vineyards in eastern Austria, the Armagnac region of France, the Mosel area of Germany, the Alto Adige and Lombardi region of Italy and the Constanta area of Romania. Resistance management guidelines recommend that no more than three applications of a strobilurin fungicide, either singly or in combination, per growing season should be used. Strobilurins used alone should be used in strict alternation with a fungicide from a non-cross-resistant class and no more than two consecutive applications of a strobilurin in a mixture with another chemical class should be made. The fungicides should be used in a preventive manner and curative applications should be avoided (www.frac.info/frac/work/FRAC_QoI_Minutes_2010_final).

3.4 Fungicide Sensitivity Monitoring

As described previously, with the exception of sulfur and meptyldinocap, the fungicide groups mentioned are, to varying degrees, prone to resistance development by *E. necator* and, as a result, a robust and consistent anti-resistance management strategy should be implemented to ensure sustainable disease control. An important aspect of a resistance management programme is to generate baseline sensitivity data prior to or at launch of

the product and then subsequently to monitor shifts in pathogen populations in vineyards where the fungicides are being used. This will allow for the determination of the prevalence and magnitude of resistance development (Brent and Hollomon, 1998).

Because *E. necator* is an obligate parasite, generating fungicide sensitivity data is more complex and cumbersome when compared to non-obligate pathogens, since both the generation of isolate inoculum and the screening assay require the use of live plant tissue. The first step in the process is to collect *E. necator* isolates, with several options available. These include collection from vineyards by transporting the infected plant material to the laboratory for further processing (Heaney, 1991; Kung, 1991) or by using an automobile rooftop-mounted spore trap. In the latter case, while passing through vineyard regions the trapped spores are deposited on to grape leaf tissue placed on water agar in a Petri dish, which is then put into a growth chamber maintained at the optimal temperature for *E. necator* growth and sporulation. The resulting powdery mildew colonies are used to generate isolates for further characterization (Genet and Jaworska, 2009). Mass transfer of sporulating fungus (Ypema *et al.*, 1997), single conidial chain isolates (Wong and Wilcox, 2002), or single spore isolates (Savocchia *et al.*, 2004) can be used as the source of initial inoculum, which is then increased for use in the sensitivity assay. Isolates derived from single spores or single chains are considered to be truly representative relative to bulk isolates, since the collection of individual DMI sensitivities comprises the population distribution (Erickson and Wilcox, 1997).

Individual isolates are evaluated for fungicide sensitivity, most commonly by use of a detached leaf or leaf-disc assay, in which grape leaves or discs are treated with various concentrations of the fungicides of interest and the amount of leaf area colonized by *E. necator* is visually determined. In contrast to many other fungicide groupings, sensitivity monitoring for DMI fungicides can use a germ tube/hyphal growth assay instead of colonization of the leaf disc, since DMI fungicides do not inhibit spore germination but instead have a strong effect on hyphal growth

(Halleen *et al.*, 2000; Petsikos Panagiotarou *et al.*, 2001). However, growth of *E. necator* is affected significantly by the leaf material used for the assay, with higher growth rates and less variability between experimental replicates when using immature, rapidly expanding leaves rather than fully expanded mature leaves (Erickson and Wilcox, 1997). Other considerations for a robust sensitivity monitoring programme are that the sample size of the isolates collected must be sufficiently large to represent the population(s) being evaluated and the assay method must provide for reproducible results.

Dose–response curves are generated and regression analyses performed to generate ED_{50}, ED_{90} or other relevant values such as the minimum inhibitory concentration (MIC) for each isolate. A single discriminatory rate is used in some cases instead of a multiple rate response, especially if there is a large number of isolates to evaluate (Shabi and Gaunt, 1992). The distribution of fungicide sensitivity of the population of isolates can then be analysed to determine the degree of cross-resistance among fungicides and the magnitude of the population shift towards increased resistance relative to a baseline population which has not yet been exposed to the fungicides of interest (Erickson and Wilcox, 1997).

3.5   Resistance Management

As described previously, disease control strategies should be focused on reducing initial inoculum and protection of foliage from subsequent infection. Early preventive fungicide treatment to control powdery mildew is crucial to keep disease levels low and reduce the fungicide resistance selection pressure on the pathogen. Key aspects to consider when designing a viable anti-resistance strategy are: to limit the use of site-specific fungicides in the application programme; to alternate or tank mix fungicides from different groupings; and to adhere to label recommendations in regards to spray interval and use rates. Also, where feasible, use fungicides or other disease control technology where resistance development is not an issue. This can include

the use of biological control agents, potassium salts, various oils, plant defence activators and fungicides which have no resistance issues, such as sulfur and dinocap (Gubler *et al.*, 2009). One example is a biofungicide comprised of the hyperparasite *Ampelomyces quisqualis*, which is commercially available for GPM control. After application to grapevines, germinating spores of *A. quisqualis* form hyphae that penetrate the hyphae of *E. necator* via a specific host–parasite interaction. GPM can be controlled effectively with *A. quisqualis* if applied to vines at stages where disease incidence is very low and good coverage of the entire vine canopy is attained (Daoust and Hofstein, 1996).

Campbell and Latorre report that the plant defence activator acibenzolar-*S*-methyl enhances vine resistance to GPM, reducing both disease incidence and severity. The infection rates on plants treated with either acibenzolar-*S*-methyl or the strobilurin fungicide kresoxim-methyl were significantly lower than those from untreated plants, with similar efficacy of acibenzolar-*S*-methyl to a single application of kresoxim-methyl (Campbell and Latorre, 2004). Since acibenzolar-*S*-methyl is not fungitoxic but instead acts on the plant to increase its natural defence mechanisms, no selection pressure is applied to the pathogen to become resistant.

The fungicide meptyldinocap is particularly efficacious very early in the season to eradicate mildew in infected buds and delay the onset of the disease epidemic by reducing initial inoculum. It is used either by itself or in mixtures with other fungicides in a GPM programme. Steva (1994) investigated strategies for controlling *E. necator* resistance to DMIs in French vineyards with fungicide programmes that included the following treatments: sulfur alone; triadimenol at full and half-label rates; triadimenol tank mixed with either sulfur or dinocap; and alternation of triadimenol at full rate with either sulfur or dinocap. Applications with the low rate of triadimenol resulted in a rapid increase in resistant isolates when compared to applications with the full rate. Applications with a mixture of triadimenol and sulfur did not slow resistance development and the recommended strategy was to reduce the number of DMI treatments and to use sulfur alone. Velusceck (2001) used dino-cap in a fungicide programme to control both triazole-sensitive and -resistant *E. necator* populations. Dinocap was applied to vines at early or late season (between flowering and fruit set or after fruit set). For the early infection period, dinocap alone or dinocap mixed with a half rate of sulfur was more effective than sulfur alone. For late season applications, dinocap alternated with triazoles at full rate or full rate of dinocap mixed with triazoles led to the prevention of *E. necator* resistance development to triazole fungicides.

In New York State, USA, Wilcox *et al.* (1998) evaluated three resistance management strategies in a grape vineyard with *E. necator* resistance to DMI fungicides. Myclobutanil was chosen as the experimental DMI fungicide, and the components for resistance management strategies included application frequency, rate and disease pressure. Lower rates (half-label rate) resulted in inferior disease control and enhanced fungicide resistance development. Disease control was 35% greater with the six high-rate sprays compared to the low-rate sprays, and resistant populations were fourfold higher, with six low-rate applications compared to three high-rate sprays followed by sulfur.

3.6 Summary

GPM can inflict extensive damage to grape vineyards and can result in significant reduction in grape quality and quantity. An integrated approach to disease control is critical, with the use of fungicides providing a valuable tool for combating disease. A number of chemical fungicide groups are available for disease control, but the majority are prone to resistance development by *E. necator*. To provide a sustainable level of efficacy, a robust anti-resistance management strategy must be implemented. This can include an ongoing monitoring and detection programme of *E. necator* populations to determine the prevalence and magnitude of population shifts towards increased fungicide resistance. For site-specific fungicides, practices such as restricting their use where practical by relying more heavily on agronomic cultural practices and use of fungicides such as sulfur and meptyldinocap can delay resistance development. Also of high importance is to adhere strictly to manufacturer label requirements for use of proper rates and application timings, as well as the effective use of alternations and mixtures with chemical groups with different modes of action.

References

Aloi, C., Gullino, M.L. and Garibaldi, A. (1990) Sensitivity of grape powdery mildew isolates from California towards fenarimol, myclobutanil and triadimefon. *Phytopathology* 80(10), 975.

Ari, M. and Delen, N. (1988) Studies on the fungicide sensitivity of vine mildew (*Uncinula necator* (Schwein) Burr.) in Aegean region of Turkey. *Journal of Turkish Phytopathology* 17(1), 19–30.

Bartlett, D.W., Clough, J.M., Godwin, R., Hall, A.A., Hamer, M. and Parr-Dobrzanski, B. (2002) The strobilurin fungicides. *Pest Management Science* 58(7), 649–662.

Brent, K.J. and Hollomon, D.W. (1998) Fungicide resistance: the assessment of risk. *FRAC Monograph No. 2*. Global Crop Protection Federation, Brussels.

Campbell, P.A. and Latorre, B.A. (2004) Suppression of grapevine powdery mildew (*Uncinula necator*) by acibenzolar-S-methyl. *Vitis* 43(4), 209–210.

Colcol, J. (2008) Fungicide sensitivity of *Erysiphe necator* and *Plasmopora viticola* from Virginia and nearby states. MSc thesis, Virginia Polytechnic Institute and State University, Blacksburg, Virginia.

Crane, V., Beatty, M., Zeka, B., Armstrong, R., Geddens, R. and Sweigard, J. (2008) Proquinazid activates host defense gene expression in *Arabidopsis thaliana*. In: *Modern Fungicides and Antifungal Compounds*. Proceedings of the 15th International Rheinhardsbrunn Symposium, 6–10 May 2007, Friedrichroda, Germany. BCPC Publications, Alton, UK, pp. 19–26.

Daoust, R.A. and Hofstein, R. (1996) *Ampelomyces quisqualis*, a new biofungicide to control powdery mildew in grapes. *Proceedings of the British Crop Protection Conference, Pests and Diseases* 1, 33–40.

Delye, C. and Corio-Costet, M. (1994) Resistance of grape powdery mildew (*Uncinula necator*) to triadimenol, a sterol biosynthesis inhibitor: biochemical characterisation of sensitive and resistant strains. *British Crop Protection Council Monograph 60, Fungicide Resistance*. BCPC, Alton, Hampshire, UK, pp. 87–92.

Delye, C., Laigret, F. and Corio-Costet, M. (1997) New tools for studying epidemiology and resistance of grape powdery mildew to DMI fungicides. *Pesticide Science* 51(3), 309–314.

De Waard, M.A. (1994) Resistance to fungicides which inhibit sterol 14alpha-demethylation, an historical perspective. In: *British Crop Protection Council Monograph 60, Fungicide Resistance*. BCPC Publications, Surrey, UK, pp. 3–10.

Erickson, E.O. and Wilcox, W.F. (1997) Distributions of sensitivities to three sterol demethylation inhibitor fungicides among populations of *Uncinula necator* sensitive and resistant to triadimefon. *Phytopathology* 87(8), 784–791.

Fourie, P.J. and Zahn, K. (2001) Prosper® and Falcon®-spiroxamine based new products for control of powdery mildew in grape vine. *Pflanzenschutz-Nachrichten* 54(3), 399–412.

Genet, J. and Jaworska, G. (2009) Baseline sensitivity to proquinazid in *Blumeria graminis* f. sp. *tritici* and *Erysiphe necator* and cross-resistance with other fungicides. *Pest Management Science* 65(8), 878–884.

Gilbert, S.R., Cools, H.J., Fraaije, B.A., Bailey, A.M. and Lucas, J.A. (2009) Impact of proquinazid on appressorial development of the barley powdery mildew fungus *Blumeria graminis* f. sp. *hordei*. *Pesticide Biochemistry and Physiology* 94, 127–132.

Gisi, U., Chin, K.M., Knapova, G., Kueng Faerber, R., Mohr, U., Parisi, S., *et al.* (2000) Recent developments in elucidating modes of resistance to phenylamide, DMI and strobilurin fungicides. *Crop Protection* 19(810), 863–872.

Gisi, U., Sierotzki, H., Cook, A. and McCaffery, A. (2002) Mechanisms influencing the evolution of resistance to Qo inhibitor fungicides. *Pest Management Science* 58(9), 859–867.

Grasso, V., Palermo, S., Sierotzki, H., Garibaldi, A. and Gisi, U. (2006) Cytochrome b gene structure and consequences for resistance to Qo inhibitor fungicides in plant pathogens. *Pest Management Science* 62(6), 465–472.

Green, E. and Duriatti, A. (2005) Sensitivity of *Uncinula necator* isolates to quinoxyfen: baseline studies, validation of baseline method and targeted sensitivity monitoring after several years of commercial use. In: *Proceedings of the BCPC International Congress. Crop Science and Technology*. BCPC Alton, Hants, UK, pp. 163–168.

Green, E.A. and Gustafson, G.D. (2006) Sensitivity of *Uncinula necator* to quinoxyfen: evaluation of isolates selected using a discriminatory dose screen. *Pest Management Science* 62(6), 492–497.

Gubler, W.D., Ypema, H.L., Ouimette, D.G. and Bettiga, L.J. (1994) Resistance of *Uncinula necator* to DMI fungicides in California vines. *British Crop Protection Council Monograph 60, Fungicide Resistance*. BCPC Alton, Hants, UK, pp. 19–26.

Gubler, W.D., Ypema, H.L., Ouimette, D.G. and Bettiga, L.J. (1996) Occurrence of resistance in *Uncinula necator* to triadimefon, myclobutanil, and fenarimol in California grapevines. *Plant Disease* 80(8), 902–909.

Gubler, W.D., Smith, R.J., Varela, L.G., Vasquez, S., Stapleton, J.J. and Purcell, A.H. (2009) *UC IPM Pest Management Guidelines: Grape*. UC ANR Publication 3448, University of California, California.

Gustafson, G.D., Green, E.A., Keeler, L.C. and Henry, M.J. (2000) Resistance profiling of quinoxyfen on grape powdery mildew. *Phytopathology* 90(6), 32.

Halleen, F., Holz, G. and Pringle, K.L. (2000) Resistance in *Uncinula necator* to triazole fungicides in South African grapevines. *South African Journal of Enology and Viticulture* 21(2), 71–80.

Heaney, S.P. (1991) A method for the examination of the sensitivity of *Uncinula necator* to DMI fungicides. *European and Mediterranean Plant Protection Organization Bulletin* 21(2), 319–321.

Hewitt, H.G. (1998) *Fungicides in Crop Protection*. CAB International, New York.

Hufnagl, A. (2007) Meptyldinocap: a new active substance for powdery mildew control. In: *XVth International Plant Protection Congress*, Glasgow, Scotland 15–18 October 2007. British Crop Production Council, UK, Volume 1, pp. 32–39.

Kuck, K.H. (1997) KWG 4168 (spiroxamine): baseline sensitivity and cross-resistance with other fungicides. *Pflanzenschutz-Nachrichten Bayer* 50(1), 17–18.

Kuck, K.H. and Mehl, A. (2003) Trifloxystrobin: resistance risk and resistance management. *Pflanzenschutz-Nachrichten Bayer* 56(2), 313–325.

Kumar, J.H., Rao, K.C., Reddy, D.R. and Babu, T.R. (2003) Resistance development in grapevine powdery mildew *Uncinula necator* (Schwein) Burrill to carbendazim in Andhra Pradesh. *Indian Journal of Plant Protection* 31(2), 66–68.

Kung, R. (1991) Test method for examination of the sensitivity of *Uncinula necator* to pyrifenox. *European and Mediterranean Plant Protection Organization Bulletin* 21(2), 317–319.

Lee, S., Gustafson, G., Skamniot, P., Baloch, R. and Gurr, S. (2008) Host perception and signal transduction studies in wild-type *Blumeria graminis* f. sp. *hordei* and a quinoxyfen-resistant mutant implicate quinoxyfen in the inhibition of serine esterase activity. *Pest Management Science* 64(5), 544–555.

Miller, T.C. and Gubler, W.D. (2003) *Uncinula necator* retains high resistance levels to triadimefon in a survey of California populations despite product absence for 14 years. *Phytopathology* 93(6), 113.

Miller, T.C. and Gubler, W.D. (2004) Sensitivity of California isolates of *Uncinula necator* to trifloxystrobin and spiroxamine, and update on triadimefon sensitivity. *Plant Disease* 88(11), 1205–1212.

Northover, J. and Homeyer, C.A. (2001) Detection and management of myclobutanil-resistant grapevine powdery mildew (*Uncinula necator*) in Ontario. *Canadian Journal of Plant Pathology* 23(4), 337–345.

Opalski, K.S., Tresch, S., Kogel, K.H., Grossman, K., Kohle, H. and Huckelhoven, R. (2006) Metrafenone: studies on the mode of action of a novel cereal powdery mildew fungicide. *Pest Management Science* 62(5), 393–401.

Pasche, J.S., Wharam, C.M. and Gudmestad, N.C. (2004) Shift in sensitivity of *Alternaria solani* in response to QoI fungicides. *Plant Disease* 88, 181–187.

Pearson, R.C. and Goheen, A.C. (1988) *Compendium of Grape Diseases*. American Phytopathological Society, St Paul, Minnesota.

Pearson, R.C. and Taschenberg, E.F. (1980) Benomyl-resistant strains of *Uncinula necator* on grapes. *Plant Disease* 64(7), 677–680.

Pearson, R.C., Riegel, D.G. and Gadoury, D.M. (1994) Control of powdery mildew in vineyards using single-application vapor-action treatments of triazole fungicides. *Plant Disease* 78, 164–168.

Petsikos Panagiotarou, N., Markellou, A., Kalamarakis, A., Konstantinidou Doltsinis, S. and Ziogas, B.N. (2001) Differences in sensitivity to the DMI fungicide triadimenol among populations of *Uncinula necator* in Greece. *Phytopathologia Mediterranea* 40(2), 211–212.

Santomauro, A., Tauro, G., Dongiovanni, C., Giampaolo, C., Abbatecola, A., Miazzi, M., *et al.* (2003) A 4-year experience with trifloxystrobin against powdery mildew on table grape in southern Italy. *Pflanzenschutz-Nachrichten Bayer* 56(2), 373–386.

Savocchia, S., Stummer, B.E., Wicks, T.J., Van Heeswijck, R. and Scott, E.S. (2004) Reduced sensitivity of *Uncinula necator* to sterol demethylation inhibiting fungicides in southern Australian vineyards. *Australasian Plant Pathology* 33(4), 465–473.

Schmitt, M.R., Carzaniga, R., Cotter, H. van T., O'Connell, R.O. and Hollomon, D. (2006) Microscopy reveals disease control through novel effects on fungal development: a case study with an early-generation benzophenone fungicide. *Pest Management Science* 62, 383–392.

Shabi, E. and Gaunt, R.E. (1992) Measurement of sensitivity to DMI fungicides in populations of *Uncinula necator*. In: *Proceedings of the 45th New Zealand Plant Protection Conference*. New Zealand Plant Protection Society, Christchurch, New Zealand, pp. 126–128.

Steden, C., Forster, B. and Steva, H. (1994) Sensitivity of *Uncinula necator* to penconazole in European countries. *British Crop Protection Council Monograph 60, Fungicide Resistance*. BCPC Alton, Hants, UK, pp. 97–102.

Steinkellner, S. and Redl, H. (2002) Sensitivity of *Uncinula necator* populations following DMI-fungicide usage in Austrian vineyards. *Bodenkultur* 52(4), 213–219.

Steva, H. (1994) Evaluating anti-resistance strategies for control of *Uncinula necator*. *British Crop Protection Council Monograph 60, Fungicide Resistance*. BCPC Alton, Hants, UK, pp. 59–66.

Steva, H. and Cazenave, C. (1996) Evolution of grape powdery mildew insensitivity to DMI fungicides. *Brighton Crop Protection Conference – Pests and Diseases*, Volume 2. BCPC Alton, Hants, UK, pp. 725–730.

Steva, H., Clerjeau, M. and Silva, M.T.G. da (1989) Reduced sensitivity to triadimenol in Portuguese field populations of *Uncinula necator*. *International Society of Plant Pathology Chemical Control Newsletter* 12, 30–31.

Velusceck, S. (2001) Karathane: product positioning in spraying programs to prevent development of resistance to powdery mildew in grapes. Selectivity on predatory mites and other benefits. *Phytopathologia Mediterranea* 40(2), 212.

Wheeler, I., Holloman, D.W., Gustafson, G., Mitchell, J., Longhurst, C., Zhang, Z., *et al.* (2003) Quinoxyfen perturbs signal transduction in barley powdery mildew (*Blumeria graminis* f. sp. *hordei*). *Molecular Plant Pathology* 4(3), 177–186.

Wilcox, W.F., Riegel, D.M., Erickson, E.O. and Burr, J.A. (1998) Vineyard evaluation of resistance-management strategies for DMI fungicides and *Uncinula necator*. *Phytopathology* 88(9), 97.

Wong, F.P. and Wilcox, W.F. (2002) Sensitivity to azoxystrobin among isolates of *Uncinula necator*: baseline distribution and relationship to myclobutanil sensitivity. *Plant Disease* 86(4), 394–404.

Ypema, H.L., Ypema, M. and Gubler, W.D. (1997) Sensitivity of *Uncinula necator* to benomyl, triadimefon, myclobutanil, and fenarimol in California. *Plant Disease* 81(3), 293–297.

4   Fungicide Resistance in *Pseudoperonospora cubensis*, the Causal Pathogen of Cucurbit Downy Mildew

Aleš Lebeda[1] and Yigal Cohen[2]

[1]*Department of Botany, Faculty of Science, Palacký University, Olomouc,
Czech Republic;* [2]*Faculty of Life Sciences, Bar Ilan University, Ramat Gan, Israel*

4.1   Introduction

Cucurbitaceae is a very large and heterogeneous family of plants, originating from America, Africa and Asia (Bates *et al.*, 1990; Lebeda *et al.*, 2007). Cucurbits are grown worldwide, mainly in temperate regions. This family includes economically important species, particularly those having edible fruit (Robinson and Decker-Walters, 1997). Cucurbit production in various parts of the world is seriously limited due to epidemics of downy mildew caused by the oomycete *Pseudoperonospora cubensis* (Lebeda and Cohen, 2011). Such epidemics cause severe damage, ending with death of the crop plants at the adult stage. The disease therefore plays a crucial role in determining cucurbit production in certain regions (Cohen, 1981; Perchepied *et al.*, 2005; Velichi, 2009; Lebeda and Cohen, 2011).

P. cubensis was first recorded in Cuba in 1868 (Berkeley and Curtis, 1868) and currently is distributed in the north and south hemispheres where cucurbits are cultivated. It occurs in warm, temperate, subtropic and tropic regions, on field as well as on protected (glasshouse, plastic house and shade house) crops (Lebeda and Cohen, 2011), especially in areas with annual precipitation of >300 mm

(Lebeda, 1990). Although distributed worldwide, the occurrence of *P. cubensis* and the damage it causes to various host species may differ among various geographic regions (Palti and Cohen, 1980; Cohen, 1981; Lebeda, 1990).

In the 20th century, it spread rapidly throughout most European countries; in Austria and Hungary in 1904, in Yugoslavia in 1952, in Russia in 1963 and in Bulgaria, Romania, Switzerland, Germany, The Netherlands, Greece, France and the UK after 1970 (Lebeda, 1990). Since 1984 it has been considered in Czechoslovakia and in Central Europe as a major disease with high economic impact (Lebeda, 1990). In 1985, *P. cubensis* spread from Czechoslovakia (Lebeda, 1986) to Poland (Rondomanski, 1988) and, by air currents, sporangia migrated to Sweden and Finland (Tahvonen, 1985; Forsberg, 1986). Cucurbit downy mildew has been reported recently in more than 80 countries on more than 60 species (Lebeda and Cohen, 2011). Except for cucumber (*Cucumis sativus*), the most often and severely affected hosts are muskmelon (*Cucumis melo*), watermelon (*Citrullus lanatus*), pumpkin (*Cucurbita maxima*) and squash (*Cucurbita pepo*) (Lebeda and Widrlechner, 2003; Lebeda and Cohen, 2011).

In Europe, severe outbreaks of cucurbit downy mildew have been reported repeatedly on cucumber (Lebeda and Cohen, 2011). Yield reductions on field-grown cucumbers have been recorded frequently during the past 20 years (Doruchowski and Lakowska-Ryk, 1992), but no epidemics have been observed on *C. melo* or *Cucurbita* spp. under either field or glasshouse conditions (Lebeda, 1999; Lebeda *et al.*, 2010; Lebeda and Cohen, 2011). In the Czech Republic, repeated downy mildew infections have been recorded recently (2009/2010) on *C. pepo, C. maxima* and *Cucurbita moschata* (Lebeda, unpublished data; Pavelkova *et al.*, 2011), indicating a dramatic change in the population structure of the pathogen. Similar change occurred in Israel in 2002, when *P. cubensis* suddenly appeared on *C. moschata* (Cohen *et al.*, 2003). Since 2004, *P. cubensis* has re-emerged as a major problem on cucumbers in eastern USA (Holmes *et al.*, 2004; Coluci *et al.*, 2006; Holmes and Ojiambo, 2009; Holmes and Thomas, 2009). *P. cubensis* is potentially a very dangerous, devastating pathogen which annually causes a serious threat to cucurbit crops (e.g. melon, cucumber, pumpkin, squash, watermelon and luffa) grown around the world (Lebeda and Cohen, 2011).

Disease symptoms (Fig. 4.1a) are confined to the leaves and adversely affect the quality and yield of fruit (Lebeda and Cohen, 2011). Development of *P. cubensis* is dependent on host susceptibility and favourable environmental conditions (Iwata, 1942, 1953a,b; Palti, 1974; Palti and Cohen, 1980). It develops rapidly and the disease spreads quickly in the presence of free water on the leaf surface (Lebeda and Cohen, 2011). The pathogen induces chlorotic lesions on the leaves, which gradually turn necrotic, mainly in older leaves. Sporulation (Fig. 4.1b) occurs on the adaxial leaf surface.

Despite the importance of *P. cubensis*, progress in resistance breeding, fungicide control and disease management is rather slow (Lebeda and Cohen, 2011). Knowledge on the population structure of the pathogen and epidemiology of the disease is still scarce (Lebeda *et al.*, 2006, 2010; Sarris *et al.*, 2009). Limited information is available in Europe, the USA and Asia on spatial distribution, host range, temporal changes in disease prevalence in various geographic regions, virulence variation and fungicide resistance (Lebeda *et al.*, 2010; Lebeda and Cohen, 2011).

The objectives of this paper are to review the literature on fungicide resistance in *P. cubensis* and to highlight future research needs for this pathogen.

(b)

(a)

Fig. 4.1. (a) Heavy infection in cucumbers with downy mildew in the field; (b) symptoms of downy mildew caused by *P. cubensis* on a cucumber leaf (adaxial leaf surface).

4.2 Fungicide Resistance in *P. cubensis* – Historical Overview and Current Status

Fungicides remain the preferred method for the control of plant diseases, including diseases caused by Oomycetes (Cohen and Coffey, 1986; De Waard *et al.*, 1993; Lebeda and Schwinn, 1994; Knight *et al.*, 1997; Gisi, 2002). The discovery of systemic fungicides active against Oomycetes greatly improved the control of foliar downy mildews (Schwinn and Staub, 1995).

P. cubensis is highly variable in pathogenicity (Lebeda and Widrlechner, 2003, 2004; Lebeda *et al.*, 2006) and its control by using resistant cultivars is, as yet, ineffective (Lebeda and Cohen, 2011). *P. cubensis* belongs to the group of 'highest risk pathogens' with high evolutionary potential (McDonald and Linde, 2002; Lebeda and Urban, 2004a,b; Lebeda *et al.*, 2006, 2010). Deployment of resistance genes has to be combined with additional practices of integrated disease management, such as fungicides, to minimize the high risk of the pathogen overcoming such genes (Lebeda and Cohen, 2011).

According to Gisi (2002), the sales value of fungicides against downy mildews amounted to SFr 1.2 billion in 1996, of which 10% were used mainly to fight *P. cubensis* on cucurbit crops. For many decades, copper formulations were the only contact fungicides available for the control of *P. cubensis*. During the past several decades, new contact and systemic fungicides became available for disease control (fosetyl-Al in 1977; phenylamides in 1977–1983; propamocarb in 1978; dimethomorph in 1988; cyazofamid in 2001). Systemic fungicides have a specific, single-site mode of action and therefore provide a high risk for resistance to develop. Several reports (Samoucha and Cohen, 1985; O'Brien and Weinert, 1995; Ishii *et al.*, 2002) show the appearance and build-up of subpopulations resistant to single-site fungicides shortly after their introduction to the market as a result of continuous, exclusive use under high disease pressure. On the other hand, contact fungicides are multi-site inhibitors and therefore provide low resistance. For this reason, integrated disease management should use weather forecasts, scouting for initial foci, resistant cultivars and diverse fungicides. When used combined and coordinated, the number of fungicide applications may be reduced (Lebeda and Schwinn, 1994).

4.3 Methods for Detecting Fungicide Resistance

Laboratory methods

Sampling and establishing P. cubensis *isolates*

Details of methods of laboratory work with *P. cubensis* have been described recently (Lebeda and Urban, 2010). Infected leaves are collected from infected cucurbit crops, placed adaxial surface uppermost on wet filter paper in sealed plastic containers at *c*.20°C, and the sporangia produced on the leaf surface after 1–2 days of incubation are removed by brush or shaking a single sporulating lesion in cold distilled water. The sporangial suspension is drop-inoculated or atomized on to the adaxial surface of a leaf of a highly susceptible genotype laid in a Petri dish on wet filter paper. Inoculated leaves are incubated in a growth chamber at 12 h light/dark photoperiod, 18°C during light/15°C during darkness. The pathogen produces sporangiophores bearing sporangia at 6–8 dpi (Lebeda and Urban, 2010).

Plant material

Highly susceptible cultivars of cucumber ('Marketer 430', Bet-Alpha, Dalila) are used for pathogen multiplication and bioassays. Plants are grown in the greenhouse (25–30°C/ 15–20°C day/night, daily watering and weekly fertilization) and used at 3–6 weeks after planting (2–6 true-leaf stage) (Lebeda and Widrlechner, 2003; Lebeda and Urban, 2010).

Maintenance of P. cubensis *isolates*

Methods for preserving obligate plant parasites have been summarized by Lebeda and Bartoš (1988). Sporulating detached leaves are stored in Petri dishes on dry filter paper at –20°C (2 days) and then at –80°C. Sporangia remain viable for ≥6 months (Lebeda and Urban, 2010). Isolates can also be maintained by weekly transfers to detached cotyledons, leaves or leaf discs on wet filter paper (Lebeda, 1992; Lebeda and Urban, 2010), or water agar in Petri dishes (O'Brien and Weinert, 1995; Ishii *et al.*, 2002), or by transfer inoculation of whole plants in separate growth chambers (Samoucha and Cohen, 1984a; Ishii *et al.*, 2001).

Leaf-disc bioassay for determining fungicide sensitivity

Various methods are available to assess the sensitivity of *P. cubensis* isolates to fungicides under laboratory conditions (Urban and Lebeda, 2006). One method is floating leaf discs on fungicide solutions (Anon., 1982; Urban and Lebeda, 2004a,b, 2006, 2007). Fungicides are each used in five concentrations, one as recommended by the producer and two below and two above that optimum (Urban and Lebeda, 2007; Hübschová and Lebeda, 2010). Leaf discs (15 mm in diameter, $n = 4$/concentration) are floated, abaxial surface upward (Fig. 4.2), on fungicide solutions in 12-well titre plates (Anon., 1982). Control

leaf discs are floated on distilled water. The test is replicated three times. After 24 h of incubation, leaf discs are spray inoculated with a sporangial suspension (1×105 spores/ml) of the test isolate and plates incubated as described above (Urban and Lebeda, 2007). A slightly different method was used by O'Brien and Weinert (1995) to test the efficacy of metalaxyl. Leaf discs taken from cucumber cotyledons were inoculated by placing a single 10 µl droplet of sporangial suspension (1×104/ml) on each disc. Discs were incubated at 20°C and rated at 12 dpi for sporulation intensity: 0, no symptoms; 1, reaction spot with no sporulation; 2, sporulation low-moderate covering <50% of the disc surface; 3, sporulation covering >50% of the disc.

Rating fungicide resistance in leaf-disc bioassays

Sporulation intensity is estimated visually at 2-day intervals between 6 and 16 dpi using a 0–4 scale (Lebeda, 1992; Lebeda and Widrlechner, 2003; Lebeda and Urban, 2010). In control discs, sporulation is seen initially at 5–6 dpi. Sporulation intensity is expressed as

Fig. 4.2. A leaf-disc floating bioassay for fungicide resistance screening in *P. cubensis* (left – control leaf discs without inoculation/on solution of propamocarb/; right – 14 days after inoculation, profuse sporulation of *P. cubensis* on leaf discs, resistant isolate to propamocarb/Previcur 607 SL/607 µg ai/ml).

a percentage of the maximum scores according to Towsend and Heuberger (1943):

$$P = \sum \frac{(n \times v) \times 100}{x \times N},$$

where P is the total score of sporulation (%), n = the number of discs in each category (0–4), v = sporulation intensity, x = the extent of a used scale (4) and N = total number of leaf discs evaluated in three replications.

Isolates were identified as sensitive (total score of sporulation $P \leq 10\%$), tolerant ($10 < P \leq 35\%$) and resistant ($P > 35\%$) (Urban and Lebeda, 2004a,b, 2007; Lebeda and Urban, 2007). ED_{50} values (fungicide concentration providing 50% inhibition of fungal growth) are calculated for each isolate for comparative purposes (Urban and Lebeda, 2007; Hübschová and Lebeda, 2010).

Bioassay with cotyledons

Two methods of fungicide application are used: (i) soil drench; and (ii) foliar spray (for details see Urban and Lebeda, 2006). Soil drench is used with systemic fungicides such as metalaxyl, propamocarb, cyprofuram + folpet, SAN 371 F and fosetyl-Al (Cohen and Samoucha, 1984; Samoucha and Cohen, 1985). Foliar spray is appropriate for all types of fungicides, including contact fungicides such as mancozeb (Samoucha and Cohen, 1984b). Spray droplets are allowed to dry for about 2 h before inoculation.

Bioassay with whole plants

Again, two methods of fungicide application are used: (i) soil drench and (ii) foliar spray. Cohen (1979) and Samoucha and Cohen (1985) applied propamocarb (or previcur) and metalaxyl, respectively, to the soil to two-leaf plants (3 weeks old) 1 day before inoculation. Disease control by propamocarb (or previcur) was estimated according to the number of sporangia produced, and frequency of sporangia resistant to metalaxyl was calculated according to the ratio between disease severity in treated and untreated plants.

These authors used foliar spray to test the efficacy of mancozeb. Spray droplets were allowed to dry for about 2 h before inoculation. Disease data were recorded at 7 dpi. The leaf area showing downy mildew symptoms on two inoculated leaves was measured by tracing on a transparent paper and weighing the cut-outs. This method was also used by O'Brien and Weinert (1995) and Mitani et al. (2001).

Efficacy tests in the field

Field trials are rarely used for evaluating the resistance of *P. cubensis* to fungicides (Urban and Lebeda, 2006). Mitani et al. (2003) examined the efficacy of the systemic fungicide cyazofamid in the field: plots with 5–7 plants per plot were arranged randomly (two replicates). Plants were sprayed with the fungicide and allowed to be infected naturally. Disease intensity on individual leaves was evaluated using a 0–4 scale, where 0 is no disease, 1 is 1–5% of the leaf area infected, 2 is 6–25%, 3 is 26–49% and 4 is more than 50%. The disease severity of a plot was expressed by the following formula:

$$\text{Disease severity} = 100 \times (B + 2C + 3D + 4E)/4(A + B + C + D + E),$$

where A, B, C, D and E represent the number of leaves rated at disease index 0, 1, 2, 3 and 4, respectively (Urban and Lebeda, 2006).

Molecular tools

Conventional methods for detecting fungicide resistance are generally labour-intensive and time-consuming when large numbers of isolates are tested. Molecular tools enable rapid detection of fungicide-resistant genotype once the mechanisms of resistance are known. Molecular techniques such as PCR, PCR-restriction fragment length polymorphism (PCR-RFLP), allele-specific PCR and allele-specific real-time PCR have been used successfully to detect fungicide-resistant genotypes in various plant pathogens (Sierotzki and Gisi, 2003; Ma and Michailides, 2005; Zhang *et al.*, 2008). In *P. cubensis*, Ishii (2001) and Ishii *et al.* (2002) reported detecting resistance to strobilurins by using PCR-RFLP. A leaf disc bearing *P. cubensis* spores was used for DNA extraction with a mortar and pestle. A fragment of the pathogen cytochrome *b* gene was amplified by PCR. The products were then treated with the restriction enzyme ItaI, which specifically recognized the mutated sequence in the fragment (Urban and Lebeda, 2006). Sierotzki *et al.* (2011) have developed a PCR technique to detect mutations in *CesA3* of *P. cubensis*, which are responsible for resistance against CAA fungicides. DNA was extracted from sporangia of the pathogen taken from the leaf surface by vacuum suction.

4.4 Fungicides Effective Against *P. cubensis*

Fungicides active against *P. cubensis* are distinguished according to the developmental stage they inhibit in the disease cycle, the biochemical process they affect and their mobility in the plant (Table 4.1). Fungicides that lack any penetration ability into the plant (contact, e.g. mancozeb, inhibits the pathogen prior to its penetration into the tissue) are used preventively. Others that have translaminar mobility (e.g. mandipropamid) are also used preventively. Still others that show systemic translocation in the plant, acropetally and/or basipetally (e.g. phenylamides, fosetyl-Al), may be used preventively or curatively.

Protection of the foliage with contact fungicides in field-grown cucurbits is only partially effective because of the difficulty in applying them to the lower leaf surfaces. There are only a few reports on infection of *P. cubensis* on plant parts other than leaves (Van Haltern, 1933; D'Ercole, 1975).

Systemic fungicides exhibit various levels of systemicity in plants. They can be fully systemic (e.g. metalaxyl) or partially systemic. Partially systemic fungicides are either locally systemic (e.g. dimethomorph) or mesostemic (trifloxystrobin; Margot *et al.*, 1998). A mesostemic fungicide has a high affinity with the

plant surface, is absorbed by the waxy layers of the plant and is redistributed at the plant surface by superficial vapour movement and re-deposition. The term 'quasi-systemic' or 'surface systemic' has been used for kresoxim-methyl (Ypema and Gold, 1999). Systemic fungicides can exhibit preventive, curative or eradicative activity. Curative compounds affect pathogen stages between the penetration and appearance of the first symptoms, while eradicative compounds affect later stages of pathogen development (hyphal growth in plant tissues and sporulation) (Urban and Lebeda, 2006). Mandipropamid penetrates plant tissue and exhibits translaminar activity but no transport within the vascular system of the plant (Cohen and Gisi, 2007).

There are differences in modes of translocation of various systemic fungicides in the plant. It can be apoplastic (acropetal) mobility, i.e. translocation within the free space, cell walls and xylem elements of the plant tissue governed by diffusion and the rate of transpiration. Or, mobility is symplastic (basipetal), i.e. translocation through plasmodesmata from cell to cell, involving uptake and distribution from source to sink via the phloem (Neumann and Jacob, 1995).

4.5 Mechanisms of *P. cubensis* Resistance to Fungicides

Fungicide resistance can be conferred by various mechanisms (Gisi *et al.*, 2000; Ma and Michailides, 2005; Gisi and Sierotzki, 2008), including: an altered target site which reduces the binding of the fungicide; the synthesis of an alternative enzyme capable of substituting the target enzyme; the overproduction of the fungicide target; an active efflux or reduced uptake of the fungicide; and a metabolic breakdown of the fungicide. In addition, some unrecognized mechanisms could also be responsible for fungicide resistance (Urban and Lebeda, 2006).

Since resistance of *P. cubensis* to phenylamide fungicide metalaxyl first appeared, there have been many reports of failure of metalaxyl to control other oomycete pathogens such as *Phytophthora infestans* (e.g. Fisher and

Table 4.1. Fungicides used against *P. cubensis* and several other oomycete pathogens.

Level of systemicity	Cross-resistance group	Common name of compound	Type of activity/ translocation behaviour in plants	Biochemical and physiological mode of action
Fully systemic	Phenylamides	Metalaxyl Mefenoxam Benalaxyl Ofurace	Preventive, curative, eradicative/apoplastic, symplastic, translaminar	Inhibition of rRNA synthesis
	Phosphonates	Fosetyl-Al	Preventive, curative/ apoplastic, symplastic	Inhibition of spore germination, retardation of mycelia development and sporulation, induction of host resistance
	Carbamates	Propamocarb Prothiocarb	Preventive, eradicative/ apoplastic	Multi-site inhibitor, affect membrane permeability
	Cyano-acetamide oximes	Cymoxanil	Preventive, curative/ apoplastic, symplastic, translaminar	(?)
	Benzamide	Fluopicolide	Preventive, curative/ apoplastic, symplastic, translaminar	Delocalization of spectrin-like proteins
Partially systemic	Complex III respiration inhibit (QoI)	Azoxystrobin Fenamidone Trifloxystrobin Kresoxim-methyl Pyraclostrobin	Preventive/translaminar apoplastic (azoxystrobin, fenamid), 'mesostemic' (trifloxystrobin), 'quasi-systemic' (kresoxim-methyl)	Inhibit mitochondrial respiration at the enzyme complex III (Qo site)
	Complex III respiration inhibit (QiI)	Cyazofamid	Preventive, curative, eradicative/ translaminar	Inhibit mitochondrial respiration at the enzyme complex III (Qi site)
Non-systemic	Carboxylic acid amides (CAA)	Dimethomorph Mandipropamid Iprovalicarb Bethiavalicarb Flumorph	All are preventive, translaminar	Cell wall synthesis, Ces3A cellulose synthase inhibitors
	Dinitroanilines	Fluazinam	Preventive/–	Inhibit ATP production
	Miscellaneous	Zoxamide	Preventive/–	(?)
	Multi-site inhibitors	Inorganic copper fungicides (Cu-oxychloride, Cu-hydroxide)	Preventive/–	Multi-site inhibitor
		Organic dithiocarbamate fungicides (e.g. mancozeb)	Preventive/–	Multi-site inhibitor
		Chlorothalonil	Preventive/–	Multi-site inhibitor
		Folpet Other multi-sites	Preventive/–	Multi-site inhibitor

Note: (?) not well and/or exactly known. (Information is modified according to Urban and Lebeda, 2006; Lebeda and Cohen, 2011.)

Hayes, 1984; Deahl *et al.*, 1993; Daayf and Platt, 1999; Shattock, 2002; Stein and Kirk, 2003), *Plasmopara viticola* (Clerjeau and Simone, 1982; Moreau *et al.*, 1987) and *Bremia lactucae* (Crute, 1987). The genetics of metalaxyl resistance in *P. cubensis* is not known. Inheritance studies of metalaxyl resistance in *P. infestans* suggested that a single, partially dominant gene controlled resistance (Shattock, 1988; Shaw and Shattock, 1991). However, Lee *et al.* (1999) reported that one dominant gene plus minor genes were involved in phenylamide resistance. Resistance to metalaxyl is also controlled by a single, incompletely dominant gene in *Phytophthora capsici* (Lucas *et al.*, 1990) and *Phytophthora sojae* (Bhat *et al.*, 1993). Some authors have identified, by using bulk segregant analysis, molecular markers linked to the loci controlling resistance in *P. infestans* (Fabritius *et al.*, 1997; Judelson and Roberts, 1999). So far, six random amplified polymorphic DNA (RAPD) markers have been identified, and mapping assays indicate that one or two semi-dominant loci, MEX, are involved. The data further suggested that not all alleles were functionally equivalent and, in addition, epistatic minor genes interacted with the MEX loci. In biochemical studies, endogenous RNA polymerase II activity of isolated nuclei of *Phytophthora megasperma* and *P. infestans* was highly sensitive to metalaxyl in sensitive isolates but insensitive in resistant isolates, suggesting that a target site mutation was responsible for resistance (Davidse, 1988).

The mechanism of resistance of *P. cubensis* against the strobilurin fungicides azoxystrobin and kresoxim-methyl is well known. In strobilurin-resistant isolates of *P. cubensis*, a single point mutation leading to an amino acid change from glycine to alanine at codon 143 (G143A) of the mitochondrial cytochrome *b* gene conferred resistance (Heaney *et al.*, 2000; Ishii *et al.*, 2000, 2001, 2002). The G143A mutation has been correlated with strobilurin resistance in a wide variety of plant pathogens, e.g. *Mycosphaerella fijiensis* (Sierotzki *et al.*, 2000a), *Blumeria graminis* f. sp. *tritici* (Sierotzki *et al.*, 2000b) and *P. viticola* (Gisi, 2002). Mutation at different codons of the cytochrome *b* gene (Kim *et al.*, 2003) and various substitutions at codon 143 (Avila-Adame and Köller, 2003) were also reported to be

responsible for strobilurin resistance. Miguez *et al.* (2004) reported an additional resistance mechanism of *Mycosphaerella graminicola* to azoxystrobin involving activation of an alternative oxidase, which increased flexibility in respiration and allowed resistant strains to survive in the presence of the fungicide.

Mancozeb, chlorothalonil, fluazinam and zoxamide are multi-site inhibitors that interact non-specifically with many biochemical steps in the metabolism of the pathogen (Urban and Lebeda, 2006). Therefore, the risk is low that resistance to these fungicides will develop.

4.6 Spatial and Temporal Aspects of *P. cubensis* Fungicide Resistance

Fungicides are a major tool for managing cucurbit downy mildew. Unfortunately, single-site inhibitory fungicides are highly effective but bear a high risk of resistance build-up (Gisi, 2002), especially in *P. cubensis*, which bears a high intrinsic potential for developing such resistances. Contact fungicides have a considerably lower risk of resistance build-up due to their multi-site inhibitory nature (McGrath, 2001), but they are less effective because they do not reach the abaxial leaf surfaces of the treated plants (Urban and Lebeda, 2006). The occurrence of field isolates of *P. cubensis* resistant to fungicides is given in Table 4.2.

Metalaxyl resistance

The first failure of metalaxyl was observed in the winter of 1979, 2 years after its commercialization, in cucumbers grown in plastic houses in the Hadera district in Israel. Isolates collected from these plastic houses could not be controlled with a metalaxyl dose 300 times higher than the rate recommended for the control of sensitive strains (Reuveni *et al.*, 1980). In spite of metalaxyl being abandoned in mid-1981, 78 and 92% of the isolates of *P. cubensis* isolates collected during 1983 and 1984, respectively, from many sites in Israel were resistant (Samoucha and Cohen, 1985).

Table 4.2. Occurrence of fungicide resistance/tolerance in *Pseudoperonospora cubensis*.

Chemical group/ chemical class	Common name	Countries where resistant/ tolerant strains occurred
Phenylamides	Metalaxyl	Israel (1980)[a] Greece (1981) Italy (1985) USA (1987) USSR (1992) Australia (1995) Czech Republic (1990/2004)
Strobilurins	Azoxystrobin, kresoxim-methyl	Japan (1999) Taiwan (2001)
	Pyraclostrobin	USA (2004)
Phosphonates	Fosetyl-Al	Israel (1984) Czech Republic (2004)
Carbamates	Propamocarb	Israel (1984)
Phthalimides	Folpet	Israel (1984)
Dithiocarbamates	Mancozeb	Israel (1984)
Carboxyic acid amides	Dimethomorph	Czech Republic (2005), Israel (2006)
	Mandipropamid Iprovalicarb Benthiavalicarb	USA (2007)

Note: [a]Year of first described occurrence. (Information is modified according to Urban and Lebeda, 2006; Olaya *et al.*, 2009; Lebeda and Cohen, 2011; Cohen, unpublished.)

The results indicated that the initial rapid build-up of resistance was due to the strong selection pressure imposed by metalaxyl, but later, after metalaxyl was withdrawn from the market, resistant isolates continued to increase due to their higher fitness. Resistant isolates showed higher fitness than sensitive isolates in competition experiments in the absence of metalaxyl (Cohen *et al.*, 1983). The resistant (R) isolates predominated after a single sporulation cycle, even at an unfavourable initial ratio of 1:20 to the sensitive (S) isolates. This and more sensitive isolates increased (3–6×) the infectivity of resistant isolates when the mixtures of the two were inoculated on to metalaxyl-treated cucumber plants. Metalaxyl-resistant isolates tolerated higher doses of the unrelated fungicides propamocarb, fosetyl-Al and mancozeb, but not cymoxanil, as compared to metalaxyl-sensitive isolates (Cohen and Samoucha, 1984; Cohen *et al.*, 1985). A study by Samoucha and Cohen (1984b) showed that higher doses of mancozeb were required for controlling metalaxyl-resistant than metalaxyl-sensitive isolates of *P. cubensis*

on intact cucumber plants: the ED_{50} of metalaxyl for S and R isolates ranged between 3.12–6.25 and 125–750 µg/ml, respectively (40–120×) and the ED_{50} of mancozeb for S and R isolates ranged between 2–8 and 24–40, respectively (5–12×). It should be noted that since the first appearance of resistant isolates in 1979 until 2010 (31 years), all isolates collected in Israel were resistant to metalaxyl or mefenoxam (the active enantiomer of metalaxyl). In mid-2010, one isolate with intermediate resistance was discovered in *C. moschata* (Y. Cohen, unpublished data).

The worldwide distribution of metalaxyl-resistant isolates of *P. cubensis* has since been confirmed in Greece (Georgopoulos and Grigoriu, 1981), Italy (D'Ercole and Nipoti, 1985), the USA (Moss, 1987) and Russia (Grin'ko, 1992). In Australia, O'Brien and Weinert (1995) reported three levels of sensitivity to metalaxyl of *P. cubensis* isolates collected in the Murrumbidgee Irrigation Area (MIA), New South Wales, and the Burdekin district, Queensland. In floating disc experiments, the EC_{50} values of sensitive, intermediate and resistant isolates were in the ranges of >0.01 to <0.1, >1 to <10 and >10 µg metalaxyl/ml, respectively. They postulated that the reduced (intermediate) sensitivity of *P. cubensis* isolates from MIA probably represented a transitory stage and later collections would show the presence of resistant types.

The occurrence of metalaxyl-resistant isolates of *P. cubensis* and its historical development was also explored in the Czech Republic (Urban and Lebeda, 2004a,b; Hübschová and Lebeda, 2010; Lebeda *et al.*, 2010). A large number of isolates were collected during 2001–2010 and 15 isolates were collected in the former Czechoslovakia during 1985–2000 (Lebeda, 1992; Lebeda and Gadasová, 2002) and screened for reaction to metalaxyl. The results showed (see examples in Table 4.3) a different pattern of resistance build-up compared to Israel. Whereas in Israel build-up was abrupt, expanding all over the country within a year or two, a continuous selection of highly resistant isolates occurred among the Czech populations of *P. cubensis*. Until 2000, metalaxyl-sensitive isolates predominated, although the existence of some resistant isolates has already been proved in 1995. Since 2000, resistant isolates have become predominant, and a gradual selection for higher resistance has been observed during 2000–2009 (Urban and

Lebeda, 2004a,b; Hübschová and Lebeda, 2010; Lebeda *et al.*, 2010).

Strobilurin resistance

Field resistance to strobilurin fungicides was first reported in wheat powdery mildew (*B. graminis* f. sp. *tritici*) in northern Germany in 1998 (Erichsen, 1999). Resistance in cucumber downy mildew was first reported in Japan and Taiwan (Ishii, 2001). In Japan, kresoxim-methyl and azoxystrobin were officially registered in December 1997 and April 1998, respectively. Most cucumber growers followed the manufacturers' recommended rate and applied their products only twice per crop in alternation with other fungicides with different modes of action. However, a reduced efficacy and build-up of highly resistant subpopulations of *P. cubensis* had already been observed in 1999 (Takeda *et al.*, 1999). Heaney *et al.* (2000) stated that resistant subpopulations of *P. cubensis* were present in almost all locations sampled in Japan. Ishii *et al.* (2002) reported that resistant *P. cubensis* isolates persisted in the absence of strobilurins in several greenhouses in Tsukuba and Akeno, Ibaraki. Strobilurin fungicides were removed from practice in 1999. Nevertheless, a high proportion of the isolates were still resistant in 2000

Table 4.3. Temporal shift in occurrence of metalaxyl-resistant isolates of *Pseudoperonospora cubensis* collected in the Czech Republic in 1985–2002.

Isolate[a] (year)	Metalaxyl concentration (µg ai/ml)[b]					
	0	50	100.0	200.0[c]	400.0	800.0
3/85 (1985)	+	–	–	–	–	–
1/88 (1988)	+	–	–	–	–	–
4/95 (1995)	+	–	–	–	–	–
1/98 (1998)	+	–	–	–	–	–
35/01 (2001)	+	+	(–)	(–)	–	–
28/01 (2001)	+	+	+	(–)	–	–
60/02 (2002)	+	+	+	+	(–)	–
43/02 (2002)	+	+	+	+	+	+

Note: [a]For origin and pathogenicity characteristics see Lebeda (1992), Lebeda and Gadasová (2002), Lebeda and Widrlechner (2003, 2004), Lebeda and Urban (2004a,b); [b]leaf-disc bioassay with Ridomil 48 WP (AgroBio Opava Ltd, Opava, Czech Republic); [c]the concentration recommended by the producer (Kužma, 2001, 2005). Reaction (Urban and Lebeda, 2004a,b) of *P. cubensis* isolates to metalaxyl: – = sensitive (no sporulation), (–) = tolerant (limited sporulation), + = resistant (profuse sporulation).
Source: Urban and Lebeda (2004a,b, 2006).

and 2001, suggesting that resistant isolates were not necessarily less fit than sensitive isolates. In *Magnaporthe grisea*, several parameters tested to measure fitness penalties inherent to the mutational changes in the cytochrome *b* gene responsible for quinone outside inhibitor (QoI) resistance revealed that the G143A mutant was not compromised (Avila-Adame and Köller, 2003). Also, cross-resistance between azoxystrobin and kresoxym-methyl was confirmed in *P. cubensis* isolates.

Resistance to fosetyl-Al and other fungicides

Despite the good efficacy of fosetyl-Al in controlling *P. cubensis*, there are reports of highly resistant isolates of *P. cubensis* in the Czech Republic and Israel. Cohen and Samoucha (1984) reported that metalaxyl-resistant isolates of *P. cubensis* collected in 1979 and 1982 in Israel were also resistant to a soil drench of fosetyl-Al at a concentration of 3200 µg ai/ml. In the Czech Republic, the efficacy of fosetyl-Al against *P. cubensis* was monitored in detail during 2001–2004 (Urban and Lebeda, 2004a,b, 2007). In 2001, a high proportion of isolates had a certain level of resistance/tolerance to this fungicide. However, resistance

did not build up in 2002 and 2004. The loss of resistance to fosetyl-Al is probably related to the fact that the epidemics in the Czech Republic are caused by inoculum transported annually by air flows coming from southern and south-eastern areas of Europe (Lebeda, 1990; Lebeda and Cohen, 2011). Such transports allow for annual restoration and extinction of isolates. It might also result from the reduced fitness of the resistant subpopulations relative to the wild-type population.

Cohen and Samoucha (1984) described propamocarb resistance of metalaxyl-resistant *P. cubensis* isolates. That fungicide gave no control as a soil drench at a concentration of 420 g ai/ml.

The efficacy of cymoxanil in controlling *P. cubensis* in intact cucumber and melon plants was studied by Cohen and Grinberger (1987). They showed that the dose required to achieve 90% control of the disease (ED_{90}) ranged between 197–647 and 201–878 µg/ml for metalaxyl-sensitive and metalaxyl-resistant isolates.

Samoucha and Cohen (1988a,b) discovered that fungicidal mixtures, especially those containing cymoxanil, were highly synergistic in controlling downy mildew in intact cucumber plants (Table 4.4). They showed that such mixtures were about twice as effective (relative to their components combined)

Table 4.4. Efficacy of cymoxanil mixtures in controlling downy mildew in cucumber plants.

Fungicides	Mixture ratio	MS isolate ED_{90}, mg/l	SF	MR isolate ED_{90}, mg/l	SF	Resistance factor
Oxadixyl		24		8659		361
Metalaxyl		22		7524		342
Cymoxanil		573		599		1
Mancozeb		28		49		1.8
Oxadixyl + mancozeb	1 : 7	9	3	40	1.4	4.4
Metalaxyl + mancozeb	1 : 7.5	7	3.9	38	1.4	5.4
Cymoxanil + mancozeb	1 : 4	16	2.2	17	3.5	1
Oxadixyl + cymoxanil + mancozeb	1 : 0.4 : 7	15	1.9	18	3.2	1.2
Oxadixyl + cymoxanil + mancozeb	1 : 2 : 7	14	2.4	17	4	1.2

Note: ED_{90} = the dose in mg ai/ml required to reduce the disease by 90% relative to control inoculated untreated plants. SF = synergy factor. Fold increase in efficacy of the mixture relative to its expected efficacy. Expected efficacy of the mixture is calculated according to Wadely (see Samoucha and Cohen, 1988a,b). Resistance factor = the ratio between ED_{90} of MR and ED_{90} of MS.

in controlling a metalaxyl-sensitive isolate and 3.2–4 times more effective in controlling a metalaxyl-resistant isolate.

4.7 Fungicide Resistance in *P. cubensis* in the Czech Republic

Origin and characterization of cucurbit downy mildew isolates

Distribution, host range and damage caused by *P. cubensis* to cucurbits were evaluated repeatedly (at yearly intervals) at *c.*80–100 locations/year in the Czech Republic over the period 2001–2008 (130 locations in 2001, 109 in 2002, 107 in 2003, 110 in 2004, 96 in 2005, 105 in 2006, 91 in 2007 and 76 in 2008). Monitoring was undertaken when plants reached maturity and at harvest time (end of July and August), and sites included domestic vegetable gardens, small private fields and larger commercial production areas in the main cucurbitaceous vegetable production areas (Lebeda and Urban, 2007; Urban and Lebeda, 2007; Hübschová and Lebeda, 2010; Lebeda *et al.*, 2010). Some marginal cucurbit production areas were also surveyed. To assess damage caused by *P. cubensis*, disease incidence and disease prevalence were determined. Disease incidence was expressed as the percentage of surveyed localities and host crops at which *P. cubensis* occurred. Disease prevalence (intensity) was assessed visually by using a graded 0–4 scale, modified for *P. cubensis* (Lebeda and Křístková, 1994).

Fungicide efficacy assays

A floating leaf-disc bioassay (Fig. 4.2) was used to screen 183 isolates of *P. cubensis* (collected in 2001–2009 in the Czech Republic) for sensitivity to the following fungicides: Ridomil Plus 48 WP (active ingredients: 40% Cu-oxychloride, 8% metalaxyl); Ridomil Gold MZ 68 WP (ai: 64% mancozeb, 4% metalaxyl-M); Acrobat MZ (ai: 600 g/l mancozeb, 90 g/l dimethomorph); Aliette 80 WP (ai: 80% fosetyl-Al); Previcur 607 SL (ai: 607 g/l propamocarb); and Curzate K (ai: 77.3%

Cu-oxychloride, 4% cymoxanil). Ridomil Gold MZ 68 WP, Curzate K and Acrobat MZ were tested from 2005. Five concentrations of each fungicide were tested. Leaf discs (15 mm in diameter) were prepared (for methodological details see the relevant section of this chapter) and inoculated 24 h later. Incubation and evaluation were carried out as described above. Three types of reactions were assigned: sensitive, $P \leq 10\%$, tolerant ($10 < P \leq 35\%$) and resistant ($P > 35\%$) (Lebeda and Urban, 2007; Urban and Lebeda, 2007).

Changes in fungicide efficacy

A total of 183 Czech isolates of *P. cubensis* were screened for sensitivity to six major fungicides. The frequency of sensitive/tolerant/resistant isolates was then compared among the fungicides (Lebeda *et al.*, 2010). The fungicides tested could be divided into three groups: propamocarb and fosetyl-Al, towards which no resistance has developed; metalaxyl, metalaxyl-M and cymoxanil, towards which a high resistance has developed (Fig. 4.3); and the CAA fungicide dimethomorph (DMM), towards which a shift from resistance to tolerance has developed (Lebeda and Urban, 2007; Urban and Lebeda, 2007; Hübschová and Lebeda, 2010; Lebeda *et al.*, 2010). Detailed data related to the temporal shift of resistance/tolerance of *P. cubensis* (in total 107 isolates) to metalaxyl, metalaxyl-M, cymoxanil and dimethomorph in the years 2005–2009 are summarized in Fig. 4.3.

In Israel, DMM has been used for 15 years for the control of *P. cubensis*. Isolates resistant to DMM were discovered in 2006, and resistant to other members of the group, iprovalicarb, benthiavalicarb and mandipropamide, in 2008 (Y. Cohen, unpublished; Fig. 4.4). In the USA, mandipropamide-resistant isolates were detected in 2008 in Florida, North Carolina and Michigan (Sierotzki *et al.*, 2011). Resistant isolates showed a resistance factor of >100. Sierotzki *et al.* (2011) showed that a single mutation in codon 1105 in the cellulose synthase *CesA3* gene of the pathogen, from glycine to valine in Israeli isolates, or from glycine to tryptophane in US isolates, was

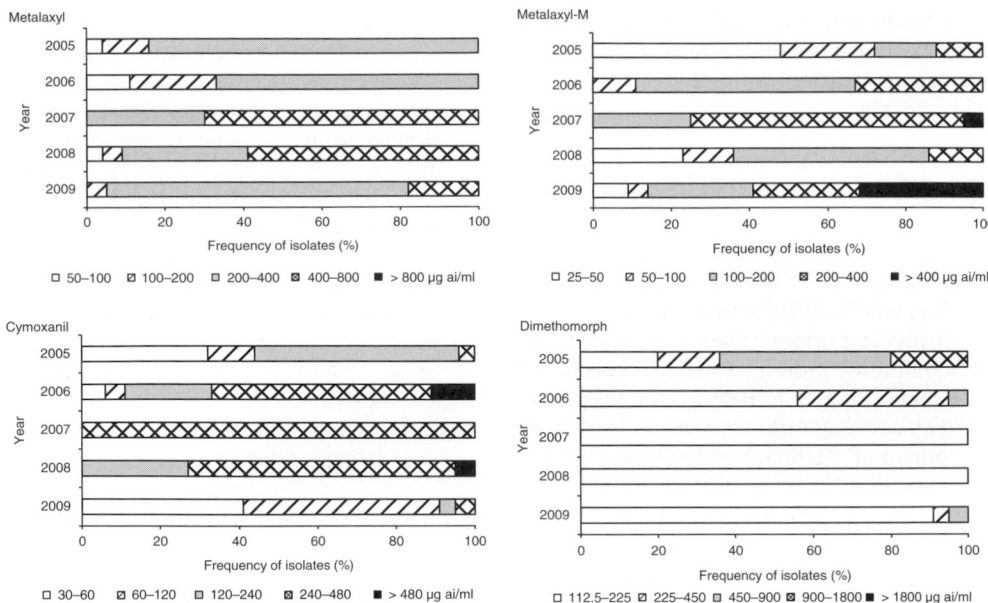

Fig. 4.3. Structure of *P. cubensis* populations in relation to frequency of ED$_{50}$ values for metalaxyl, metalaxyl-M, cymoxanil and dimethomorph in 2005–2009 in the Czech Republic (Lebeda and Pavelkova, unpublished).

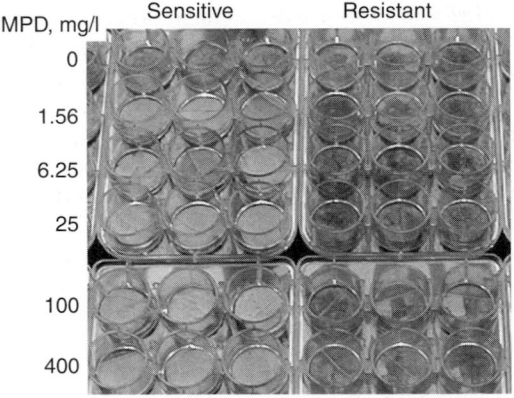

Fig. 4.4. Downy mildew development on leaf discs of cucumber treated with the CAA fungicide mandipropamide (MPD). Leaf discs were floated on MPD solutions of various concentrations and thereafter inoculated with a CAA-sensitive isolate or a CAA-resistant isolate of *P. cubensis*. Photo was taken at 7 dpi. Note that MPD is most effective in controlling the sensitive isolate but not the resistant isolate. The resistant factor is >100.

responsible for resistance. A study by Gisi *et al.* (2007) revealed that in *P. viticola* resistance against CAAs was controlled by two recessive genes. Ojiambo *et al.* (2010) conducted a meta-analysis of 105 reports dealing with chemical control of *P. cubensis* published in the USA during 2000–2008. They showed that fluopicolide was most effective, carbamates and quinone inside inhibitors (QiIs) were moderately effective, phenylamides and

QoIs lowly effective and mandipropamide least effective. *P. cubensis* seems to be capable of overcoming the control efficacy of fungicides quickly.

4.8 Conclusions and Future Prospects

- Cucurbit downy mildew caused by the oomycete *P. cubensis* is a most devastating, worldwide distributed and economically important disease of almost all commonly grown cucurbits (Lebeda and Cohen, 2011). It belongs to 'high risk pathogens' with a high evolutionary potential (Urban and Lebeda, 2006; Lebeda and Cohen, 2011).
- *P. cubensis* is highly variable in host specificity, pathogenicity, fungicide resistance and genetic polymorphism (Lebeda and Cohen, 2011).
- Fungicides are the major tool in protecting cucurbits against downy mildew. Resistant cultivars are available in some but not all cucurbits. When available, they are often used in combination with fungicides to minimize selection of isolates able to overcome genetic resistance (Urban and Lebeda, 2006; Lebeda and Cohen, 2011).
- Fungicide resistance in *P. cubensis* was first detected in 1979 (Reuveni *et al.*, 1980) and since then has occurred worldwide (Gisi, 2002; Urban and Lebeda, 2006, 2007; Gisi and Sierotzki, 2008; Hübschová and Lebeda, 2010; Lebeda *et al.*, 2010; Lebeda and Cohen, 2011).
- Many highly effective fungicides have been introduced to the market in the past three decades (Gisi, 2002; Tomlin, 2003; Kuck and Russell, 2006; Urban and Lebeda, 2006; Gisi and Sierotzki, 2008; Leadbeater and Gisi, 2010). They are mostly single-site inhibitors which bear a high risk of resistance development (Gisi, 2002; Urban and Lebeda, 2006, 2007; Gisi and Sierotzki, 2008).
- The mechanism of resistance is known for strobilurins (Takeda *et al.*, 1999; Ishii *et al.*, 2002) and CAAs (Sierotzki *et al.*, 2011).

- In some countries, an effort was made to obtain comprehensive data on fungicide resistance, its geographic distribution, spatial and temporal variability and changes in resistance frequencies: in Central Europe (Czech Republic; Urban and Lebeda, 2004a,b, 2007; Lebeda and Urban, 2007; Hübschová and Lebeda, 2010; Lebeda *et al.*, 2010); the USA (Holmes and Oijambo, 2009; Olaya *et al.*, 2009); China (Zhang *et al.*, 2008; Zhu *et al.*, 2008; Wang *et al.*, 2009); and Japan (Okada and Furukawa, 2008).
- Various methods of investigating fungicide resistance in *P. cubensis* have been described in this paper.
- More studies on the control of downy mildew are needed, especially on the mechanisms of fungicide resistance, its genetic background and inheritance, persistence, fitness of resistant versus wild-type populations and epidemiology and spatial and temporal changes in resistance frequencies. International cooperation and coordination are required for significant progress in this field and for more efficient ways to protect cucurbits against *P. cubensis*.
- Systemic plant resistance induced by chemicals or biotic agents may be incorporated into integrated pest management (IPM) programmes to combat *P. cubensis* in cucurbits and thus reduce fungicide use (Lebeda and Cohen, 2011).

Acknowledgements

This work was supported by grants NAZV QH 71229, MSM 6198959215 and PrF_2011_001. Some isolates of cucurbit downy mildew used in this chapter are maintained in the Czech National Collection of Microorganisms (http://www.vurv.cz/collections/vurv.exe/) at the Department of Botany, Palacký University in Olomouc (http://botany.upol.cz).

References

Anon. (1982) FAO Method No 30. *FAO Plant Protection Bulletin* Volume 30/2.

Avila-Adame, C. and Köller, W. (2003) Characterization of spontaneous mutants of *Magnaporthe grisea* expressing stable resistance to the Qo-inhibiting fungicide azoxystrobin. *Current Genetics* 42, 332–338.

Bates, D.M., Robinson, R.W. and Jeffrey, C. (eds) (1990) *Biology and Utilization of the Cucurbitaceae*. Comstock Publishers Association, Ithaca, New York.

Berkeley, M.S. and Curtis, A. (1868) *Peronospora cubensis. Journal of the Linnean Society of Botany* 10, 363.

Bhat, R.G., McBlain, B.A. and Schmitthenner, A.F. (1993) The inheritance of resistance to metalaxyl and to fluorophenylalanine in matings of homothalic *Phytophthora sojae. Mycological Research* 97, 865–870.

Clerjeau, M. and Simone, H. (1982) Apparition en France de souches de mildiou (*Plasmopara viticola*) résistantes aux fongicides de la famille des anilides (métalaxyl, milfurame). *Progrés Agricole et Viticole* 99, 59–61.

Cohen, Y. (1979) A new systemic fungicide against the downy mildew disease of cucumbers. *Phytopathology* 69, 433–436.

Cohen, Y. (1981) Downy mildew of cucurbits. In: Spencer, D.M. (ed.) *The Downy Mildews*. Academic Press, London, pp. 341–354.

Cohen, Y. and Coffey, M.D. (1986) Systemic fungicides and the control of Oomycetes. *Annual Review of Phytopathology* 24, 311–338.

Cohen, Y. and Gisi, U. (2007) Differential activity of carboxylic acid amide (CAA) fungicides against various developmental stages of *Phytophthora infestans. Phytopathology* 97, 1274–1283.

Cohen, Y. and Grinberger, M. (1987) Control of the metalaxyl-resistant causal agents of late blight in potato and tomato and downy mildew in cucumber by cymoxanil. *Phytopathology* 77, 1283–1288.

Cohen, Y. and Samoucha, Y. (1984) Cross-resistance to four systemic fungicides in metalaxyl-resistant strains of *Phytophthora infestans* and *Pseudoperonospora cubensis. Plant Disease* 68, 137–139.

Cohen, Y., Reuveni, M. and Samoucha, Y. (1983) Competition between metalaxyl-resistant and sensitive strains of *Pseudoperonospora cubensis* on cucumber plants. *Phytopathology* 73, 1516–1520.

Cohen, Y., Eyal, H. and Sheinboim, Y. (1985) Efficacy of cymoxanil in controlling metalaxyl-resistant isolates of *Phytophthora infestans* and *Pseudoperonospora cubensis. 1985 Fungicides for Crop Protection*. BCPC Monograph No. 31. British Crop Protection Council, Croydon, UK, pp. 307–310.

Cohen, Y., Meron, I., Mor, N. and Zuriel, S. (2003) A new pathotype of. *Pseudoperonospora cubensis* causing downy mildew in cucurbits in Israel. *Phytoparasitica* 31, 458–466.

Colucci, S.J., Wehner, T.C. and Holmes, G.J. (2006) The downy mildew epidemic of 2004 and 2005 in the eastern United States. In: Holmes, G.J. (ed.) *Proceedings Cucurbitaceae 2006*. Universal Press, Raleigh, North Carolina, pp. 403–411.

Crute, I.R. (1987) The occurrence, characteristics, distribution, genetics and control of a metalaxyl-resistant genotype of *Bremia lactucae* in the United Kingdom. *Plant Disease* 71, 763–767.

Daayf, F. and Platt, H.W. (1999) Assessment of mating types and resistance to metalaxyl of Canadian populations of *Phytophthora infestans* in 1997. *American Journal of Potato Research* 76, 287–295.

Davidse, L.C. (1988) Phenylamide fungicides: mechanism of action and resistance. In: Delp, C.J. (ed.) *Fungicide Resistance in North America*. APS Press, St Paul, Minnesota, pp. 63–65.

Deahl, K.L., Inglis, D.A. and DeMuth, S.P. (1993) Testing for resistance to metalaxyl in *Phytophthora infestans* isolates from Northwestern Washington. *American Potato Journal* 70, 779–795.

D'Ercole, N. (1975) La peronoospora del cetriolo in coltura protetta. *Informatore Fitopatologia* 25, 11–13.

D'Ercole, N. and Nipoti, P. (1985) Downy mildew of melon due to *Pseudoperonospora cubensis* (Berk. et Curt.) Rostow. *Informatore Fitopatologie* 35, 61–63.

De Waard, M.A., Georgopoulos, S.G., Hollomon, D.W., Ishii, H., Leroux, P., Ragsdale, N.N., *et al.* (1993) Chemical control of plant diseases: problems and prospects. *Annual Review of Phytopathology* 31, 403–421.

Doruchowski, R. and Lakowska-Ryk, E. (1992) Inheritance of resistance to downy mildew (*Pseudoperonospora cubensis* Berk. & Court.) in *Cucumis sativus*. In: Doruchowski, R.W., Kozik, E. and Niemirowicz-Szczytt, K. (eds) *Proceedings of Cucurbitaceae 1992, The 5th EUCARPIA Cucurbitaceae Symposium*. Vegetable Research Institute, Skierniewice, Poland, pp. 66–69.

Erichsen, E. (1999) Problems in mildew control in northern Germany. *Getreide* 5, 44–46.

Fabritius. A.L., Shattock, R.C. and Judelson, H.S. (1997) Genetic analysis of metalaxyl insensitivity loci in *Phytophthora infestans* using linked DNA markers. *Phytopathology* 87, 1034–1040.

Fisher, D.J. and Hayes, A.L. (1984) Studies of mechanism of metalaxyl fungitoxicity and resistance to metalaxyl. *Crop Protection* 3, 177–185.

Forsberg, A.S. (1986) Downy mildew – *Pseudoperonospora cubensis* in Swedish cucumber fields. *Växtskyddsnotiser* 50, 17–19.

Georgopoulos, S.G. and Grigoriu, A.C. (1981) Metalaxyl-resistant strains of *Pseudoperonospora cubensis* in cucumber greenhouses of southern Greece. *Plant Disease* 65, 729–731.

Gisi, U. (2002) Chemical control of downy mildews. In: Spencer-Phillips, P.T.N., Gisi, U. and Lebeda, A. (eds) *Advances in Downy Mildew Research*. Kluwer Academic Publishers, Dordrecht, The Netherlands, pp. 119–159.

Gisi, U. and Sierotzki, H. (2008) Fungicide modes of action and resistance in downy mildews. *European Journal of Plant Pathology* 122, 157–167.

Gisi, U., Chin, K.M., Knapova, G., Färber, R.K., Mohr, U., Parisi, S., *et al.* (2000) Recent developments in elucidating modes of resistance to phenylamide, DMI and strobilurin fungicides. *Crop Protection* 19, 863–872.

Gisi, U., Waldner, M., Kraus, N., Dubuis, P.H. and Sierotzki, H. (2007) Inheritance of resistance to carboxylic acid amide (CAA) fungicides in *Plasmopara viticola*. *Plant Pathology* 56, 199–208.

Grin'ko, N.N. (1992) Tolerance of the causal agent of downy mildew of cucumber. *Zashchita Rastenii (Moskva)*, 12, 14. [Abstract in *Review of Plant Pathology* (1993) 72, 440. Abstract No. 3805.]

Heaney, S.P., Hall, A.A., Davies, S.A. and Olaya, G. (2000) Resistance to fungicides in the QoI-STAR cross-resistance group: current perspectives. In: *Brighton Crop Protection Conference*. BCPC Publications, Croydon, UK, pp. 755–762.

Holmes, G.J. and Ojiambo, P. (2009) Chemical control of cucurbit downy mildew: a summary of field experiments in the US. *Phytopathology* 99, S171.

Holmes, G.J. and Thomas, C. (2009) The history and re-emergence of cucurbit downy mildew. *Phytopathology* 99, S171.

Holmes, G.J., Main, C.E. and Keever, Z.T. III (2004) Cucurbit downy mildew: a unique pathosystem for disease forecasting. In: Spencer-Phillips, P.T.N. and Jeger, M. (eds) *Advances in Downy Mildew Research*, Volume 2. Kluwer Academic Publishers, Dordrecht, The Netherlands, pp. 69–80.

Hübschová, J. and Lebeda, A. (2010) Fungicide effectiveness on Czech populations of *Pseudoperonospora cubensis*. *Acta Horticulturae* 871, 457–463.

Ishii, H. (2001) DNA-based approaches for diagnosis of fungicide resistance. In: Clark, J. and Yamaguchi, I. (eds) *Agrochemical Resistance-extent, Mechanism and Detection*. ACS Symposium Series 808, American Chemical Society, Washington, DC, pp. 242–259.

Ishii, H., Fraaije, B.A., Noguchi, K., Nishimura, K., Takeda, T., Amano, T., *et al.* (2000) Cytochrome *b* genes in phytopathogenic fungi and their involvement in strobilurin resistance. *Japanese Journal of Phytopathology* 66, 183 (Abstract).

Ishii, H., Fraaije, B.A., Sugiyama, T., Noguchi, K., Nishimura, K., Takeda, T., *et al.* (2001) Occurrence and molecular characterization of strobilurin resistance in cucumber powdery mildew and downy mildew. *Phytopathology* 91, 1166–1171.

Ishii, H., Sugiyama, T., Nishimura, K. and Ishikawa, Y. (2002) Strobilurin resistance in cucumber pathogens: persistence and molecular diagnosis of resistance. In: Dehne, H.W., Gisi, U., Kuck, K.H., Russell, P.E. and Lyr, H. (eds) *Modern Fungicides and Antifungal Compounds III, The 13th International Reinhardsbrunn Symposium*. AgroConcept, Bonn, Germany, pp. 149–159.

Iwata, Y. (1942) Specialization of *Pseudoperonospora cubensis* (Berk. et Curt.) Rostow. II. Comparative studies of the morphologies of the fungi from *Cucumis sativus* L. and *Cucurbita moschata* Duchesne. *Annals of the Phytopathological Society of Japan* 11, 172–185 [in Japanese].

Iwata, Y. (1953a) Specialization in *Pseudoperonospora cubensis* (Berk. et Curt.) Rostow. IV. Studies on the fungus from Oriental pickling melon (*Cucumis melo* var. *conomon* Makino). *Bulletin of the Faculty of Agriculture, Mie University* 6, 30–35.

Iwata, Y. (1953b) Specialization in *Pseudoperonospora cubensis* (Berk. et Curt.) Rostow. V. on the fungus from Calabash gourd *Lagenaria vulgaris* Ser. var. *clavata* Ser. *Bulletin of the Faculty of Agriculture, Mie University* 6, 32–36.

Judelson, H.S. and Roberts, S. (1999) Multiple loci determining insensitivity to phenylamide fungicides in *Phytophthora infestans*. *Phytopathology* 89, 754–760.

Kim, Y.S., Dixon, E.W., Vincelli, P. and Farman, M.L. (2003) Field resistance to strobilurin (QoI) fungicides in *Pyricularia grisea* caused by mutations in the mitochondrial cytochrome *b* gene. *Phytopathology* 93, 891–900.

Knight, S.C., Anthony, V.M., Brady, A.M., Greenland, A.J., Heaney, S.P., Murray, D.C., *et al.* (1997) Rationale and perspectives on the development of fungicides. *Annual Review of Phytopathology* 35, 349–372.

Kuck, K.H. and Russell, E.P. (2006) FRAC: combined resistance risk assessment. *Aspects of Applied Biology* 78, 3–10.

Kužma, Š. (ed.) (2001) Seznam registrovaných přípravků na ochranu rostlin 2001 (List of Registered Compounds for Plant Protection 2001). Státní rostlinolékařská správa (State Phytosanitary Administration), Praha, Czech Republic.

Kužma, Š. (ed.) (2005) Seznam registrovaných přípravků na ochranu rostlin 2005 (List of Registered Compounds for Plant Protection 2005). Státní rostlinolékařská správa (State Phytosanitary Administration), Praha, Czech Republic.

Leadbeater, A. and Gisi, U. (2010) The challenges of chemical control of plant diseases. In: Gisi, U., Chet, I. and Gullino, M.L. (eds) *Recent Developments in Management of Plant Diseases. Plant Pathology in the 21st Century, 1.* Springer, Dordrecht, The Netherlands, pp. 3–17.

Lebeda, A. (1986) Epidemic occurrence of *Pseudoperonospora cubensis* in Czechoslovakia. *Temperate Downy Mildews Newsletter* 4, 15–17.

Lebeda, A. (1990) Biology and ecology of cucurbit downy mildew In: Lebeda, A. (ed.) *Cucurbit Downy Mildew.* Czechoslovak Scientific Society for Mycology by Czechoslovak Academy of Sciences, Praha, pp. 13–46.

Lebeda, A. (1992) Screening of wild *Cucumis* species against downy mildew (*Pseudoperonospora cubensis*) isolates from cucumbers. *Phytoparasitica* 20, 203–210.

Lebeda, A. (1999) *Pseudoperonospora cubensis* on *Cucumis* spp. and *Cucurbita* spp. – resistance breeding aspects. *Acta Horticulturae* 492, 363–370.

Lebeda, A. and Bartoš, P. (1988) Methods of maintenance of biotrophic plant pathogenic fungi. *Sborník ÚVTIZ – Ochrana rostlin* (Proceedings ÚVTIZ – Plant Protection) 24, 155–159.

Lebeda, A. and Cohen, Y. (2011) Cucurbit downy mildew (*Pseudoperonospora cubensis*) – biology, ecology, epidemiology, host–pathogen interactions and control. *European Journal of Plant Pathology* 129, 157–192 (doi: 10.1007/s10658-010-9658-1).

Lebeda, A. and Gadasová, V. (2002) Pathogenic variation of *Pseudoperonospora cubensis* in the Czech Republic and some other European countries. *Acta Horticulturae* 588, 137–141.

Lebeda, A. and Křístková, E. (1994) Field resistance of *Cucurbita* species to powdery mildew (*Erysiphe cichoracearum*). *Journal of Plant Diseases and Protection* 101, 598–603.

Lebeda, A. and Schwinn, F.J. (1994) The downy mildews – an overview of recent research progress. *Journal of Plant Diseases and Protection* 101, 225–254.

Lebeda, A. and Urban, J. (2004a) Distribution, harmfulness and pathogenic variability of cucurbit downy mildew in the Czech Republic. *Acta fytotechnica et zootechnica* 7, 170–173.

Lebeda, A. and Urban, J. (2004b) Disease impact and pathogenicity variation in Czech populations of *Pseudoperonospora cubensis*. In: Lebeda, A. and Paris, H.S. (eds) *Progress in Cucurbit Genetics and Breeding Research*, Proceedings of Cucurbitaceae 2004, the 8th EUCARPIA Meeting on Cucurbit Genetics and Breeding. Palacký University in Olomouc, Olomouc, Czech Republic, pp. 267–273.

Lebeda, A. and Urban, J. (2007) Temporal changes in pathogenicity and fungicide resistance in *Pseudoperonospora cubensis* populations. *Acta Horticulturae* 731, 327–336.

Lebeda, A. and Urban, J. (2010) Screening for resistance to cucurbit downy mildew (*Pseudoperonospora cubensis*). In: Spencer, M.M. and Lebeda, A. (eds) *Mass Screening Techniques for Selecting Crops Resistant to Disease.* International Atomic Energy Agency (IAEA), Vienna, Chapter 18, pp. 285–294.

Lebeda, A. and Widrlechner, M.P. (2003) A set of Cucurbitaceae taxa for differentiation of *Pseudoperonospora cubensis* pathotypes. *Journal of Plant Diseases and Protection* 110, 337–349.

Lebeda, A. and Widrlechner, M.P. (2004) Response of wild and weedy *Cucurbita* L. to pathotypes of *Pseudoperonospora cubensis* (Berk. et Curt.) Rostov. (Cucurbit downy mildew). In: Spencer-Phillips, P. and Jeger, M. (eds) *Advances in Downy Mildew Research*, Volume 2. Kluwer Academic Publishers, Dordrecht, The Netherlands, pp. 203–210.

Lebeda, A., Widrlechner, M.P. and Urban, J. (2006) Individual and population aspects of interactions between cucurbits and *Pseudoperonospora cubensis*: pathotypes and races. In: Holmes, G.J. (ed.) *Proceedings of Cucurbitaceae 2006.* Universal Press, Raleigh, North Carolina, pp. 453–467.

Lebeda, A., Widrlechner, M.P., Staub, J., Ezura, H., Zalapa, J. and Křístková, E. (2007) Cucurbits (Cucurbitaceae; *Cucumis* spp., *Cucurbita* spp., *Citrullus* spp.). In: Singh, R. (ed.) *Genetic Resources, Chromosome Engineering, and Crop Improvement Series, Volume 3 – Vegetable Crops.* CRC Press, Boca Raton, Florida, Chapter 8, pp. 271–376.

Lebeda, A., Hübschová, J. and Urban, J. (2010) Temporal population dynamics of *Pseudoperonospora cubensis.* In: Thies, J.A., Kousik, S. and Levi, A. (eds) *Cucurbitaceae 2010 Proceedings.* American Society for Horticultural Science, Alexandria, Virginia, pp. 240–243.

Lee, T.Y., Mizubuti, E.S. and Fry, W.E. (1999) Genetics of metalaxyl resistance in *Phytophthora infestans. Fungal Genetics and Biology* 26, 118–130.

Lucas, J.A., Greer, G., Oudemans, P.V. and Coffey, M.D. (1990) Fungicide sensitivity in somatic hybrids of *Phytophthora capsici* by protoplast fusion. *Physiological and Molecular Plant Pathology* 36, 175–187.

Ma, Z. and Michailides, T.J. (2005) Advances in understanding molecular mechanisms of fungicide resistance and molecular detection of resistant genotypes in phytopathogenic fungi. *Crop Protection* 24, 853–863.

McDonald, B.A. and Linde, C. (2002) Pathogen population genetics, evolutionary potential and durable resistance. *Annual Review of Phytopathology* 40, 349–379.

McGrath, M.T. (2001) Fungicide resistance in cucurbit powdery mildew: experiences and challenges. *Plant Disease* 85, 236–245.

Margot, P., Huggenberger, F., Amrein, J. and Weiss, B. (1998) CGA 279202: a new broad-spectrum strobilurin fungicide. In: *Brighton Crop Protection Conference.* BCPC Publications, Croydon, UK, pp. 375–382.

Miguez, M., Reeve, C., Wood, P.M. and Hollomon, D.W. (2004) Alternative oxidase reduces the sensitivity of *Mycosphaerella graminicola* to QoI fungicides. *Pest Management Science* 60, 3–7.

Mitani, S., Araki, S., Yamaguchi, T., Takii, Y., Ohshima, T. and Matsuo, N. (2001) Biological properties of the novel fungicide cyazofamid against *Phytophthora infestans* on tomato and *Pseudoperonospora cubensis* on cucumber. *Pest Management Science* 58, 139–145.

Mitani, S., Kamachi, K., Sugimoto, K., Araki, S. and Yamaguchi, T. (2003) Control of cucumber downy mildew by cyazofamid. *Journal of Pesticide Science* 28, 64–68.

Moreau, C., Clerjeau, M. and Morziéres, J.P. (1987) Bilan des essais détection de souches de *Plasmopara viticola* résistantes aux anilides anti-oomycetes (métalaxyl, ofurace, oxadixyl et benalaxyl) dans le vignoble français en 1987. Rapport G.R.I.S.P. de Bordeaux, Pont de la Maye, France.

Moss, M.A. (1987) Resistance to metalaxyl in the *Pseudoperonospora cubensis* population causing downy mildew of cucumber in South Florida. *Plant Disease* 71, 1045.

Neumann, S. and Jacob, F. (1995) Principles of uptake and systemic transport of fungicides within plants. In: Lyr, H. (ed.) *Modern Selective Fungicides, 2nd edn.* Gustav Fischer, Jena, Germany, pp. 53–73.

O'Brien, R.G.O. and Weinert, M.P. (1995) Three metalaxyl sensitivity levels in Australian isolates of *Pseudoperonospora cubensis* (Berk. et Curt.) Rost. *Australian Journal of Experimental Agriculture* 35, 543–546.

Ojiambo, P.S., Paul, P.A. and Holmes, G.J. (2010) A quantitative review of fungicide efficacy for managing downy mildew in cucurbits. *Phytopathology* 100, 1066–1076.

Okada, K. and Furukawa, M. (2008) Occurrence and countermeasure of fungicide-resistant pathogens in vegetable field of Osaka prefecture. *Journal of Pesticide Science* 33, 326–329.

Olaya, G., Kuhn, P., Hert, A., Holmes, G. and Colucci, S. (2009) Fungicide resistance in cucurbit downy mildew. *Phytopathology* 99, S171.

Palti, J. (1974) The significance of pronounced divergences in the distribution of *Pseudoperonospora cubensis* on its crop hosts. *Phytoparasitica* 2, 109–115.

Palti, J. and Cohen, Y. (1980) Downy mildew of cucurbits (*Pseudoperonospora cubensis*). The fungus and its hosts, distribution, epidemiology, and control. *Phytoparasitica* 8, 109–147.

Pavelkova, J., Lebeda, A. and Sedláková, B. (2011) First report of *Pseudoperonospora cubensis* on *Cucurbita moschata* in the Czech Republic. *Plant Disease* 95, 878–879.

Perchepied, L., Bardin, M., Dogimont, C. and Pitrat, M. (2005) Relationship between loci conferring downy mildew and powdery mildew in melon assessed by quantitative trait loci mapping. *Phytopathology* 95, 556–565.

Reuveni, M., Eyal, H. and Cohen, Y. (1980) Development of resistance to metalaxyl in *Pseudoperonospora cubensis. Plant Disease* 64, 1108–1109.

Robinson, R.W. and Decker-Walters, D.S. (1997) *Cucurbits.* Crop Production Science in Horticulture Series. CAB International, Wallingford, UK.

Rondomanski, W. (1988) Downy mildew on cucumber – a serious problem in Poland. Abstracts of papers 5th International Congress of Plant Pathology, Kyoto, Japan 1988. Poster P.VIII-2-48.

Samoucha, Y. and Cohen, Y. (1984a) Differential sensitivity to mancozeb of metalaxyl-sensitive and metal-axyl-resistant isolates of *Pseudoperonospora cubensis*. *Phytopathology* 74, 1437–1439.

Samoucha, Y. and Cohen, Y. (1984b) Differential sensitivity of mancozeb to metalaxyl-sensitive and -resist-ant isolates of *Pseudoperonospora cubensis*. *Phytopathology* 74, 1437–1439.

Samoucha, Y. and Cohen, Y. (1985) Occurrence of metalaxyl-resistant isolates of *Pseudoperonospora cubensis* in Israel: a 5-year survey. *Bulletin OEPP/EPPO* 15, 419–422.

Samoucha, Y. and Cohen, Y. (1988a) Synergistic interactions of cymoxanil mixtures in the control of meta-laxyl-resistant *Phytophthora infestans* of potato. *Phytopathology* 78, 636–640.

Samoucha, Y. and Cohen, Y. (1988b) Synergism in fungicide mixtures against *Pseudoperonospora cubensis*. *Phytoparasitica* 16, 337–342.

Sarris, P.F., Abdelhalim, M., Kitner, M., Skandalis, N., Panopoulos, N.J., Doulis, A.G., *et al.* (2009) Molecular polymorphism between populations of *Pseudoperonospora cubensis* from Greece and the Czech Republic and their phytopathological and phylogenetic implications. *Plant Pathology* 58, 933–943.

Schwinn, F.J. and Staub, T. (1995) Oomycetes fungicides: phenylamides and other fungicides against Oomycetes. In: Lyr, H. (ed.) *Modern Selective Fungicides, 2nd edn*. Gustav Fischer, Jena, Germany, pp. 323–346.

Shattock, R.C. (1988) Studies on the inheritance of resistance to metalaxyl in *Phytophthora infestans*. *Plant Pathology* 37, 4–11.

Shattock, R.C. (2002) *Phytophthora infestans*: populations, pathogenicity and phenylamides. *Pesticide Management Science* 58, 944–950.

Shaw, D.S. and Shattock, R.C. (1991) Genetics of *Phytophthora infestans*: the Mendelian approach. In: Lucas, J.A., Shattock, R.C., Shaw, D.S. and Cooke, L.R. (eds) *Phytophthora*. Cambridge University Press, Cambridge, UK, pp. 218–230.

Sierotzki, H. and Gisi, U. (2003) Molecular diagnostics for fungicide resistance in plant pathogens. In: Voss, G. and Ramos, G. (eds) *Chemistry of Crop Protection – Progress and Perspectives in Science and Regulation*. 10th IUPAC International Congress on the Chemistry of Crop Protection, 4–9 August 2002, Basel, Switzerland. Wiley-VCH, New York, pp. 71–88.

Sierotzki, H., Parisi, S., Steinfeld, U., Tenzer, I., Poirey, S. and Gisi, U. (2000a) Mode of resistance to respira-tion inhibitors at the cytochrome bc$_1$ enzyme complex of *Mycosphaerella fijiensis* field isolates. *Pest Management Science* 56, 833–841.

Sierotzki, H., Wullschleger, J. and Gisi, U. (2000b) Point mutation in cytochrome *b* gene conferring resist-ance to strobilurin fungicides in *Erysiphe* f.sp. *tritici* field isolates. *Pesticide Biochemistry and Physiology* 68, 107–112.

Sierotzki, H., Blum, M., Olaya, G., Waldner, Cohen, Y. and Gisi, U. (2011) Sensitivity to CAA fungicides and frequency of mutations in cellulose synthase *CesA3* gene of oomycete pathogen populations. In: Dehne, H.W., Gisi, U., Kuck, K.H., Russell, P.E. and Lyr, H. (eds) *Modern Fungicides and Antifungal Compounds VI*. DPG-Verlag, Braunschweig, pp. 151–154.

Stein, J.M. and Kirk, W.W. (2003) Variations in the sensitivity of *Phytophthora infestans* isolates from differ-ent genetic backgrounds to dimethomorph. *Plant Disease* 87, 1283–1289.

Tahvonen, R. (1985) Downy mildew of cucurbits found for the first time in Finland. *Växtskyddsnotiser* 49, 42–44.

Takeda, T., Kawagoe, Y., Uchida, K., Fuji, M. and Amano, T. (1999) The appearance of resistant isolates to strobilurins (Abstr.). *Annales Phytopathological Society of Japan* 65, 655.

Tomlin, S.D.C. (ed.) (2003) *The Pesticide Manual: A World Compendium*. 13th edn. British Crop Protection Council, Alton, Hampshire, UK.

Towsend, G.R. and Heuberger, W. (1943) Methods for estimating losses caused by diseases in fungicide experiments. *Plant Disease Reporter* 27, 340–343.

Urban, J. and Lebeda, A. (2004a) Differential sensitivity to fungicides in Czech population of *Pseudoperonospora cubensis*. In: Lebeda, A. and Paris, H.S. (eds) *Progress in Cucurbit Genetics and Breeding Research. Proceedings of Cucurbitaceae 2004, the 8th EUCARPIA Meeting on Cucurbit Genetics and Breeding*. Palacký University in Olomouc, Olomouc, Czech Republic, pp. 275–280.

Urban, J. and Lebeda, A. (2004b) Resistance to fungicides in population of cucurbit downy mildew in the Czech Republic. *Acta fytotechnica et zootechnica* 7, 327–329.

Urban, J. and Lebeda, A. (2006) Fungicide resistance in cucurbit downy mildew – methodological, biologi-cal and population aspects. *Annals of Applied Biology* 149, 63–75.

Urban, J. and Lebeda, A. (2007) Variation for fungicide resistance in Czech populations of *Pseudoperonospora cubensis. Journal of Phytopathology* 155, 143–151.

Van Haltern, F. (1933) Spraying cantaloups for the control of downy mildew and other diseases. *Georgia Agricultural Experiment Station Bulletin 175.* Agricultural Experiment Station, Griffin, The University of Georgia, Athens, pp. 1–53.

Velichi, E. (2009) Dynamics of appearance and evolution to the watermelon (*Citrullus lanatus* L.), of downy mildew [*Pseudoperonospora cubensis* (Berk. et Curt.) Rostow.], in the rainy years 2004, 2005, in Baragan field, (Braila area). *Research Journal of Agricultural Science* 41, 345–350.

Wang, H.C., Zhou, M.G., Wang, J.X., Chen, C.J., Li, H.X. and Sun, H.Y. (2009) Biological mode of action of dimethomorph on *Pseudoperonospora cubensis* and its systemic activity in cucumber. *Agricultural Sciences in China* 8, 172–181.

Ypema, H.L. and Gold, R.E. (1999) Kresoxim-methyl: modification of a naturally occurring compound to produce a new fungicide. *Plant Disease* 83, 4–19.

Zhang, X., Chen, Y., Zhang, Y.J. and Zhou, M.G. (2008) Occurrence and molecular characterization of azoxystrobin resistance in cucumber downy mildew in Shandong province in China. *Phytoparasitica* 36, 136–143.

Zhu, S.S., Liu, P.F., Liu, X.L., Li, J.Q., Yuan, S.K. and Sil, N.G. (2008) Assessing the risk of resistance in *Pseudoperonospora cubensis* to the fungicide flumorph *in vitro. Pest Management Science* 64, 255–261.

5 Resistance to Azole Fungicides in *Mycosphaerella graminicola*: Mechanisms and Management

Hans J. Cools and Bart A. Fraaije

Department of Plant Pathology and Microbiology, Rothamsted Research, Harpenden, UK

5.1 Introduction

Azole (imidazole and triazole) fungicides are the most widely used class of antifungal agents for the control of fungal diseases of humans, animals and plants. Azoles have a single-site mode of action, binding to the sterol 14α-demethylase (CYP51) enzyme, thereby preventing the production of ergosterol, an essential component of fungal cell membranes. In agriculture, the first azoles, the triazole triadimefon and the imidazole imazalil, were introduced in the early 1970s (Russell, 2005). During the same period, the imidazole clotrimazole was introduced into the clinic, primarily for the topical control of the opportunistic pathogen, *Candida albicans*. Interestingly, despite their long-term use and single biochemical target, resistance to azoles in agriculture remains relatively rare. In contrast, resistance to azoles in human pathogens, including *Candida* spp. and *Aspergillus fumigatus*, is more common and a problem in the treatment of systemic mycosis, particularly in immunocompromised patients. The genetic changes underlying alterations in azole sensitivity are therefore well defined in human pathogens, particularly *C. albicans*, whereas in plant pathogens the molecular mechanisms conferring azole resistance are less well understood. Recently, however, there has been a progressive decline in the field performance of some azole fungicides against the wheat pathogen, *Mycosphaerella graminicola*. Consequently, current research efforts, both in academia and industry, have focused on understanding the mechanisms responsible for this recent change in sensitivity, the current status of azole resistance in the European *M. graminicola* population and whether strategies to manage the ongoing evolution of the pathogen population in response to azole fungicide use can be developed.

5.2 Decline in the Effectiveness of Azole Fungicides Against *M. graminicola*

M. graminicola (Fuckel) J Schroeter in Cohn (anamorph: *Septoria tritici* Roberge in Desmaz), causes septoria leaf blotch, currently the most important foliar disease of winter wheat in Western Europe (Hardwick *et al.*, 2001). None of the available commercial wheat cultivars is fully resistant to the disease. Therefore, control is mainly by the programmed application of fungicides, with an estimated value of £80–90 million/year sprayed to control *M. graminicola* in the UK alone. However, the fungus has adapted and developed resistance to single-site systemic fungicides introduced for its control. Resistance to the methyl benzimidazole carbamates (MBCs) emerged in the 1980s (Griffin and Fisher, 1985) and the quinone outside inhibitors (QoIs; strobilurins) in 2002 (Fraaije *et al.*, 2005). Therefore, the control of *M. graminicola* for the past 30 years has relied to a large extent on the application of azole fungicides. Monitoring studies undertaken in the 1990s did not find evidence of shifts in azole sensitivity in the UK *M. graminicola* population that could compromise disease control (Hollomon *et al.*, 2002), although comparisons with strains isolated in the 1980s did reveal a reduction in azole sensitivity. After QoI resistance became widespread in the early 2000s, the increased reliance on azoles again prompted concerns over the development of resistance to these compounds. In 2003, isolates of *M. graminicola* from a site in Kent, UK, with previously unseen reductions in sensitivity to some azole fungicides, particularly tebuconazole, were found (Cools *et al.*, 2005a). Subsequently, after reports of poor performance of some products in the field in 2005, a series of trials funded by the Pesticide Safety Directorate of the Department of Environment and Rural

Affairs of the UK confirmed that several azoles marketed for use in cereals no longer gave adequate control of *M. graminicola* (Clark, 2006). Compounds particularly affected included flusilazole, propiconazole and tebuconazole. However, several azoles, notably epoxiconazole and prothioconazole, remained very effective. Comparison of the sensitivities of isolates of *M. graminicola* obtained from untreated fields at Rothamsted Research in 2003 and 2010 confirmed differences in sensitivity shifts to different azoles (Fig. 5.1), with the population becoming resistant to tebuconazole, less sensitive to epoxiconazole and, interestingly, more sensitive to prochloraz.

5.3 Mechanisms Conferring Resistance to Azole Fungicides in *M. graminicola*

Studies of azole-resistant strains of human and plant pathogenic fungi have associated four mechanisms with an azole-resistant

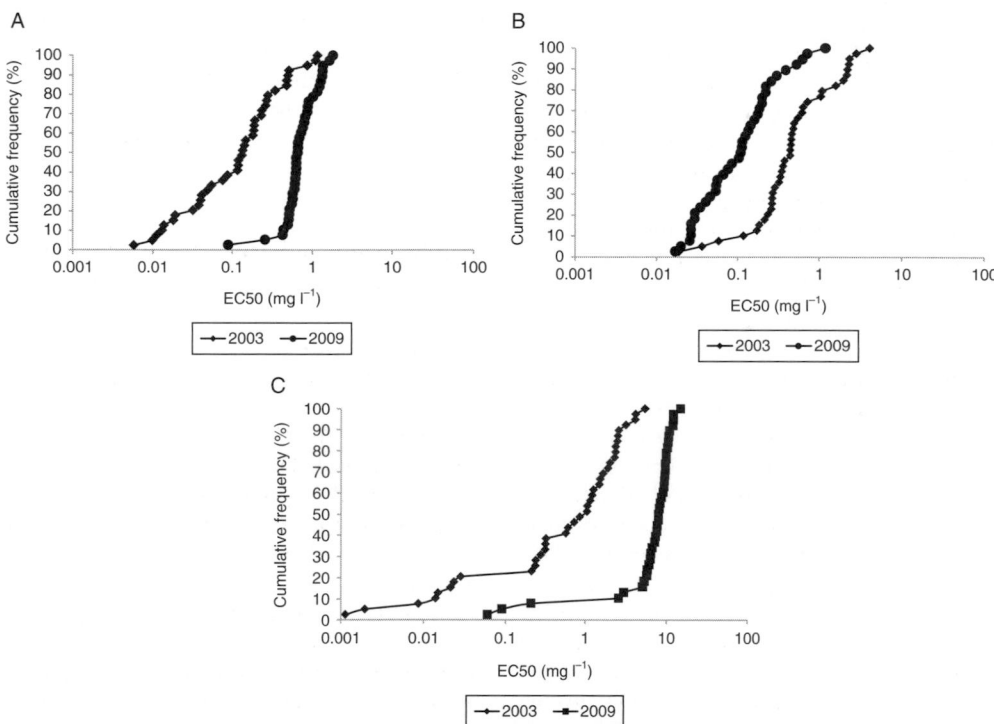

Fig. 5.1. Sensitivity of *M. graminicola* isolates to tebuconazole (A), epoxiconazole (B) and prochloraz (C) at Rothamsted in 2003 and 2009. Isolates are ranked according to increasing EC_{50} values (cumulative).

phenotype. These are: (i) alteration of the target CYP51 enzyme by mutation; (ii) over-expression of *CYP51*; (iii) enhanced active efflux of azoles; and (iv) altered sterol biosynthesis. With the exception of changes in sterol biosynthesis, all these mechanisms have been suggested to contribute to azole resistance in *M. graminicola*.

Mutation of the target-encoding *CYP51* gene

Resistance to most single-site fungicides is conferred by point mutations in target genes encoding single amino acid substitutions which prevent inhibitor binding while maintaining protein activity. In populations of *M. graminicola*, examples of fungicide classes affected by single target site mutations include the MBCs (E198A of β-tubulin) and the QoIs (F129L and G143A of cytochrome *b*) (Fraaije *et al.*, 2005; Lucas and Fraaije, 2008). However, unlike most single-site fungicides, resistance to azoles is conferred by the accumulation of multiple mutations in the target-encoding *CYP51* gene. This phenomenon is best studied in azole-resistant strains of *C. albicans* (Sanglard *et al.*, 1998; Marichal *et al.*, 1999; Perea *et al.*, 2001; Chau *et al.*, 2004; Morio *et al.*, 2010). *CYP51* changes have also accumulated in populations of *M. graminicola* (Zhan *et al.*, 2006; Leroux *et al.*, 2007; Cools and Fraaije, 2008; Stammler *et al.*, 2008) and substitutions are often at residues equivalent to known changes in *C. albicans* (Table 5.1), with positions considered important for azole sensitivity located in putative substrate recognition sites (SRS, Fig. 5.2; Lepesheva and Waterman, 2004).

Three substitutions in SRS-1 have been identified in resistant isolates of *M. graminicola*. These include Y137F (Fig. 5.2), a residue commonly identified in fluconazole-resistant isolates of *C. albicans* and also some plant pathogens, including *Uncinula necator* (Delye *et al.*, 1997), *Blumeria graminis* f. sp. *hordei* and *tritici* (Delye *et al.*, 1998) and, more recently, *Puccinia triticina* (Stammler *et al.*, 2009). In current *M. graminicola* populations, Y137F is rare or even absent (Stammler *et al.*, 2008). However, evidence suggests Y137F confers the greatest

reductions in sensitivity to triadimenol, an azole fungicide introduced in the late 1970s and now no longer used to control cereal diseases (Cools and Fraaije, 2008). In *C. albicans*, alterations at the residue equivalent to Y137 (Y136H/F) are often found in combination with other CYP51 changes, for example S405F (Sanglard *et al.*, 1998; Perea *et al.*, 2001). Similarly, Y137F has been found in combination with S524T in isolates of *M. graminicola* obtained in the late 1990s. We have demonstrated that this combination of changes further reduces sensitivity to all azoles, not only triadimenol (Cools *et al.*, 2011).

Also located in SRS-1 are substitutions D134G and V136A/C (Fig. 5.2). The equivalent residue to D134, G129, has been identified as altered (G129A) in fluconazole-resistant isolates of *C. albicans*, although only in combination with other substitutions (Sanglard *et al.*, 1998). Similarly, in *M. graminicola* D134G has only been found combined with other CYP51 changes, in particular V136A. Substitution V136A, an alteration unique to *M. graminicola*, confers reduced sensitivity to epoxiconazole and prochloraz, in particular, while having little effect on sensitivity to tebuconazole (Leroux *et al.*, 2007). Substitution V136C is rare, causing decreased sensitivity to prochloraz, epoxiconazole and tebuconazole. Similar to other CYP51 substitutions, V136A/C has been identified only in *M. graminicola* isolates in combination with alterations between Y459 and Y461. Using a heterologous expression system, we have demonstrated that V136A alone destroys CYP51 function in yeast, but this is partially rescued by combining V136A with changes between Y459 and Y461 (Cools *et al.*, 2011).

Located in SRS-5, substitutions I381V and A379G (Fig. 5.2) have been associated with the recent decline in the effectiveness of a number of azoles in controlling *M. graminicola*, particularly tebuconazole (Fraaije *et al.*, 2007; Leroux *et al.*, 2007). To date, A379G has been found only in isolates carrying I381V, suggesting these changes can only occur sequentially and isolates carrying CYP51 variants with both these alterations currently dominate the European population (Stammler *et al.*, 2008). In comparison with I381V alone, isolates carrying the combination of A379G

Table 5.1. Amino acid substitutions identified in isolates of *M. graminicola* and alterations at the equivalent residues in *C. albicans*.

Amino acid substitutions in CYP51 of *M. graminicola*	Reference(s)	Substitution at equivalent residue in CYP51 of *C. albicans*
L50S	Leroux *et al.* (2007)	None
D107V[a]	Stammler *et al.* (2008)	None
D134G	Stammler *et al.* (2008)	G129A
V136A	Leroux *et al.* (2007)	None
V136C	Leroux *et al.* (2007)	None
Y137F	Cools *et al.* (2005b); Leroux *et al.* (2007)	Y137F
		Y137H
S188N	Leroux *et al.* (2007)	None
G312A	This paper	None
A379G	Leroux *et al.* (2007)	M374V
I381V	Cools *et al.* (2005b); Fraaije *et al.* (2007); Leroux *et al.* (2007)	L376V
Y459C	Stammler *et al.* (2008)	Y447G
Y459D	Cools *et al.* (2005a); Zhan *et al.* (2006); Leroux *et al.* (2007)	Y447H
Y459N	Stammler *et al.* (2008)	
Y459S	Leroux *et al.* (2007); Stammler *et al.* (2008)	
G460D	Cools *et al.* (2005a); Leroux *et al.* (2007)	G448E
		G448R
		G448V
Y461H	Cools *et al.* (2005a); Leroux *et al.* (2007)	F449L
Y461S	Leroux *et al.* (2007)	F449S
		F449V
		F449Y
ΔY459/G460	Cools *et al.* (2005a); Leroux *et al.* (2007)	None
V490L	This paper	None
G510C	Leroux *et al.* (2007)	None
N513K	Cools *et al.* (2005b); Leroux *et al.* (2007)	None
S524T	This chapter	None

Note: [a]bold, only identified in azole-resistant isolates.

and I381V have reduced sensitivities to tebuconazole and epoxiconazole, whereas sensitivity to prochloraz is increased. In *C. albicans*, amino acid changes at residues equivalent to A379 (M374V) and I381 (L379V) (Table 5.1) have been identified, although the contribution of these alterations to azole resistance is unclear (see Morio *et al.*, 2010). Expression of *M. graminicola* CYP51 variants carrying I381V alone in *S. cerevisiae* perturbs the function of the enzyme. However, like V136A, the lethality of I381V can be partially rescued by combining with changes between Y459 and Y461 (Cools *et al.*, 2010).

The recent emergence of substitution S524T has caused concern due to reports that it contributes to decreased *in vitro* sensitivity to prothioconazole, one of the two remaining azole fungicides effective against *M. graminicola*. In recent isolates, particularly from Ireland, S524T is often found in combination with V136A. However, retrospective analysis of isolates obtained in the early 2000s identified S524T combined with Y137F. Analysis of the azole sensitivities of isolates with S524T, combined with either Y137F or V136A, reveals this substitution further decreases sensitivity to all azoles (Cools *et al.*, 2011). Expression of an *M. graminicola* CYP51 variant carrying S524T alone in *S. cerevisiae*, a variant not present in the *M. graminicola* population, confers decreased prothioconazole sensitivity. Combining S524T with Y137F not only decreases sensitivity to different azoles,

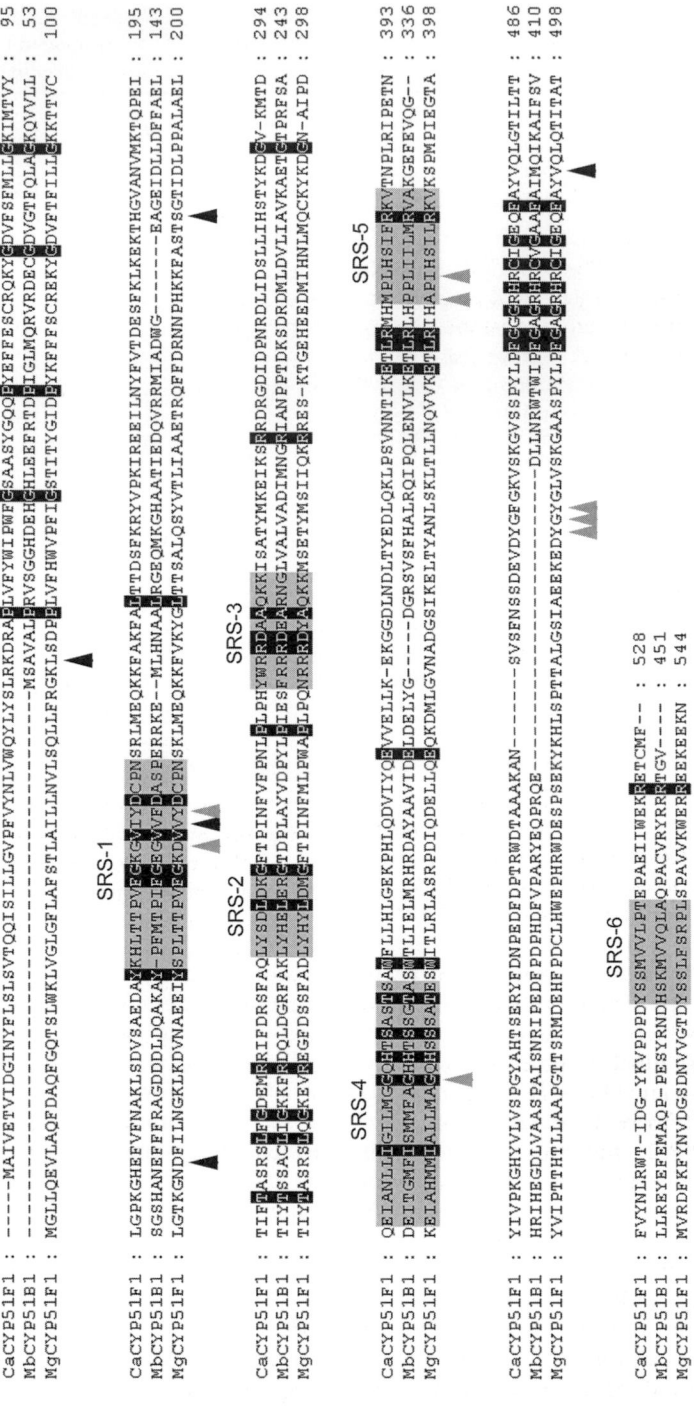

Fig. 5.2. Sequence alignment of the CYP51 from *Mycosphaerella graminicola* (MgrCYP51F1), *Candida albicans* (CaCYP51F1) and *Mycobacterium tuberculosis* (MtCYP51B1). Grey triangles indicate alterations identified in azole-resistant isolates of both *M. graminicola* and *C. albicans*. Black triangles indicate alterations unique to *M. graminicola*. Shaded areas indicate putative substrate recognition sites (SRS).

including epoxiconazole, triadimenol and propiconazole, but also rescues the lethality in yeast of an *M. graminicola* CYP51 variant carrying Y137F alone; suggesting the combination of Y137F and S524T confers not only increased resistance to azoles but also reduces the impact on enzyme activity of Y137F alone. The prevalence of isolates carrying both substitutions in the early 2000s, superseding isolates with Y137F alone, common in the early to mid-1990s, is consistent with these findings. Located in SRS-6 (Fig. 5.2), a substitution at the residue equivalent to S524 has not been identified in azole resistance strains of *C. albicans*. However, a change at this residue (S505P) has been identified in the CYP51B of an itraconazole resistant strain of *A. fumigatus* (Ferreira *et al.*, 2005).

Isolates carrying alterations at residues Y459–Y461, including Y459D, Y461H and ΔY459/G460, currently dominate the European *M. graminicola* population, suggesting that these changes confer an adaptive advantage in the presence of azoles. The region of the *M. graminicola* CYP51 encompassing residues Y459–Y461 is unique to fungi (Fig. 5.2), preventing prediction of the proximity of this region to the active site by homology to CYP51 crystal structures of bacteria (Podust *et al.*, 2001) or other eukaryotes (Lepesheva *et al.*, 2010; Strushkevich *et al.*, 2010). However, heterologous expression of *M. graminicola* CYP51 variants in *S. cerevisiae* has confirmed that changes at Y459–Y461 reduce sensitivity to all azoles, and furthermore, alterations at these residues are required to accommodate other amino acid substitutions which, when introduced alone, perturb CYP51 function in *S. cerevisiae*, for example V136A and I381V. The direct impact of changes at Y459–Y461 on azole binding and CYP51 function suggests an important role for this region in the function of the active site and, as a consequence, the recent evolution of the *M. graminicola* population in response to exposure to azole fungicides. The finding that these residues are also changed in azole-resistant strains of other pathogens, including *C. albicans* (Table 5.1) and the banana pathogen, *M. fijiensis* (Cañas-Gutiérrez *et al.*, 2009), supports this conclusion.

CYP51 overexpression

Constitutive upregulation of the *CYP51* gene encoding the azole target has been suggested to contribute to an azole-resistant phenotype in a number of human and plant pathogenic fungi. In *C. albicans* strains, increased *CYP51* expression was correlated with increasing azole resistance (White, 1997), and in *A. fumigatus* a tandem repeat in the promoter of *CYP51A* causing overexpression, together with substitution L98A, has been shown to confer cross-resistance to different azoles (Mellado *et al.*, 2007). In plant pathogens, *CYP51* overexpression, also associated with promoter insertions, has been correlated with azole resistance in *Penicillium digitatum* (Hamamoto *et al.*, 2000), *Venturia inaequalis* (Schnabel and Jones, 2000), *Blumeriella jaapii* (Ma *et al.*, 2006) and *Monilinia fructicola* (Luo and Schnabel, 2008).

The contribution of *CYP51* overexpression to azole resistance in *M. graminicola* isolates is unclear. A field isolate moderately resistant to cyproconazole with higher *CYP51* expression has been identified previously (Stergiopoulos *et al.*, 2003b), and an insertion in the *CYP51* promoter has been found in isolates carrying a CYP51 variant with substitution I381V (Chassot *et al.*, 2008), although the effect of this insertion on *CYP51* expression is uncertain. Studies of isolates carrying different CYP51 variants, and consequently with differing azole sensitivities, have demonstrated variation in constitutive *CYP51* expression levels. However, in this study *CYP51* expression could be linked to enzyme activity (eburicol demethylation). Therefore, modification of *CYP51* expression may also compensate for changes in protein activity caused by the accumulation of mutations which confer azole resistance but also affect enzyme function (Bean *et al.*, 2009).

Enhanced active efflux

The most common mechanism identified in clinical isolates of *C. albicans* resistant to azole fungicides is the overexpression of genes encoding energy-dependent efflux proteins. Membrane-spanning transporter proteins capable of exporting different xenobiotics, including fungicides, are from two families, ATP-binding cassette transporters (ABC transporters) or major facilitators (MF transporters). In *C. albicans*, efflux protein-encoding genes overexpressed in azole-resistant strains include the ABC transporter genes, *CDR1* (Sanglard *et al.*, 1995) and *CDR2* (Sanglard *et al.*, 1997), and the MF gene, *MDR1* (White, 1997). Strains overexpressing both *CDR1* and *CDR2* display a multidrug-resistant (MDR) phenotype, cross-resistant to the unrelated fungicides, terbinafine and caspofungin (Schuetzer-Muehlbauer *et al.*, 2003).

In *M. graminicola*, five genes encoding ABC transporters (*MgAtr1–5*) have been functionally characterized. Heterologous expression of *MgAtr1*, *MgAtr2*, *MgAtr4* and *MgAtr5* in a hypersensitive yeast mutant demonstrated these genes encoded proteins that could transport a range of chemically unrelated compounds, including azoles, with overlapping substrate specificity (Zwiers *et al.*, 2003). In addition, *MgAtr1* overexpression in a laboratory-derived *M. graminicola* mutant conferred cyproconazole resistance, a phenotype that could be reverted by disrupting *MgAtr1* (Zwiers *et al.*, 2002). Interestingly, deletion of *MgAtr4* reduced virulence of a wild-type isolate (Stergiolopoulos *et al.*, 2003a). Therefore, laboratory mutants and functional characterization by heterologous expression and gene deletion have demonstrated the capacity of ABC transporters of *M. graminicola* to export azoles and confer resistance.

Studies of field isolates of *M. graminicola*, however, have failed so far to correlate *MgAtr1–5* expression with azole sensitivity (Stergiopolous *et al.*, 2003b; Cools *et al.*, 2005a). In addition, transcriptional analysis of the response of *M. graminicola* to an azole fungicide using a microarray covering approximately one-quarter of the genome did not identify a transporter gene either responsive to azole treatment or constitutively more highly expressed in a less sensitive isolate (Cools *et al.*, 2007). Studies of both *in vitro* and *in planta* expression of ABC transporter encoding genes, characterized by genome analysis as candidates for azole export, also failed to identify constitutively overexpressed genes

in a resistant isolate (Bean *et al.*, 2008). Therefore, although the biological potential for efflux protein-mediated azole resistance exists, the genes responsible have not been identified in field isolates of *M. graminicola*.

More convincing evidence for a contribution of active efflux to azole resistance in *M. graminicola* field isolates comes primarily from the use of putative efflux pump inhibitors as azole synergists. Although earlier studies failed to identify modulators of ABC transporter activity with clear synergistic activity when combined with azole fungicides (Stergiopolous and De Waard, 2002), analysis of more recent isolates suggests that some compounds, for example fluphenazine, can increase sensitivity to azoles. In addition, recent analysis of some azole-resistant isolates from France, cross-resistant to other fungicides including thiolcarbamate, terbinafine and boscalid, revealed a synergistic effect of a number of modulators. These isolates have been suggested to have an MDR phenotype (Leroux and Walker, 2011).

5.4 Current Status of Azole Resistance in European *M. graminicola* Populations

The current European *M. graminicola* population is diverse, both in azole sensitivity phenotype and CYP51 variants. Detailed studies by Stammler *et al.* (2008) have shown that this diversity can be represented in a single region, or even a single field. However, broad regional differences do exist. For example, isolates carrying CYP51 variants with I381V, and therefore resistant to tebuconazole, dominate the *M. graminicola* population in most European countries. The exceptions are Ireland, Norway, Sweden, the Ukraine and Finland. In the latter two countries, wild-type and Y137F CYP51 variants dominate. In Ireland, the most frequently identified CYP51 variants carry the prochloraz resistance conferring substitution, V136A (Stammler *et al.*, 2008). This west-to-east divide, with higher frequencies of azole-resistant CYP51 variants in Western populations and lower in Eastern, has been suggested to reflect a Western European origin of substitutions, such as I381V, and subsequent gene flow by ascospore dispersal along the prevailing north-west to south-east wind across Europe (Brunner *et al.*, 2008). The absence of the I381V substitution in Ireland therefore suggested variants carrying this alteration arose in the UK, probably as a consequence of the high level of tebuconazole use (Stammler *et al.*, 2008).

In 2008 and 2009, there were reports of isolates from Ireland less sensitive to prothioconazole and, to a lesser extent, epoxiconazole, the two most effective azoles against *M. graminicola*. The appearance of these isolates was correlated with the emergence of the S524T substitution, usually identified in combination with L50S, V136A and Y461S. Although S524T had been identified in UK isolates in combination with Y137F, a variant with L50S, V136A, Y461S and S524T had not. Therefore, should new CYP51 variants carrying S524T confer increased levels of resistance to either prothioconazole or epoxiconazole, an increase in the prevalence of this alteration, caused by the influx of ascospores from Ireland along the prevailing west-to-east wind direction, would be expected to occur in UK populations in the near future. In fact, a few CYP51 variants recently detected in the UK, for example V136A, S188N, ΔY459/G460, S524T and L50S, D134G, V136A, Y461S and S524T, carry S524T.

5.5 Management of Azole Resistance in *M. graminicola*

The primary strategy recommended to manage resistance to any fungicide class is to mix or alternate with another fungicide with a different mode of action while ensuring sufficient disease control (Brent and Hollomon, 2007). In general, mixing is preferred as it minimizes the number of fungicide applications required in cereals and can provide a wider spectrum of disease control. The partner compound can be a multi-site inhibitor, presumed to have a low resistance risk, or an unrelated single-site fungicide. Currently, azoles are the most effective single-site inhibitors available for the control of *M. graminicola*. Therefore, mixing is generally with the multi-site inhibitor, chlorothalonil, the use of which has increased substantially in

recent years. A new generation of succinate dehydrogenase inhibitor (SDHI) fungicides (e.g. bixafen, isopyrozam and penthiopyrad) will also soon be introduced into the cereal market. However, the chemical diversity of azoles and their differential effects on CYP51 changes can also be exploited to improve Septoria control and to provide a wider spectrum of disease management. This is reflected by the recent successful introduction of highly effective formulated azole mixtures into the UK market.

Use of azole mixtures and/or alternations

To exploit the differential effects of target site changes on sensitivity to different azoles, a number of azole mixture products are currently available to cereal growers for the control of *M. graminicola*. These include Brutus® (BASF, epoxiconazole and metconazole), Ennobe® (BASF, epoxiconazole and prochloraz), Foil® (BASF, fluquinconazole and prochloraz), Menara® (Syngenta, cyproconazole and propiconazole), Prosaro® (Bayer Crop Science, prothioconazole and tebuconazole) and Vareon® (Dupont, prochloraz, tebuconazole and proquinazid). Early performance studies suggest these products provide improved disease control compared to their component fungicides. However, it is unclear so far whether this effect is caused by simply increasing the total amount of azole applied, improved formulation of the product, or the complementary activity of the azole components against less sensitive pathogen isolates. It has also been suggested that alternations of different azoles could be used to select against particular CYP51 changes, thereby posing the *M. graminicola* population an evolutionary conundrum (Cools and Fraaije, 2008). Therefore, the use of prochloraz in mixtures, which is unaffected by variants carrying substitutions A379G and I381V, has become more common. However, the recent emergence of novel CYP51 variants carrying previously unseen combinations of alterations that affect sensitivity to prochloraz as well as other azoles, for example variants carrying V136A and I381V combined, suggests pathogen evolution is rapidly finding ways of confounding the challenges imposed by recent resistance management strategies.

Use of high doses

Cereal growers are advised to maintain recommended dose rates that should kill all individuals in a fungal population, including the most resistant ones, thereby curtailing the emergence of resistance while also ensuring fungicide performance (Brent and Hollomon, 2007). This approach is clearly inappropriate when a single gene change confers high resistance levels, such as with the QoI fungicides, as a high dose quickly will select for the resistant individuals that survive. The experimental evidence describing the effect of dose rate on the selection for resistance to fungicides, such as the azoles, conferred by multiple genetic changes, each contributing to a continuous shift in sensitivity, suggests that

lower doses reduce selection for resistance, for example in powdery mildew (Hunter *et al.*, 1984; Porras *et al.*, 1990) and in *M. graminicola* (Metcalfe *et al.*, 2000; Mavroeidi and Shaw, 2006). Therefore, current theory supports the use of the lowest effective dose (Shaw, 2006). However, as the lowest effective dose is difficult, if not impossible, to quantify due to the ongoing evolution of CYP51 variants and differences in varietal resistance ratings for septoria leaf blotch, the recommendation to use the most effective compounds at higher doses remains the best strategy for disease control, although selection for the most resistant phenotypes might be stronger.

Use of diverse chemistry

Despite the number of different chemical classes used on cereals (Fig. 5.3), the azoles are currently the predominant group of

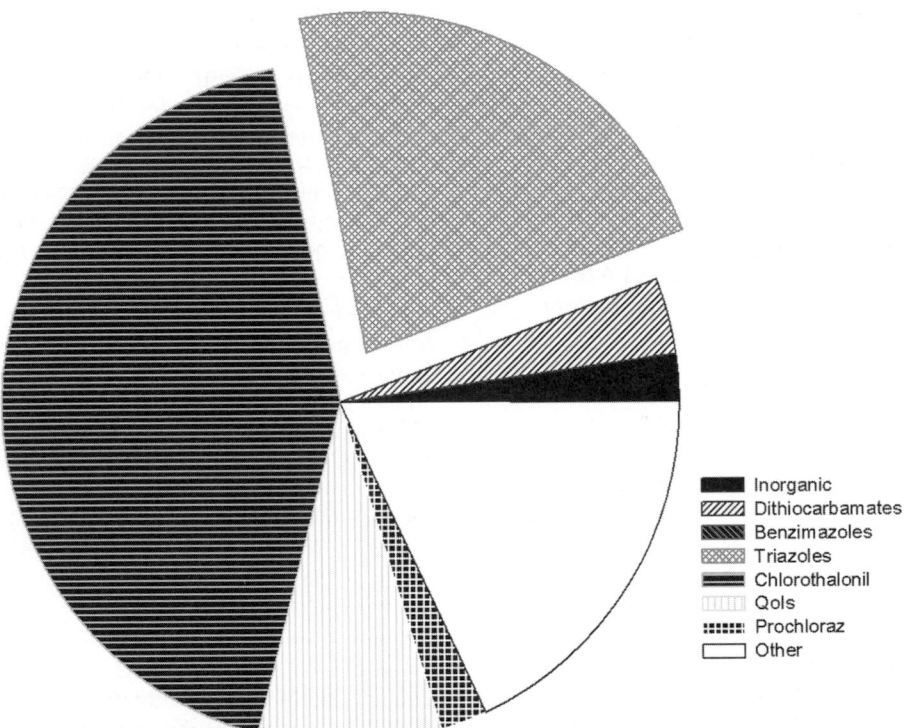

Fig. 5.3. Total weight (kg) of fungicides applied on UK cereals, excluding seed treatments, in 2008. Adapted from pesticide usage statistics (http://pusstats.csl.gov.uk).

single-site fungicides effective against *M. graminicola*. This is likely to change in the near future with the forthcoming introduction of further SDHI fungicides, but azoles will continue to form the basis for cereal crop protection strategies as they are the preferred mixing partner for most SDHIs. Therefore, exploiting the range of activities of compounds within the azoles class against different *M. graminicola* CYP51 variants will become more important, both for future management of azole resistance and to maintain disease control. However, protecting the diversity of compounds within the azole class will become more difficult under new European pesticide legislation aiming to minimize the risk to health and the environment from the use of pesticides (Sustainable Use Directive 2009/128/EC and Plant Protection Products Regulation (EC) 1107/2009). Azoles may be affected if categorized as potential endocrine disrupters. In addition, the introduction of comparative assessment criteria, where only the compound deemed the most effective within a chemical class will be retained, could also reduce the number of different azoles available. A key principle of sustainable crop protection, the need for a diversity of solutions, is hence at risk. The example of differential effects of different azoles against modern *M. graminicola* variants and other cereal diseases should be used to illustrate the advantages of maintaining chemical diversity within a single mode of action.

5.6   Summary and Future Perspectives

Azole fungicides have been used to control diseases of cereals for over 30 years, without, until recently, the development of widespread resistance in a pathogen population. Therefore, the progressive development of azole resistance in the European *M. graminicola* population is exceptional, comparable only to the evolution of azole resistance in the human opportunistic pathogen *C. albicans* in the 1990s (Kontoyiannis and Lewis, 2002). The decline in the effectiveness of azoles against *M. graminicola* has been accompanied by the accumulation of

mutations in the target-encoding *CYP51* gene. Particular mutations have been shown to affect sensitivity to various azoles differentially (mutation encoding substitution I381V, for example, conferring resistance to tebuconazole, while V136A causes resistance to prochloraz). In addition, site-directed mutagenesis and heterologous expression studies have demonstrated CYP51 variants resistant to azoles carry combinations of changes which enhance protein activity, thereby increasing tolerance for alterations important for azole sensitivity that individually impact on protein function (Cools *et al.*, 2010). Consequently, in most European countries the *M. graminicola* populations are comprised of a mixture of CYP51 variants, each conferring a different azole sensitivity phenotype.

Potential strategies to curtail the widespread development of resistance to all azoles have sought to exploit CYP51 diversity in the *M. graminicola* population by using azole mixtures and alternations to counterselect different CYP51 variants, thereby posing the pathogen population an evolutionary puzzle. For some products, this appears to have been moderately successful, with azole mixtures often performing better than their component fungicides in field trials. However, CYP51 evolution is ongoing and novel variants, selected for by recent azole treatment regimes, are emerging. The appearance of variants carrying both V136A and I381V is a good example.

Despite the forthcoming introduction to the cereal fungicide market of a new generation of highly effective single-site compounds, the SDHIs, azoles will remain the mainstay of control programmes for the foreseeable future. The risk of resistance development to the SDHIs has been classified as medium to high, and there have been recent reports of resistance in a number of different plant pathogen species (Avenot and Michailides, 2010). It is therefore possible that azole fungicides may again be relied on solely for the control of *M. graminicola*, particularly as new legislation and the change from risk- to hazard-based pesticide safety assessment criteria will increase markedly the cost of bringing new chemistry to the market. Therefore, one of the principal strategies for managing resistance to all fungicide groups in the future will be to exploit the diversity of activity within the azole class. It is therefore imperative that this diversity is maintained or extended by biochemical design to develop new azoles able to act against known or predicted changes in the protein target site.

Acknowledgements

The authors would like to thank John Lucas for critical reading of the manuscript. Rothamsted Research receives grant-aided support from the Biotechnology and Biological Sciences Research Council (BBSRC) of the UK (project BBE02257X1/BBE0218321).

References

Avenot, H.F. and Michailides, T.J. (2010) Progress in understanding molecular mechanisms and evolution of resistance to succinate dehydrogenase inhibiting (SDHI) fungicides in phytopathogenic fungi. *Crop Protection* 29, 643–651.

Bean, T.P., Cools, H.J., Lucas, J.A. and Fraaije, B.A. (2008) Expression of genes encoding efflux proteins in *Mycosphaerella graminicola* isolates with sensitive or reduced triazole sensitivity phenotypes. In: Dehne, H.W., Deising, H.B., Gisi, U., Kuck, K.H., Russell, P.E. and Lyr, H. (eds) *Modern Fungicides and Antifungal Compounds V: Proceedings 15th International Reinhardsbrunn Symposium on Modern Fungicides and Antifungal Compounds, 2007.* Deutsche Phytomedizinische Gesellschaft, Braunschweig, Germany, pp. 93–99.

Bean, T.P., Cools, H.J., Lucas, J.A., Hawkins, N.D., Ward, J.L., Shaw, M.W., *et al.* (2009) Sterol content analysis suggests altered eburicol 14α-demethylase (CYP51) activity in isolates of *Mycosphaerella graminicola* adapted to azole fungicides. *FEMS Microbiology Letters* 296, 266–273.

Brent, K.J. and Hollomon, D.W. (2007) Fungicide resistance in crop pathogens: how can it be managed. *FRAC Monograph No.2*, 2nd revised edition. Croplife International, Brussels.

Brunner, P.C., Stefanato, F.L. and McDonald, B.A. (2008) Evolution of the *CYP51* gene in *Mycosphaerella graminicola*: evidence for intragenic recombination and selective replacement. *Molecular Plant Pathology* 9, 305–316.

Cañas-Gutiérrez, G.P., Angarita-Velásquez, M.J., Restrepo-Flórez, J.M., Rodríguez, P., Moreno, C.X. and Arango, R. (2009) Analysis of the *CYP51* gene and encoded protein in propiconazole-resistant isolates of *Mycosphaerella fijiensis*. *Pest Management Science* 65, 892–899.

Chassot, C., Hugelshofer, U., Sierotzki, H. and Gisi, U. (2008) Sensitivity of *Cyp51* genotypes to DMI fungicides in *Mycosphaerella graminicola*. In: Dehne, H.W., Gisi, U., Kuck, K.H., Russell, P.E. and Lyr, H. (eds) *Modern Fungicides and Antifungal Compounds V*. DPG Seebsterverlag, Braunschweig, Germany, pp. 129–136.

Chau, A.S., Mendrick, C.A., Sabatelli, F.J., Loebenberg, D. and McNicholas, P.M. (2004) Application of real-time quantitative PCR to molecular analysis of *Candida albicans* strains exhibiting reduced susceptibility to azoles. *Antimicrobial Agents and Chemotherapy* 48, 2124–2131.

Clark, W.S. (2006) *Septoria tritici* and azole performance. Fungicide resistance: are we winning the battle but losing the war? *Aspects of Applied Biology* 78, 127–132.

Cools, H.J. and Fraaije, B.A. (2008) Spotlight: are azole fungicides losing ground against Septoria wheat disease? Resistance mechanisms in *Mycosphaerella graminicola*. *Pest Management Science* 64, 681–684.

Cools, H.J., Fraaije, B.A. and Lucas, J.A. (2005a) Molecular examination of *Septoria tritici* isolates with reduced sensitivities to triazoles. In: Dehne, H.-W., Gisi, U., Kuck, K.H., Russell, P.E. and Lyr, H. (eds) *Modern Fungicides and Antifungal Compounds IV: Proceedings of the 14th International Reinhardsbrunn Symposium, Friedrichroda, 25–29 April 2004.* BCPC, Alton, UK, pp. 103–114.

Cools, H.J., Fraaije, B.A. and Lucas, J.A. (2005b) Molecular mechanisms correlated with changes in triazole sensitivity in isolates of *Mycosphaerella graminicola*. In: *Proceedings BCPC International Congress – Crop Science and Technology, Glasgow, 31 October–2 November 2005.* BCPC, Alton, UK, pp. 267–274.

Cools, H.J., Fraaije, B.A., Bean T.P., Antoniw, J. and Lucas J.A. (2007) Transcriptome profiling of the response of *Mycosphaerella graminicola* isolates to an azole fungicide using cDNA microarrays. *Molecular Plant Pathology* 8, 639-651.

Cools, H.J., Parker, J.E., Kelly, D.E., Lucas, J.A., Fraaije, B.A. and Kelly, S.L. (2010) Heterologous expression of mutated eburicol 14α-demethylase (CYP51) proteins of *Mycosphaerella graminicola* to assess effects on azole fungicide sensitivity and intrinsic protein function. *Applied Environmental Microbiology* 76, 2866–2872.

Cools, H.J., Mullins, J.G.L., Fraaije, B.A., Parker, J.E., Kelly, D.E., Lucas, J.A., *et al.* (2011) Impact of recently emerged sterol 14α-demethylase (CYP51) variants of *Mycosphaerella graminicola* on azole fungicide sensitivity. *Applied and Environmental Microbiology* 77, 3830–3837.

Delye, C., Laigret, F. and Corio-Costet, M.F. (1997) A mutation in the sterol 14α- demethylase gene in *Uncinular necator* that correlates with resistance to sterol biosynthesis inhibitors. *Applied Environmental Microbiology* 63, 2966–2970.

Delye, C., Bousset, L. and Corio-Costet, M.F. (1998) PCR cloning and detection of point mutations in the eburicol 14α-demethylase (CYP51) gene from *Erysiphe graminis* f. sp. *hordei*, a 'recalcitrant' fungus. *Current Genetics* 34, 399–403.

Favre, B., Didmon, M. and Ryder, N.S. (1999) Multiple amino acid substitutions in lanosterol 14α-demethylase contribute to azole resistance in *Candida albicans*. *Microbiology* 145, 2715–2725.

Ferreira, M.E.D., Colombo, A.L., Paulsen, I., Ren, Q., Wortman, J., Huang, J., *et al.* (2005) The ergosterol biosynthesis pathway, transporter genes, and azole resistance in *Aspergillus fumigatus*. *Medical Mycology* 43, S313–S319.

Fraaije, B.A., Cools, H.J., Fountaine, J., Lovell, D.J., Motteram, J., West, J.S., *et al.* (2005) The role of ascospores in further spread of QoI-resistant cytochrome *b* alleles (G143A) in field populations of *Mycosphaerella graminicola*. *Phytopathology* 95, 933–941.

Fraaije, B.A., Cools, H.J., Kim, S.-H., Motteram, J., Clark, W.S. and Lucas, J.A. (2007) A novel substitution I381V in the sterol 14α-demethylase (CYP51) of *Mycosphaerella graminicola* is differentially selected by azole fungicides. *Molecular Plant Pathology* 8, 245–254.

Griffin, M.J. and Fisher, N. (1985) Laboratory studies on benzimidazole resistance in *Septoria tritici*. *EPPO Bulletin* 15, 505–511.

Hamamoto, H., Hasegawa, K., Nakaune, R., Lee, Y.J., Makizumi, Y., Akutsu, K., *et al.* (2000) Tandem repeat of a transcriptional enhancer upstream of the sterol 14 alpha-demethylase gene (*CYP51*) in *Penicillium digitatum*. *Applied and Environmental Microbiology* 66, 3421–3426.

Hardwick, N.V., Jones, D.R. and Royle, D.J. (2001) Factors affecting diseases of winter wheat in England and Wales, 1989–98. *Plant Pathology* 50, 453–462.

Hollomon, D., Cooke, L. and Locke, T. (2002) Maintaining the effectiveness of DMI fungicides in cereal disease control strategies. HGCA Project Report 275. HGCA, London.

Hunter, T., Brent, K.J. and Carter, G.A. (1984) Effects of fungicide regimes on sensitivity and control of barley mildews. In: *Proceedings 1984 British Crop Protection Conference – Pests and Diseases*. BCPC, Alton, UK, pp. 471–482.

Kontoyiannis, D.P. and Lewis, R.E. (2002) Antifungal drug resistance of pathogenic fungi. *The Lancet* 359, 1135–1144.

Lepesheva, G.I. and Waterman, M.R. (2004) CYP51 – the omnipotent P450. *Molecular and Cellular Endocrinology* 215, 165–170.

Lepesheva, G.I., Park, H.W., Hargrove, T.Y., Vanhollebeke, B., Wawrzak, Z., Harp, J.M., *et al.* (2010) Crystal structures of *Trypanosoma brucei* sterol 14α-demethylase and implications for selective treatment of human infections. *Journal of Biological Chemistry* 284, 1773–1780.

Leroux, P. and Walker, A.S. (2011) Multiple mechanisms account for resistance to sterol 14 alpha-demethylation inhibitors in field isolates of *Mycosphaerella graminicola*. *Pest Management Science* 67, 44–59.

Leroux, P., Albertini, C., Gautier, A., Gredt, M. and Walker, A.S. (2007) Mutations in the *cyp51* gene correlated with changes in sensitivity to sterol 14α-demethylation inhibitors in field isolates of *Mycosphaerella graminicola*. *Pest Management Science* 63, 688–699.

Loffler, J., Kelly, S.L., Hebart, H., Schumacher, U., LassFlorl, C. and Einsele, H. (1997) Molecular analysis of *cyp51* from fluconazole-resistant *Candida albicans* strains. *FEMS Microbiology Letters* 151, 263–268.

Lucas, J.A. and Fraaije, B.A. (2008) QoI resistance in *Mycosphaerella graminicola*: What have we learned so far? In: Dehne, H.W., Deising, H.B., Gisi, U., Kuck, K.-H., Russell, P. and Lyr, H. (eds) *Modern Fungicides and Antifungal Compounds V*. DPG Spectrum Phytomedizin, Braunschweig, Germany, Chapter 9, pp. 71–77.

Luo, C.X. and Schnabel, G. (2008) The cytochrome p450 lanosterol 14 alpha-demethylase gene is a demethylation inhibitor fungicide resistance determinant in *Monilinia fructicola* field isolates from Georgia. *Applied and Environmental Microbiology* 74, 359–366.

Ma, Z.H., Proffer, T.J., Jacobs, J.L. and Sundin, G.W. (2006) Overexpression of the 14 alpha-demethylase target gene (*CYP51*) mediates fungicide resistance in *Blumeriella jaapii*. *Applied and Environmental Microbiology* 72, 2581–2585.

Marichal, P., Koymans, L., Willemsens, S., Bellens, D., Verhasselt, P., Luyten, W., *et al.* (1999) Contribution of mutations in the cytochrome P450 14α-demethylase (Erg11p, Cyp51p) to azole resistance in *Candida albicans*. *Microbiology* 145, 2701–2713.

Mavroeidi, V.I. and Shaw, M.W. (2006) Effects of fungicide dose and mixtures on selection for triazole resistance in *Mycosphaerella graminicola* under field conditions. *Plant Pathology* 55, 715–725.

Mellado, E., Garcia-Effron, G., Alcazar-Fuoli, L., Melchers, W.J.G., Verweij, P.E., Cuenca-Estrella, A., *et al.* (2007) A new *Aspergillus fumigatus* resistance mechanism conferring *in vitro* cross-resistance to azole antifungals involves a combination of *cyp51A* alterations. *Antimicrobial Agents and Chemotherapy* 51, 1897–1904.

Metcalfe, R.J., Shaw, M.W. and Russel, P.E. (2000) The effect of dose and mobility on the strength of selection for DMI fungicide resistance in inoculated field experiments. *Plant Pathology* 49, 546–557.

Morio, F., Loge, C., Besse, B., Hennequin, C. and Le Pape, P. (2010) Screening for amino acid substitutions in the *Candida albicans* Erg11 protein of azole-susceptible and azole-resistant clinical isolates: new substitutions and a review of the literature. *Diagnostic Microbiology and Infectious Diseases* 66, 373–384.

Perea, S., Lopez-Ribot, J.L., Kirkpatrick, W.R., McAtee, R.K., Santillan, R.A., Martinez, M., *et al.* (2001) Prevalence of molecular mechanisms of resistance to azole antifungal agents in *Candida albicans* strains displaying high-level fluconazole resistance isolated from human immunodeficiency virus-infected patients. *Antimicrobial Agents Chemotherapy* 45, 2676–2684.

Podust, L.M., Poulos, T.L. and Waterman, M.R. (2001) Crystal structure of cytochrome P450 14alpha-sterol demethylase (CYP51) from *Mycobacterium tuberculosis* in complex with azole inhibitors. *Proceedings of the National Academy of Sciences USA* 98, 3068–3073.

Porras, L., Gisi, U. and Staehle-Csech, U. (1990) Selection dynamics in triazole treated populations of *Eryisphe graminis*. *Proceedings 1990 British Crop Protection Conference – Pests and Diseases*. BCPC, Alton, UK, pp. 1163–1168.

Russell, P.E. (2005) A century and fungicide evolution. *Journal of Agricultural Science* 153, 11–25.

Sanglard, D., Kuchler, K., Ischer, F., Pagani, J.L., Monod, M. and Bille, J. (1995) Mechanisms of resistance to azole antifungals in *Candida albicans* from AIDS patients involves specific multidrug resistance transporters. *Antimicrobial Agents Chemotherapy* 39, 2378–2386.

Sanglard, D., Ischer, F., Monod, M. and Bille, J. (1997) Cloning of *Candida albicans* genes conferring resistance to azole antifungal agents: characterization of *CDR2*, a new multidrug ABC transporter gene. *Microbiology* 143, 405–416.

Sanglard, D., Ischer, F., Koymans, L. and Billie, J. (1998) Amino acid substitutions in the cytochrome P450 lanosterol 14α-demethylase (*CYP51A1*) from azole-resistant *Candida albicans* clinical isolates contribute to resistance to antifungal agents. *Antimicrobial Agents Chemotherapy* 42, 241–253.

Schnabel, G. and Jones, A.L. (2000) The 14 alpha-demethylase (*CYP51A1*) gene is overexpressed in *Venturia inaequalis* strains resistant to myclobutanil. *Phytopathology* 91, 102–110.

Schuetzer-Muehlbauer, M., Willinger, B., Egner, R., Ecker, G. and Kuchler, K. (2003) Reversal of antifungal resistance mediated by ABC efflux pumps from *Candida albicans* functionally expressed in yeast. *International Journal of Antimicrobial Agents* 22, 291–300.

Shaw, M.W. (2006) Is there such a thing as a fungicide resistance strategy? A modeller's perspective. In: Fungicide resistance: are we winning the battle but losing the war? *Aspects of Applied Biology* 78, 37–44.

Stammler, G., Carstensen, M., Koch, A., Semar, M., Strobel, D. and Schlehuber, S. (2008) Frequency of different CYP51-haplotypes of *Mycosphaerella graminicola* and their impact on epoxiconazole-sensitivity and -field efficacy. *Crop Protection* 27, 1448–1456.

Stammler, G., Cordero, J., Koch, A., Semar, M. and Schlehuber, S. (2009) Role of the Y134F mutation in *cyp51* and overexpression of *cyp51* in the sensitivity response of *Puccinia triticina* to epoxiconazole. *Crop Protection* 28, 891–897.

Stergiopoulos, I. and De Waard, M.A. (2002) Activity of azole fungicides and ABC transporter modulators on *Mycosphaerella graminicola*. *Journal of Phytopathology* 150, 313–320.

Stergiopoulos, I., Zwiers, L.-H. and De Waard, M.A. (2003a) The ABC transporter MgAtr4 is a virulence factor of *Mycosphaerella graminicola* that affects colonisation of substomatal cavities in wheat leaves. *Molecular Plant Microbe Interactions* 16, 689–698.

Stergiopoulos, I., Van Nistelrooy, J.G.M., Kema, G.H.J. and De Waard, M.A. (2003b) Multiple mechanisms account for variation in base-line sensitivity to azole fungicides in field isolates of *Mycosphaerella graminicola*. *Pest Management Science* 59, 1333–1343.

Strushkevich, H., Usanov, S.A. and Park, H.-W. (2010) Structural basis of human CYP51 inhibition by antifungal azoles. *Journal of Molecular Biology* 397, 1067–1078.

White, T. (1997) Increased mRNA levels of ERG16, CDR, and MDR1 correlate with increases in azole resistance in *Candida albicans* isolates from a patient infected with human immunodeficiency virus. *Antimicrobial Agents Chemotherapy* 41, 1482–1487.

Xu, Y.H., Chen, L.M. and Li, C.Y. (2008) Susceptibility of clinical isolates of *Candida* species to fluconazole and detection of *Candida albicans ERG11* mutations. *Journal of Antimicrobial Chemotherapy* 61, 798–804.

Zhan, J., Stefanato, F. and McDonald, B.A. (2006) Selection for increased cyproconazole tolerance in *Mycosphaerella graminicola* through local adaptation and response to host resistance. *Molecular Plant Pathology* 7, 259–268.

Zwiers, L.-H., Stergiopoulos, I., Van Nistelrooy, J.G.M. and De Waard, M.A. (2002) ABC transporters and azole sensitivity in laboratory strains of the wheat pathogen *Mycosphaerella graminicola*. *Antimicrobial Agents and Chemotherapy* 46, 3900–3906.

Zwiers, L.–H., Stergiopoulos, I., Gielkens, M.M.C., Goodall, S.D. and De Waard, M.A. (2003) ABC transporters of the wheat pathogen *Mycosphaerella graminicola* function as protectants against biotic and xenobiotic toxic compounds. *Molecular Genetics and Genomics* 269, 499–507.

6 The Role of Intraspecific Parallel Genetic Adaptation to QoIs in Europe

Eva Edin[1] and Stefano Torriani[2]

[1]*Department of Forest Mycology and Plant Pathology, Swedish University of Agricultural Sciences, Uppsala, Sweden;* [2]*Plant Pathology Group, Institute of Integrative Biology, ETH Zurich, Zurich, Switzerland*

6.1   Introduction

Good agricultural management is the most effective way of controlling plant diseases. Alternation of cereal crops with various types of non-host crops, such as oilseed rape, legume crops or even oats, decreases the sources of inoculum in the field. Infested seed may be one important inoculum source which affects the seedlings from the first stage of development (Bennett *et al.*, 2007). Seeds of cereal crops in conventional farming are often treated with fungicides to reduce the impact of pathogens.

Fungicide applications have long been used as a disease control strategy to reduce crop losses. The first reports of seed treatments date back to the 17th century and followed the observation that wheat seeds recovered from the sea were free of bunt. This had occurred long before the crop protection chemical industry developed and before Benedict Prevost (1807) demonstrated that the fungus (*Tilletia caries*) that caused bunt disease of wheat could be controlled to some degree by copper sulfate. It was only later, by the mid-1940s, that the chemical crop protection industry emerged. The first agricultural fungicides were protectants, therefore limited to the external part of the plant, like the inorganic Bordeaux mixture or the organic dithiocarbamates and phthalimides. Most of the key classes of modern fungicides are, instead, systemic, thus absorbed by either the plant roots or the leaves. Among the first systemic fungicides introduced into the market in the late 1960s were the benzimidazoles and aminopyrimidine. Demethylation inhibitors (DMIs) and quinone outside inhibitors (QoIs) are other examples of systemic compounds launched in the 1970s and in 1996, respectively. Antifungal substances might be divided into different classes or families depending on their mode of action. Normally, if a pathogen develops resistance to a specific fungicide, then it becomes resistant to all fungicides enclosed within that fungicide class.

The first active substance in the fungicide QoI class was developed from the natural fungicidal derivates of β-methoxyacrylic acid produced by the Basidiomycetes *Strobilurus, Mycena* and *Oudemansiella* (Kraiczy *et al.*, 1996). The substance inhibits mitochondrial respiration by binding to the Qo site (outer quinine oxidizing pocket in complex III) at cytochrome *b*, which has the structure of a saddle (Esser *et al.*, 2004). Cytochrome *b* is a protein, encoded by the mitochondrial gene cytochrome *b* (*cytb*), located in the inner mitochondrial membrane forming the core of the mitochondrial bc_1 complex (complex III). QoIs block the electron transfer process between cytochrome *b* and cytochrome c_1, which leads to inhibition of electron transport in the energy cycle.

©CAB International 2012. *Fungicide Resistance in Crop Protection: Risk and Management* (ed. T.S. Thind)

The consequences are disruption of ATP production, and thus the metabolism of the fungus is impaired (Esser *et al.*, 2004).

The uptake of active substance of QoIs into the plant cells depends on formula, additives, mixtures with other products, application practices, weather conditions and the biological and physiological state of the crop, to mention the most important criteria (Bartlett *et al.*, 2002). Azoxystrobin and picoxystrobin are the two active substances that move systemically in the xylem and protect younger, newly developed plant parts and provide good disease control all the way to the leaf tip.

QoIs have shown activity against Ascomycetes, Basidiomycetes, Deuteromycetes and Oomycetes. The active substance interacts with the energy metabolism in order to prevent spore germination and infection, which are very energy-demanding processes (Wong and Wilcox, 2001). Strobilurins may also have a curative effect due to the decrease in mycelial growth during the latency period (time between infection and the beginning of reproduction). Eradicant effects and antisporulant effects on both sexual and asexual stages have been reported in some fungal species (Bartlett *et al.*, 2002).

QoI fungicides were first introduced into the market in 1996 with the substance azoxystrobin, which soon became a large selling product for plant protection. From the beginning, QoIs became an essential part of agricultural control programmes due to their wide-ranging efficacy against fungal pathogens of different crops and low toxicity against other organisms like mammals, insects or birds. QoIs display dose-dependent toxicity toward aquatic organisms, but since QoIs dissolve relatively quickly in the environment, long-term exposure is not a concern. Therefore, QoIs show a relatively low chronic risk (Bartlett *et al.*, 2002).

One of the reasons the QoIs have been used on such a large scale is that they have a 'greening effect' on the crop. The senescence of the leaves is delayed, even in the absence of pathogenic fungi, resulting in increased grain weight and nitrogen content of the grains (Gooding *et al.*, 2000; Ruske *et al.*, 2003), but the increase in protein content in the grain is not significant or consistent.

6.2 Reduction in Sensitivity To QoIs

Disease management in agricultural ecosystems is severely threatened by the evolution of fungicide resistance. The development of fungicide resistance is unlikely to be avoided. Therefore, in order to delay as much as possible the first appearance of resistance, it is important, besides collecting information on the molecular mechanisms characterizing it, to study the evolutionary processes that give rise to and disperse the resistance alleles. The development of fungicide resistance often involves amino acid substitutions, caused mainly by spontaneous mutations at one or several positions in the target gene. The substitution of amino acid enables alteration in the protein structure, so the active substance of the fungicide cannot attach properly, leading to failure in efficacy. The substitutions are then selected for during fungicide application, and the population of less sensitive strains is increased. The mutations may also be forced to evolve by the use of fungicides or in laboratory experiments through radiation or gene transfer (for example, Stammler *et al.*, 2007; Angelini *et al.*, 2010; Cools *et al.*, 2010).

Recent studies on *Mycosphaerella graminicola*, the causal agent of septoria tritici leaf blotch, disclosed the evolutionary patterns underlying the emergence and spread of resistance allele to QoIs and DMIs (Cools *et al.*, 2006; Brunner *et al.*, 2008; Torriani *et al.*, 2009b). Such studies represent the first step toward improved models to forecast the emergence and spread of fungicide resistance, and they might be helpful in understanding better the yet unpredictable spatial and temporal scales over which fungicide resistance develops.

6.3 Emergence of Resistance to QoIs

The natural strobilurin-producing Basidiomycetes *Strobilurus tenacellus* and *Mycena galopoda* have developed several point mutations in the mitochondrial gene *cytb* encoding cytochrome *b*, which confer natural resistance against their own metabolite (Kraiczy *et al.*, 1996). Moreover, researchers confirmed that mutations in *cytb*, leading to amino acid substitutions, are associated with loss of sensitivity to QoIs in plant pathogenic fungi (Grasso *et al.*, 2006a; Sierotzki *et al.*, 2007). These peptide changes have been identified in two domains of the cytochrome *b* protein, respectively between amino acid positions 120–155 and 255–280 (Kraiczy *et al.*, 1996; Fernandez-Ortuño *et al.*, 2008). The two domains connect closely together in the tertiary structure of the protein and contribute in binding the ligand. Alteration of amino acid sequence leads to transformation of the binding site, and the QoIs therefore cannot bind to the target site in cytochrome *b* (Esser *et al.*, 2004).

At least three amino acid substitutions have been related to QoI resistance in phytopathogenic fungi (for example, Grasso *et al.*, 2006a; www.frac.info). These are located at positions 129, 137 and 143. A point mutation at codon 129 is responsible for an amino acid change from phenylalanine to leucine (F129L). At position 137, a glycine is replaced by an arginine (G137R) and a substitution targeting codon 143 leads to the replacement of a glycine by an alanine (G143A). The strongest and most frequent mutation is G143A that confers high (complete) resistance, which is always associated with the breakdown in efficacy of QoIs to control disease. By contrast, isolates displaying F129L or G137R express only moderate (partial) resistance, which normally is controlled by recommended field levels of QoIs. In fact, based on current knowledge, G143A differentiates strongly in the resistance factor (RF = effective dose 50 [resistant strain]/effective dose 50 [sensitive wild-type strain]). RFs caused by F129L and G137R usually range between 5 and 15, and only rarely up to 50, while RFs related to G143A are commonly above 100 and usually greater than several hundred (FRAC, www.frac.info).

Shortly after the introduction of QoIs to the market in 1996, resistance occurred in *Blumeria graminis*, the causal agent of wheat powdery mildew, in 1998 (Sierotzki *et al.*, 2000), followed by *Pseudoperonospora cubensis* in 1999 (Ishii *et al.*, 2001) and *Magnaporthe grisea* in 2000 (Vincelli and Dixon, 2002), causing downy mildew in cucurbits and rice blast, respectively. The resistance to QoI fungicides was reported in several other phytopathogenic fungi, for example *Plasmopara viticola* on grape, *Sphaerotheca fuliginea* on cucumber, *Mycosphaerella fijiensis* on banana, *Venturia inaequalis* on apple, *M. graminicola* and *Phaeosphaeria nodorum* on wheat and *Rhynchosporium secalis* on barley (FRAC, www.frac.info; Blixt *et al.*, 2009). To date, the Fungicide Resistance Action Committee (FRAC, www.frac.info) has listed more than 30 different plant pathogenic species that have developed field resistance toward QoIs.

Despite the widespread use of QoI fungicides against some pathogens, such as the Basidiomycetes *Puccinia* spp. and *Uromyces* spp., there have never been problems in controlling these fungi. Sequencing analyses from *cytb* have revealed that these species carry a type I intron immediately after codon 143. Therefore, the nucleotide substitution in this codon G143 that confers QoI resistance is likely to prevent splicing of the intron, with the result that cytochrome *b* no longer functions, and the mutation presumably is lethal (Grasso *et al.*, 2006b). If and how fungal species displaying the group I intron directly after codon G143 will evolve complete QoI resistance is questionable. However, some fungal species carrying the intron after position 143, such as *Pyrenophora teres* and *Alternaria solani*, have mutated at position 129, leading to partial resistance (Grasso *et al.*, 2006a; Rosenzweig *et al.*, 2008). The sequence of cytochrome *b* cannot be retained since the characteristic of having an intron directly after position 143 is not linked to the genus, as both *Pyrenophora tritici-repentis* and *Alternia alternata* have the G143A substitution.

In 2006, putative QoI-resistant genotypes of *Puccinia horiana* were reported to show no difference in their sequences to QoI-sensitive isolates (Grasso *et al.*, 2006b). Neither G143A nor F129L mutations have been reported in the QoI-resistant genotypes of *P. horiana*, suggesting alternative mechanisms of resistance to QoI fungicides. Even though the impact on the resistance of alternative resistance mechanisms seems to be limited, the latter could not be excluded completely. Other known mechanisms of QoI resistance not related to mutations in *cytb* include alternative respiration and efflux transporters (Fernandez-Ortuño *et al.*, 2008). Alternative respiration sustained by alternative oxidase (AOX) seems only to counterbalance the QoI effects *in vitro* but not *in planta*. It offers less than half of the normal efficiency for energy conservation and therefore processes involving large amounts of energy, like spore germination or host penetration, cannot be supplied. Also, plant antioxidants might interfere with the induction of alternative respiration. On the other hand, it is possible that alternative respiration becomes more effective during the later stages of infection, such as mycelial growth and sporulation, when less energy is required (Wood and Hollomon, 2003). This has been suggested to be consistent with the observation that the control of many fungi is usually ineffective after visible symptoms have appeared (Fernandez-Ortuño *et al.*, 2008). Efflux transporters protect fungi against the accumulation of toxic compounds to toxic concentrations within the cells. Efflux transporters are plasma membrane-bound proteins that offer protection against several natural and xenobiotic chemicals, such as fungicides (Del Sorbo *et al.*, 2000). The first evidence of the role of an efflux transporter in conferring QoI resistance arose from *Aspergillus nidulans*. A transporter belonging to the family of ATP-binding cassette (ABC), encoded by the gene *AtrB*, was involved in protection against most fungicide major classes, including the QoIs (Andrade *et al.*, 2000). Instead, a major facilitator superfamily (MFS) efflux pump, encoded by gene *MgMfs1* in *M. graminicola*, was the first reported transporter involved in QoI resistance connected to the agroecosystem (Roohparvar *et al.*, 2007). However, the contribution of efflux transporters seems to be limited in the field since all QoI-resistant isolates of *M. graminicola* that were overexpressing the gene *MgMfs1* also contained the G143A mutation. Other plant pathogenic fungi, like *P. tritici-repentis* or *Colletotrichum* spp., display a similar efflux transporter-mediated mechanism of resistance to QoIs (Reimann and Deising, 2005; Mielke *et al.*, 2007). The knowledge behind the role of alternative QoI resistance mechanisms, such as alternative respiration and efflux pumps, is limited, and their role in contributing and shaping fungicide adaptation by plant pathogenic fungi should not be understated.

QoIs inhibit a protein encoded by the mitochondrial *cytb* gene. The distinctive characteristics of the mitochondrial genome, like uniparental inheritance, the near absence of genetic recombination, the small size and the fixed complement of a standard set of genes, make it possible to investigate new hypotheses about the evolution of fungicide resistance. A study on *M. graminicola* (see case study 1 below) tested the hypotheses that: (i) the QoI resistance allele G143A occurred only once in a single genetic and geographic background and subsequently was dispersed through gene flow; or that (ii) the resistance allele emerged through independent intraspecific parallel evolution in multiple geographic locations and in distinct fungal isolates (Torriani *et al.*, 2009b). At the same time, the mitochondrial origin of the QoI target site gene opens interesting evolutionary questions. In a fungal cell there could be around 100 mitochondria, each with its distinct mitochondrial DNA (mtDNA). Genes encoded by mtDNA are therefore present in large numbers per cell. After the discovery of QoIs was widely accepted, the idea that since *cytb* was a mitochondrial gene and was present in multi-copy numbers in a single fungal cell, the risk of developing full resistance would have been moderate. To date, it is not clear how fixation of the resistance allele occurs and how the transition from hetero- to homoplasmy is reached. Rapid fixation could be possible through a mechanism called mtDNA genetic bottlenecking. Genetic bottlenecking can select mitochondrial genes very rapidly because of unequal partitioning of mtDNA from one generation to the next (van Leeuwen *et al.*, 2008). The non-Mendelian mechanisms of mitochondrial inheritance and evolution are barely understood in fungi, but are probably relevant to forecasting and understanding the emergence of QoI resistance, as well as other mitochondrially-encoded adaptations.

Case study 1: *Mycosphaerella graminicola*

Before the launch of QoIs in 1996, *M. graminicola* was controlled mainly by DMI fungicides. QoIs showed a superior disease control and additional favourable effects on the physiology of the plant (Fraaije *et al.*, 2003). As for most fungi, the QoI resistance in *M. graminicola* is conferred by the amino acid substitution G143A in cytochrome *b* (Gisi *et al.*, 2002). The first *M. graminicola* field isolates resistant to QoIs were screened in 2001 in the UK (Fraaije *et al.*, 2005a) and later in 2002 in five different European countries (Gisi *et al.*, 2005). In the next 2 years, the frequency of the resistant strains increased quickly in Northern Europe. The status at the end of the season in 2009 reported QoI resistance in at least 13 European countries, with the UK, Ireland, Belgium, Germany and The Netherlands showing widespread resistance at high levels (FRAC, www.frac.info). Resistance levels in France and Germany are generally higher in the north than in the south. Differences in the intensity of QoI applications due to lower disease pressure in southern regions may explain the north–south gradient. No resistant isolates were found when screening a worldwide collection of 1000 isolates collected before 1996. The 95% confidence interval for the frequency of G143A mutation prior to the first introduction of QoIs ranged from 0 to 0.003 (Torriani *et al.*, 2009b). Analysis of the intraspecific mitochondrial diversity identified the molecular markers used to disclose the main evolutionary forces driving the rapid emergence and spread of QoI resistance in *M. graminicola* populations (Torriani *et al.*, 2008). Phylogenetic analysis, using 181 isolates of *M. graminicola* collected before (pre-QoI) and after (post-QoI) the first applications of QoIs in agriculture, revealed that QoI resistance was acquired independently through at least four recurrent mutations. Estimates of directional migration rates proved that the majority of gene flow occurred from west to east in Europe (Torriani *et al.*, 2009b). Ascospores have the potential to be dispersed over several kilometres (Fraaije *et al.*, 2005b) following wind directions. The same patterns of migration for *M. graminicola* were observed using the nuclear encoded *CYP51* locus to test azole resistance (Brunner *et al.*, 2008). The rapid emergence and spread of QoI resistance in European populations of *M. graminicola* was therefore achieved through independent parallel evolution that introduced the resistance allele into several distinct mtDNA haplotypes. The resistant isolates subsequently increased in

frequency within populations due to the strong selective pressure imposed by the large use of QoI fungicides and the spread eastward through wind dispersal of ascospores. Similar findings of parallel adaptation to fungicides were also presented for the grape pathogen *P. viticola* (Chen *et al.*, 2007). Results from *M. graminicola* and *P. viticola* suggest that the intraspecific *de novo* appearance of QoI fungicide resistance in distinct genetic backgrounds may be common for plant pathogenic fungi. Intraspecific parallel evolution of fungicide resistance may also be a common process for fungicides having nuclear encoded proteins as their target, the genes of which are subjected to recombination, like DMIs. For these resistance mechanisms, however, recombination makes it difficult to track how many times the mutation to resistance has occurred, since the same mutation could be recombined into many different genetic backgrounds.

Case study 2: *Phaeosphaeria nodorum*

Outbreaks of stagonospora nodorum blotch were rather severe in Europe during the 1980s until *M. graminicola* and *P. tritici-repentis* became more frequent during the latter half of the 1990s. However, *P. nodorum* still co-occurs with the two pathogens, even though sometimes only in small proportions (Blixt *et al.*, 2010). A survey of the strobilurin sensitivity of 231 Swedish isolates revealed that the majority of the *P. nodorum* isolates collected in four fields possessed the amino acid substitution G143A (Blixt *et al.*, 2009). The high frequency probably was caused by the selective use of strobilurins since only half of the isolates of another field, rarely treated with strobilurins, had the substitution. The origin of the substitution and evolvement was not investigated, but the causes were most likely the same as for *M. graminicola*.

Case study 3: *Alternaria solani*

The QoIs have so far been efficient in controlling early blight on potato, which results in a higher yield of tubers and starch (for example,

MacDonald *et al.*, 2006; Andersson and Wiik, 2008). Unfortunately, it has been revealed that QoIs no longer have as good an effect against species of *Alternaria* spp. in some parts of the USA (Ma *et al.*, 2003; Pasche *et al.*, 2004; Rosenzweig *et al.*, 2008). During the last few years, three different nucleotide substitutions at position 129, which all contribute to the amino acid substitution F129L, have been found, leading to reduced sensitivity to stro-bilurins. In some areas, a large proportion of isolates of *A. solani* collected in potato crops in the USA were shown to have one of the three possible substitutions at position 129. Isolates carrying the F129L substitution may have a fitness disadvantage. Greenhouse and *in vitro* experiments on *A. solani* possessing substitutions showed that the spore germination was reduced along with the disease severity of early blight on potato (Pasche and Gudmestad, 2008). So far, the substitution G143A has not been observed in *A. solani* due to the intron located directly after position 143. The European population of *A. solani* has yet not been fully investigated, but the preliminary results from a Swedish survey in 2009 showed that all 270 samples collected were of wild type at both position 129 and 143 (E. Edin, unpublished data).

Case study 4: *Rhycosporium secalis*

Rhycosporium secalis causes leaf blotch or scald disease of barley. Fungicides are used rou-tinely in Europe to control *R. secalis*. They have shown different efficacy against scald and resistance has developed for most fungi-cides (Cooke *et al.*, 2004). The performances of QoI fungicides in Europe were always good and extensive monitoring carried out in 2009 in ten European countries showed no G143A mutation (FRAC, www.frac.info). After 14 years from the introduction of QoIs, the European population of *R. secalis* still showed a fully sensitive picture. The *cytb* gene of *R. secalis* is not interspersed with introns, so that the G143A mutation will not be lethal if it occurs (Torriani *et al.*, 2009a). A total of 841 global field isolates of *R. secalis* were screened for the presence of G143A mutation, but no

mutants were found. These results indicated that the resistance allele did not exist at a detectable frequency globally. However, in 2008 one sample containing G143A mutation was collected from northern France (FRAC, www.frac.info). The mechanisms behind the long efficacy of QoIs to control *R. secalis* remain unclear. One interest would be to monitor the frequency of QoI resistance in northern France during the next few years, where a rapid fixation of the resistance allele would be expected after the screened intro-duction of G143A mutations.

6.4 Conclusions

From the beginning of agriculture about 12,000 years ago in the Fertile Crescent, humans have been faced with the problem of plant pathogens. Since the 19th century, but mainly from the 20th century, chemical dis-ease control has been routinely applied, increasing field productivity. Immediately fol-lowing the increased use and sometimes over-use of chemical compounds, pathogens developed resistance to the applied chemicals. The importance of monitoring programmes quickly became clear, as did the need to coord-inate resistance management strategies, so that in the early 1980s the Fungicide Resistance Action Committee (FRAC) was established. As well described in the welcome message on the FRAC's website home page: 'Fungicides have become an integral part of efficient food production. The loss of a fungicide to agricul-ture through resistance is a problem that affects us all.' In fact, the 'loss' of a fungicide implies critical yield losses and economic damages for farmers. Reduced yields lead to a decrease in food production, an increase in food prices and, potentially, famines. By 2050, the United Nations forecasts that the global human population will reach 9 billion. The logarithmic increase of world human popula-tion needs to be sustained by agriculture, which must nearly double global food pro-duction on a global scale.

Understanding of the evolutionary basis that controls the emergence and dispersal of fungicide resistance is of primary interest to establish the correct control strategies aimed to increase the lifespan of fungicides. As presented in this chapter, resistance to fungicides is unlikely to be avoided and is especially favoured by the large-scale growing of monocultures and by the strong selective pressure imposed by fungicide applications. Depending on the mode of action of the chemical, a pathogen can evolve resistance suddenly or gradually. In some cases, like the QoIs, resistance is 'qualitative' and extends to all compounds within the same mode of action. Fungicide selection shapes quite distinct resistant populations that are not easily set back if the treatments are stopped. In other cases, like for DMIs, resistance is 'quantitative', since it develops gradually, producing a shift toward less sensitive overlapping populations. When fungicide use is stopped, resistance tends to decline (Hollomon and Brent, 2009). 'Qualitative' resistances are, in most cases, mediated by a single amino acid substitution in the target protein. Instead, the genetic and biochemical changes behind 'quantitative' resistances are more complex, involving

generally more mutations at different points in the gene that encodes for the target site protein. A good anti-resistance strategy is to try to reduce the exposure to risk fungicides by limiting the number of applications and by using as low dose rates as possible. In order to preserve and prolong the efficacy of the different compounds, it is also important to integrate into the control strategy fungicides that display different modes of action. The use of products in mixtures or the alternation in spray programmes decreases the likelihood that a pathogen develops fungicide resistance. It is likely that fungicides will remain one of the principle control strategies for the forseeable future. For this reason, the discovery of new active antifungal molecules is crucial. But even more essential in the future will be the concurrent use of integrated disease management. Nonchemical methods like using disease-resistant varieties and biological controls, applying rotation of crops and respecting hygienic practices will be necessary for sustainable disease control.

References

Andersson, B. and Wiik, L. (2008) Betydelsen av torrfläcksjuka (*Alternaria* spp.) på potatis. Slutrapport av SLF 0455031 [in Swedish].

Andrade, A.C., del Sorbo, G., Van Nistelrooy, J.G.M and de Waard, M.A. (2000) The ABC transporter AtrB from *Aspergillus nidulans* mediates resistance to all major classes of fungicides and some natural toxic compounds. *Microbiology* 146, 1987–1997.

Angelini, R.M.D., Habib, W., Rotolo, C., Pollastro, S. and Faretra, F. (2010) Selection, characterization and genetic analysis of laboratory mutants of *Botryotinia fuckeliana* (*Botrytis cinerea*) resistant to the fungicide boscalid. *European Journal of Plant Pathology* 128, 185–199.

Bartlett, D.W., Clough, J.M., Godwin, J.R., Hall, A.A., Hamer, M. and Parr-Dobrzanski, B. (2002) The strobilurin fungicides. *Pest Management Science* 58, 649–662.

Bennett, R.S., Milgroom, M.G., Sainudiin, R., Cunfer, B.M. and Bergstrom, G.C. (2007) Relative contribution of seed-transmitted inoculum to foliar populations of *Phaeosphaeria nodorum*. *Phytopathology* 97, 584–591.

Blixt, E., Djurle, A., Yuen, J. and Olson, Å. (2009) Fungicide sensitivity in Swedish isolates of *Phaeosphaeria nodorum*. *Plant Pathology* 58, 655–664.

Blixt, E., Olson, Å., Lindahl, B., Djurle, A. and Yuen, J. (2010) Spatiotemporal variation in the fungal community associated with wheat leaves showing symptoms similar to stagonospora nodorum blotch. *European Journal of Plant Pathology* 126, 373–386.

Brunner, P.C., Stefanato, F.L. and McDonald, B.A. (2008) Evolution of the CYP51 gene in *Mycosphaerella graminicola*: evidence for intragenic recombination and selective replacement. *Molecular Plant Pathology* 9, 305–316.

Chen, W.J., Delmotte, F., Richard-Cervera, S., Douence, L., Greif, C. and Corio-Costet, M.F. (2007) At least two origins of fungicide resistance in grapevine downy mildew populations. *Applied and Environmental Microbiology* 73, 5162–5172.

Cooke, L.R., Locke, T., Lockley, K.D., Phillips, A., Sadiq, M.D.S., Coll, R., *et al.* (2004) The effect of fungicide programmes based on epoxiconazole on the control and DMI sensitivity of *Rhynchosporium secalis* in winter barley. *Crop Protection* 23, 393–406.

Cools, H.J., Fraaije, B.A., Kim, S.H. and Lucas, J.A. (2006) Impact of changes in the target P450 CYP51 enzyme associated with altered triazole-sensitivity in fungal pathogens of cereal crops. *Biochemical Society Transactions* 34, 1219–1222.

Cools, H.J., Parker, J.E., Kelly, D.E., Lucas, J.A., Fraaije, B.A. and Kelly, S.L. (2010) Heterologous expression of mutated eburicol 14 alpha-demethylase (CYP51) proteins of *Mycosphaerella graminicola* to assess effects on azole fungicide sensitivity and intrinsic protein function. *Applied and Environmental Microbiology* 76, 2866–2872.

Del Sorbo, G., Schoonbeek, H. and de Waard, M.A. (2000) Fungal transporters involved in efflux of natural toxic compounds and fungicides. *Fungal Genetics and Biology* 30, 1–15.

Esser, L., Quinn, B., Li, Y.-F., Zhang, M., Elberry, M., Yu, L., *et al.* (2004) Crystallographic studies of quinol oxidation site inhibitors: a modified classification of inhibitors for the cytochrome *bc1* complex. *Journal of Molecular Biology* 341, 281–302.

Fernandez-Ortuño, D., Tores, J.A., de Vicente, A. and Perez-Garcia, A. (2008) Mechanisms of resistance to QoI fungicides in phytopathogenic fungi. *International Microbiology* 11, 1–9.

Fraaije, B.A., Lucas, J.A., Clark, W.S. and Burnett, F.J. (2003) QoI resistance development in populations of cereal pathogens in the UK. In: *Proceedings BCPC International Congress – Crop Science and Technology*. British Crop Production Council, Alton, UK, pp. 689–694.

Fraaije, B.A., Burnett, F.J., Clark, W.S., Motteram, J. and Lucas, J.A. (2005a) Resistance development to QoI inhibitors in populations of *Mycosphaerella graminicola* in the UK. In: Dehne, H.W., Gisi, U., Kuck, K.H., Russell, P.E. and H. Lyr (eds) *Modern Fungicides and Antifungal Compounds IV*. BCPC, Alton, UK, pp. 63–71.

Fraaije, B.A., Cools, H.J., Fountaine, J., Lovell, D.J., Motteram, J., West, J.S., *et al.* (2005b) Role of ascospores in further spread of QoI-resistant cytochrome b alleles (G143A) in field populations of *Mycosphaerella graminicola*. *Phytopathology* 95, 933–941.

Gisi, U., Sierotzki, H., Cook, A. and McCaffery, A. (2002) Mechanisms influencing the evolution of resistance to Qo inhibitor fungicides. *Pest Management Science* 58, 859–867.

Gisi, U., Pavic, L., Stanger, C., Hugelshofer, U. and Sierotzki, H. (2005) Dynamics of *Mycosphaerella graminicola* populations in response to selection by different fungicides. In: Dehne, H.W., Gisi, U., Kuck, K.H., Russell, P.E. and Lyr, H. (eds) *Modern Fungicides and Antifungal Compounds IV*. BCPC, Alton, UK, pp. 73–80.

Gooding, M.J., Dimmock, J., France, J. and Jones, S.A. (2000) Green leaf area decline of wheat flag leaves: the influence of fungicides and relationships with mean grain weight and grain yield. *Annals of Applied Biology* 136, 77–84.

Grasso, V., Palermo, S., Sierotzki, H., Garibaldi, A. and Gisi, U. (2006a) Cytochrome *b* gene structure and consequences tor resistance to Qₒ inhibitor fungicides in plant pathogens. *Pest Management Science* 62, 465–472.

Grasso, V., Sierotzki, H., Garibaldi, A. and Gisi, U. (2006b) Characterization of the cytochrome *b* gene fragment of *Puccinia* species responsible for the binding site of QoI fungicides. *Pesticide Biochemistry and Physiology* 84, 72–82.

Hollomon, D.W. and Brent, K.J. (2009) Combating plant diseases – the Darwin connection. *Pest Management Science* 65, 1156–1163.

Ishii, H., Fraaije, B.A., Sugiyama, T., Noguchi, K., Nishimura, K., Takeda, T., *et al.* (2001) Occurrence and molecular characterization of strobilurin resistance in cucumber powdery mildew and downy mildew. *Phytopathology* 91, 1166–1171.

Kraiczy, P., Haase, V., Gencic, S., Flindt, S., Anke, T., Brandkt, V., *et al.* (1996) The molecular basis for the natural resistance of the cytochrome *bc1* complex from strobilurin-producing basidiomycetes to center Qo inhibitors. *European Journal of Biochemistry* 235, 54–64.

MacDonald, W., Peters, R.D., Coffin, R.H. and Lacroix, C. (2006) Suppression of potato early blight (*in vivo*) and germination of *Alternaria* spp. conidia (*in vitro*) with strobilurin fungicides. *Canadian Journal of Plant Pathology – Revue Canadienne De Phytopathologie* 28, 327.

Mielke, M.C., Melzer, E., Serfling, A., Ahmetovic, U., Horbach, R., Reimann, S., *et al.* (2007) The molecular basis of adaptation to fungicides in plant pathogenic Colletotrichum species. In: Dehne, H.W., Deising, H.B., Gisi, U., Kuck, K.H., Russell, P.E. and Lyr, H. (eds) *15th International Reinhardsbrunn Symposium on Modern Fungicides and Antifungal Compounds*, 6–10 May 2007, Friedrichroda, Germany. Deutsche Phytomedizinische Gesellschaft, Braunschweig, Germany, pp. 63–72.

Pasche, J.S. and Gudmestad, N.C. (2008) Prevalence, competitive fitness and impact of the F129L mutation in *Alternaria solani* from the United States. *Crop Protection* 27, 427–435.

Pasche, J.S., Wharam, C.M. and Gudmestad, N.C. (2004) Shift in sensitivity of *Alternaria solani* in response to Q_oI fungicides. *Plant Disease* 88, 181–187.

Prevost, B. (1807) *Memoir on the Immediate Cause of Bunt or Smut of Wheat: And of Several Other Diseases of Plants.* Translated by George Wannamaker Keitt in 2009. Kessinger Publishing, Whitefish, Montana, pp. 94.

Reimann, S. and Deising, H.B. (2005) Inhibition of efflux transporter-mediated fungicide resistance in *Pyrenophora tritici-repentis* by a derivative of 4'-hydroxyflavone and enhancement of fungicide activity. *Applied Environmental Microbiology* 71, 3269–3275.

Roohparvar, R., De Waard, M., Kema, G.H.J. and Zwiers, L.H. (2007) *MgMfs1*, a major facilitator superfamily transporter from the fungal wheat pathogen *Mycosphaerella graminicola*, is a strong protectant against natural toxic compounds and fungicides. *Fungal Genetics and Biology* 44, 378–388.

Rosenzweig, N., Atallah, Z.K., Olaya, G. and Stevenson, W.R. (2008) Evaluation of Q_oI fungicide application strategies for managing fungicide resistance and potato early blight epidemics in Wisconsin. *Plant Disease* 92, 561–568.

Ruske, R.E., Gooding, M.J. and Jones, S.A. (2003) The effects of triazole and strobilurin fungicide programmes on nitrogen uptake, partitioning, remobilization and grain N accumulation in winter wheat cultivars. *Journal of Agricultural Science* 140, 395–407.

Sierotzki, H., Wullschleger, J. and Gisi, U. (2000) Point mutation in cytochrome *b* gene conferring resistance to strobilurin fungicides in *Erysiphe graminis* f. sp. *tritici* field isolates. *Pesticide Biochemistry and Physiology* 68, 107–112.

Sierotzki, H., Frey, R., Wullschleger, J., Palermo, S., Karlin, S., Godwin, J., *et al.* (2007) Cytochrome *b* gene sequence and structure of *Pyrenophora teres* and *P. tritici-repentis* and implications for QoI resistance. *Pest Management Science* 63, 225–233.

Stammler, G., Brix, H.D., Glätti, A., Semar, M. and Schoefl, U. (2007) Biological properties of the carboxamide boscalid including recent studies on its mode of action. In: McKim, F.M. (ed.) *Proceedings of the XVI International Plant Protection Congress*, Glasgow 2007. British Crop Production Council, Alton, UK, pp. 40–45.

Torriani, S.F.F., Goodwin, S.B., Kema, G.H.J., Pangilinan, J.L. and McDonald, B.A. (2008) Intraspecific comparison and annotation of two complete mitochondrial genome sequences from the plant pathogenic fungus *Mycosphaerella graminicola*. *Fungal Genetics and Biology* 45, 628–637.

Torriani, S.F.F., Linde, C.C. and McDonald, B.A. (2009a) Lack of G143A QoI resistance allele and sequence conservation for the mitochondrial cytochrome *b* gene in a global sample of *Rhynchosporium secalis*. *Australasian Plant Pathology* 38, 202–207.

Torriani, S.F.F., Brunner, P.C., McDonald, B.A. and Sierotzki H. (2009b) QoI resistance emerged independently at least four times in European populations of *Mycosphaerella graminicola*. *Pest Management Science* 65, 155–162.

Van Leeuwen, T., Vanholme, B., Van Pottelberge, S., Van Nieuwenhuyse, P., Nauen, R., Tirry, L., *et al.* (2008) Mitochondrial heteroplasmy and the evolution of insecticide resistance: non-Mendelian inheritance in action. *Proceedings of the National Academy of Science of the United States of America* 105, 5980–5985.

Vincelli, P. and Dixon, E. (2002) Resistance to QoI (strobilurin-like) fungicides in isolates of *Pyricularia grisea* from perennial ryegrass. *Plant Disease* 86, 235–240.

Wong, F.P. and Wilcox, W.F. (2001) Comparative physical modes of action of azoxystrobin, mancozeb, and metalaxyl against *Plasmopara viticola* (grapevine downy mildew). *Plant Disease* 85, 649–656.

Wood, P.M. and Hollomon, D.W. (2003) A critical evaluation of the role of alternative oxidase in the performance of strobilurin and related fungicides acting at the Qo site of Complex III. *Pest Management Science* 59, 499–511.

7 Risk and Management of Fungicide Resistance in the Asian Soybean Rust Fungus *Phakopsora pachyrhizi*

Cláudia Vieira Godoy

*Empresa Brasileira de Pesquisa Agropecuária (Embrapa),
National Center of Soybean Research, Londrina, PR – Brazil*

7.1 Introduction

Asian soybean rust (ASR), caused by the biotrophic fungus *Phakopsora pachyrhizi* Syd. & P. Syd, is considered the most damaging foliar disease of soybean [*Glycine max* (L.) Merr.]. In the absence of control measures, yield losses of up to 80% have been reported (Bromfield, 1984; Hartman *et al.*, 1991; Yang *et al.*, 1991). It was first recorded in Japan as *Uredo sojae* Henn., 1902, and identified throughout tropical and subtropical Asia and Oceania in the early 20th century (Bromfield, 1984). Over several ensuing decades, the disease spread around the world and is now established in all major soybean producing countries (Rossi, 2003; Ivancovich, 2005; Schneider *et al.*, 2005; Yorinori *et al.*, 2005).

The typical symptom of this disease is sporulating lesions on the abaxial surface of the leaf, usually associated with leaf chlorosis. Lesions appear first in the lower canopy and then advance up to the mid and upper soybean canopy. As the disease progresses, high-density lesions can develop and this leads to premature defoliation and early maturity (Sinclair and Hartman, 1999). Besides soybean, there are more than 90 legume species reported as *P. pachyrhizi* hosts from inoculated and non-inoculated plants (Rytter *et al.*, 1984; Ono *et al.*, 1992; Slaminko *et al.*, 2008).

In general, optimum climatic conditions for the crop are considered favourable for the establishment and development of the rust. The fungus infects plants under temperatures ranging from 10°C to 27.5°C (optimum 20–23°C) and a minimum dew period of 6 h (Melching *et al.*, 1989; Alves *et al.*, 2006). Continuous leaf wetness, promoted either by dew or rain, favours disease development after its establishment, with rainfall being considered an important factor in determining epidemic levels in the field (Del Ponte *et al.*, 2006).

Breeding for resistance to *P. pachyrhizi* has been conducted by classical germplasm screens based on three infection phenotypes: Tan, RB and immune. Susceptible interactions (Tan) are characterized by tan-coloured lesions with fully sporulating uredinia, whereas resistant interactions are characterized by limited fungal growth and sporulation and the formation of reddish-brown lesions (RB). The immune reaction phenotype is an incompatible interaction without any visible disease symptoms on host leaves (Bromfield, 1984). To date, five major ASR resistance (*R*) genes, *Rpp1*, *Rpp2*, *Rpp3*, *Rpp4* and *Rpp5*, have been described (Bromfield and Hartwig, 1980; McLean and Byth, 1980; Bromfield, 1984; Hartwig, 1986; Garcia *et al.*, 2008).

The majority of commercial soybean culti-vars used are susceptible to ASR, and management of the disease has been based on the use of fungicides, although some cultural and crop management practices may also decrease disease risk at field and regional scales (Sinclair and Hartman, 1996; Yorinori *et al.*, 2005). In Brazil, three commercial resistant cultivars were released in 2009 using *R* genes strategy, TMG 801, TMG 803 and BRSGO 7560. Since the effectiveness of these *R* genes is limited by virulent isolates which are able to overcome each of them (Bonde *et al.*, 2006; Miles *et al.*, 2006), the resistant cultivars have been recommended associated with the use of fungicides.

7.2   Fungicides for Soybean Rust Control and Risk of Resistance

Many fungicides have been evaluated to control ASR. Early research in the eastern hemisphere indicated that protective compounds such as mancozeb were effective in reducing disease severity and providing some yield protection, although the results varied by test (Sinclair and Hartman, 1996). Fungicide trials in India (Patil and Anahosur, 1998) identified several triazole compounds, hexaconazole, triadimefon, propiconazole and difenoconazole, which performed better than mancozeb, tridemorph and chlorothalonil. Following the spread of soybean rust to Africa and South America in the last decade, additional triazoles as well as several strobilurin fungicides and mixtures were evaluated and disease control was improved greatly (Miles *et al.*, 2003, 2007; Scherm *et al.*, 2009). The mixtures of triazoles and strobilurin fungicides tend to be most consistent, providing less severe ASR and higher yields (Miles *et al.*, 2003, 2007; Scherm *et al.*, 2009).

The azole fungicides used for ASR control are sterol biosynthesis inhibitors (SBIs) that inhibit the C-14 demethylation step in fungal sterol biosynthesis and are commonly characterized as demethylation inhibitors (DMIs). DMI fungicides have been the leading agents for the control of fungal diseases of plants, humans and animals for over 20 years. For some time, it was thought that these com-pounds were not able to develop resistance because experimental evidence suggested that azole-resistant mutants were less fit than azole-sensitive strains (Koller and Scheinpflug, 1987). However, intensive use in several crops led to the appearance of less sensitive strains for some pathogens (Koller and Scheinpflug, 1987; Staub, 1991; Erickson and Wilcox, 1997; Schnabel *et al.*, 2004). The genetic background of resistance to DMIs is claimed to be polygenic for several plant pathogens, and high resistance levels are observed only after a stepwise adaptation (Gisi *et al.*, 2000). The development of resistance to DMIs has not led to a complete loss of disease control and is often described as continuous selection or shifting. The resistance risk of DMI fungicides is classified as medium or moderate (Brent and Hollomon, 2007).

Strobilurins belong to the chemical group classified as quinone outside inhibitor (QoI) fungicides, which inhibit electron transport in mitochondrial respiration at the Qo site of the cytochrome bc_1 enzyme complex (complex III, ubiquinol-cytochrome *c* oxidoreductase, named also coenzyme Q-cytochrome *c* reductase) (Gisi *et al.*, 2000). Resistance to QoI fungicides is associated with mutations in cytochrome *b* (*cyt b*). Eleven point mutations in two regions of the *cyt b* gene were described singly or in combination conferring QoI resistance (Gisi *et al.*, 2000). The target mutation in position G143, leading to a change in amino acid from glycine to alanine (G143A), is considered the most important, since a complete loss of control with QoI fungicide is possible (Mehl, 2009). For *P. pachyrhizi* and other rusts, Grasso *et al.* (2006) observed a type I intron directly after the codon for glycine at position 143. The authors predicted that a nucleotide substitution in codon 143 would prevent splicing of the intron, leading to a deficient cytochrome *b*, which was lethal, and therefore the resistance based on G143A was not likely to evolve. In species carrying an intron directly after codon 143, it cannot be ruled out that mechanisms other than the G143A mutation will arise, conferring resistance to QoIs, as observed for *Altenaria solani*. Resistance to QoI fungicides was detected in this species, based on a substitution of phenylalanine with leucine at position 129, F129L (Pasche *et al.*, 2004). In contrast to the G143A mutation, F129L mutants showed only moderate resistance factors, referred to as a shift in sensitivity or reduced sensitivity.

7.3 Asian Soybean Rust in Brazil

Brazil is one of the largest producers of soybean in the world, second only to the USA. The cultivated soybean area has nearly doubled in the past decade, rising from 12.9 million ha in 1998/99 to 23.4 million ha in 2009/2010 (Conab, 2010). One of the main factors for the success of soybean expansion in Brazil was the development of cultivars adapted to low latitudes (up to 0°), allowing for the expansion of the crop to the Cerrado (central Brazil) region. The regular soybean growing season goes from October to December in many parts of Brazil, but under irrigation in central Brazil, it is possible to grow soybean all year long.

In 2001, ASR was first found in South America (Yorinori *et al.*, 2002), from where it spread rapidly to the major soybean production regions in Brazil (Yorinori *et al.*, 2005). In the Americas, the greatest economic impact of soybean rust has been in Brazil, where environmental conditions are often favourable for epidemic development (Henning and Godoy, 2006). Estimates of the annual economic losses in Brazil due to ASR, both for direct yield loss and the cost of disease control, have been higher than US$2 billion over the past few years (Henning and Godoy, 2006; Soares, 2007).

The ambient conditions in much of Brazil are conducive for year-round survival of the pathogen (Pivonia and Yang, 2004; Li *et al.*, 2010) and until 2006 it was favoured by the presence of soybean sown during winter in the central Cerrado region. Disease outbreaks were observed in several regions, even in the first growing season after the initial discovery of the pathogen in 2001 (Yorinori *et al.*, 2005). Due to the survival of the fungus in soybean sown between crop seasons, in 2006 three states in Brazil started to adopt the *free host period*, a period of 60–90 days from June to September during which farmers were restricted from planting soybean except under strict controlled conditions. This measure was adopted to break the continuous cycle of the fungus and delay the beginning of the epidemic in the regular season. This measure was adopted by other states in subsequent years. In 2010, 12 states in Brazil adopted the *free host period* (Seixas and Godoy, 2007).

Soybean fungicide recommendations began in Brazil in 1996/97 to control powdery mildew (*Erysiphe diffusa*) and late season diseases (*Cercospora kikuchii* and *Septoria glycines*), mainly with benzimidazoles and triazoles (Embrapa, 1997). With the introduction of *P. pachyrhizi*, the use of fungicide was intensified and the first labelled fungicides to control ASR were those already labelled for other pathogen targets on soybean that had presented efficiency to control ASR in previous trials. Although reports from central Brazil indicated that four or more applications might be necessary to control soybean rust effectively when disease onset occurred early and disease pressure was high (Siqueri, 2005), most of the acreage was treated twice during the period of reproductive crop development (Henning and Godoy, 2006). The number of labelled fungicides to control ASR increased from 5, in 2002, to approximately 70, in 2010. Most of them belong to the DMI group, with several generic compounds. QoIs were recommended alone in the first growing seasons only and, due to lower efficiency compared to triazoles alone (Scherm *et al.*, 2009), they have been recommended only in mixtures with triazoles.

Due to the difference in efficiency among fungicides for ASR control, they have been evaluated annually since 2003/2004 in a nationwide network of standardized uniform field trials (UFTs), coordinated by Embrapa Soybean, a research unit of the Brazilian Agricultural Research Corporation. The experimental design of UFTs was a randomized complete block with four replications; each replication plot was at least six rows wide and 6 m long, with the middle four rows treated with fungicide. Applications were made with a CO_2-pressurized backpack sprayer equipped with a spray wand to deliver 150–200 l/ha, with fungicides applied at label-recommended rates. Disease and yield data were obtained from the two centre rows. General crop management practices were similar to the locally adapted commercial production practices, except that most trials were sown later in the season to increase the likelihood of epidemic development owing to inoculum build-up from earlier plantings. The number of compounds evaluated in UFTs was variable each year and included new compounds and most used fungicides.

The results from 2003/2004 to 2006/2007 were reviewed quantitatively using a meta-analytical approach (Scherm *et al.*, 2009). The results showed that, in general, triazole fungicides applied alone performed better than strobilurins alone, but there was a wide range in efficiency among triazoles, with prothioconazole and tebuconazole performing best and fluquinconazole and difenoconazole being the least efficient. Combinations of strobilurins with triazoles improved disease and yield loss control compared with either class alone. The triazolinthione prothioconazole that belonged to a new class of azole compound was not present in the Brazilian soybean market until 2010, although it has been evaluated since 2005/2006.

Comparing the average percentage of ASR control in the results of ANOVA from UFTs (Fig. 7.1) (Godoy, 2005a,b; Godoy *et al.*, 2007, 2009, 2010), it can be observed that the most widely used triazole (tebuconazole) performed similarly to a standard mixture of DMI + QoI (cyproconazole + azoxystrobin) until 2006/2007. The percentage of disease control is based on the disease severity of the treatment assessed between R5 and R6 (Fehr *et al.*, 1971) divided by the disease severity of the corresponding untreated check. In 2005/2006, the DMI + QoI mixture was not included in the UFTs. The variation in fungicides efficiency among seasons was expected since application timing in the Brazilian UFTs was crop phenology-based and with a fixed number of applications. A weaker efficiency of all fungicides is expected in growing seasons favourable to disease outbreaks.

A weaker efficiency of straight triazole compounds was observed in some regions at the end of the 2007/2008 growing season, mainly in the central Cerrado region, reflecting the control average in UFTs (Fig. 7.1). The orientation of management resistance risk, especially to triazoles that were used alone, started only after the problem was observed. Up to 2008/2009, in the southern region of Brazil, the azole fungicides performed as well

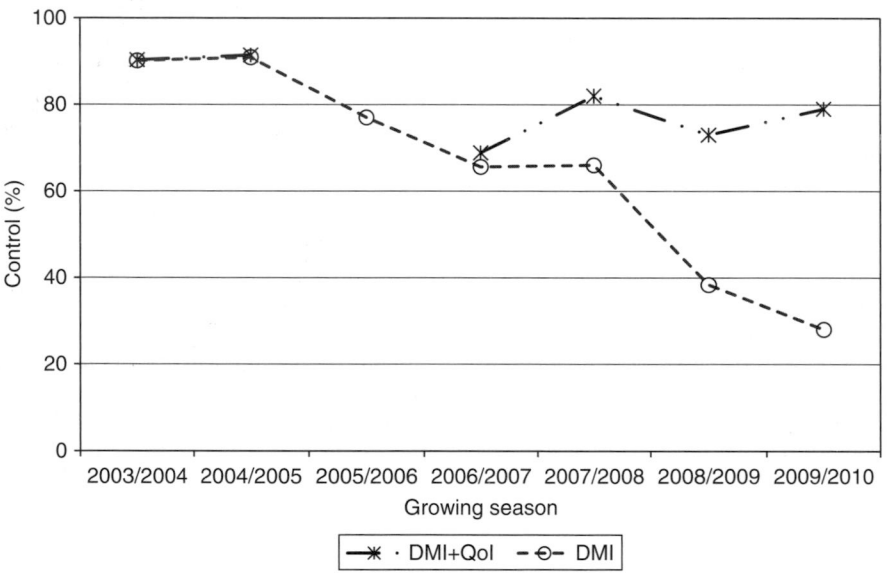

Fig. 7.1. Percentage of soybean rust severity control in uniformed field trials in the growing seasons 2003/2004 (11 trials), 2004/2005 (20 trials), 2005/2006 (15 trials), 2006/2007 (10 trials), 2007/2008 (7 trials), 2008/2009 (23 trials) and 2009/2010 (15 trials) in different producing regions in Brazil. DMI + QoI = cyproconazole + azoxystrobin; DMI = tebuconazole.
Source: Godoy (2005a,b); Godoy *et al.* (2007, 2009, 2010).

as mixtures of triazoles–strobilurins, but were recommended only at the beginning of the growing season and never in sequential application, following the general FRAC (Fungicide Resistance Action Committee) recommendations for the use of SBI fungicides (FRAC, 2010). In the Cerrado region, the orientation since 2008/2009 was to avoid straight azoles and to use preferably mixtures of triazoles–strobilurins for ASR control (Embrapa, 2008). In 2009/2010, lower efficiency with triazole alone was also observed in the southern region of Brazil, and the orientation to use preferably mixtures of triazoles–strobilurins for ASR control was extended to all growing regions in Brazil (Embrapa, 2010).

A previous problem was reported for other triazoles (flutriafol) that used to have similar efficiency compared to tebuconazole in UFTs carried out in 2003/2004 and 2004/2005 (Godoy, 2005a,b). Flutriafol was not evaluated in UFTs in 2005/2006, when some field complaints about efficiency started. In UFTs carried out in 2006/2007, tebuconazole had 61% of ASR control, while flutriafol had 42% (Godoy *et al.*, 2007). The efficiency of flutriafol remained low in subsequent years, and since this problem was isolated, questions about formulation or resistance have never been clarified. After 2007, the erosion of efficiency was general in fungicides belonging to the DMI group.

The weaker efficiency observed with DMIs 6 years after rust introduction in Brazil has been associated with the hypothesis of less sensitive population selection of the fungus. Several factors have contributed to this hypothesis. Up to 2007, due to the high level of efficiency, the presence of generic brands in the market and low cost, straight azole compounds, especially tebuconazole, were widely used in sequential applications and were recommended, under curative conditions, for ASR control. Considering that the average latent period (the time from infection to the next generation of inoculums) of *P. pachyrhizi* is about 7–9 days (Marchetti *et al.*, 1976; Alves *et al.*, 2006) and the disease tends to start at bloom stage (around 45 days after sowing), when fungicide application is started, suggests that 8–10 cycles of infection can take place in a single crop season, and they are all likely to be exposed to fungicide selection.

Another factor for selection is that, in Brazil, the window for soybean sowing is from October to December, and a double crop season is possible in some regions. Therefore, we can have a generation overlap exposed to fungicide that is spread easily by the wind from one field to another.

It was expected that the intercrop period, where the *free host period* was adopted, contributed to reducing this less sensitive population, but UFTs from 2009/2010 did not confirm this hypothesis. The efficiency of triazoles in 2009/2010 was even weaker than in 2008/2009, and it was observed in all regions of soybean production.

7.4 Sensitivity Monitoring Programmes

Measuring of sensitivity in resistance monitoring requires effective tests, which should be simple, reliable, repeatable and able to be related to sensitivity responses in the field (Russell, 2005). The monitoring programmes for *P. pachyrhizi* in Brazil started in 2005, 4 years after rust introduction when fungicides were already used intensively on soybean. Bayer CropScience started the first monitoring programme in 2005. Since strobilurins inhibit spore germination, the method for this group consists in evaluating this component. For DMIs, radioactive label incorporation studies have demonstrated that within the first 6 h of *P. pachyrhizi* germination, no significant sterol biosynthesis can be detected. As germinating uredospores of *P. pachyrhizi* only synthesize sterols almost after completion of germ tube development, spore germination tests are not recommended (Mehl, 2009). The same pattern was observed for *Puccinia graminis* f. sp. *tritici* (Pontzen and Scheinpflug, 1989). The suitable method in this case was the detached leaf method. Both methodologies are published on the FRAC website. For resistance testing, soybean leaves with rust symptoms are collected from fields during the growing season and sent to laboratories for analysis.

According to Calegaro *et al.* (2009), Bayer CropScience sensitivity monitoring data up to 2008/2009 for tebuconazole and generated with *P. pachyrhizi* samples from different Brazilian states, including the central Cerrado region, did not show a significant increase in EC_{50} (half maximal effective concentration) values. Higher EC_{50} values were detected in late season samples (Table 7.1), especially in the central Cerrado region, returning to normal values in the subsequent growing season. No sensitivity change was detected for strobilurin.

Syngenta and BASF started *P. pachyrhizi* monitoring programmes for other compounds in subsequent years, but data have not been published. Embrapa started the bioassays for tebuconazole, metconazole, cyproconazole and prothioconazole in the 2008/2009 growing season. Embrapa's data did not show a large increase in EC_{50} values, and the higher EC_{50} values were even lower in the 2009/2010 growing season. The EC_{50} values observed in sensitivity monitoring tests carried out by Embrapa do not explain the erosion of triazole efficiency observed in the field (Table 7.2). It is important to emphasize that all *P. pachyrhizi* monitoring programmes were started after the intensive use of fungicides on soybean.

Table 7.1. Range of EC_{50} values for tebuconazole in different growing seasons, from Bayer Crop Science's sensitivity monitoring data (Singer *et al.*, 2008).

Growing season	Tebuconazole EC_{50} range
2005/2006	0.016–0.52
2006/2007	0.08–1.85
2007/2008	0.04–3.9

Table 7.2. Range of EC_{50} values for cyproconazole, metconazole, tebuconazole and prothioconazole in the 2008/2009 and 2009/2010 growing seasons, from Embrapa's sensitivity monitoring data.

| DMI fungicides | EC_{50} range | |
	2008/2009	2009/2010
Cyproconazole	0.06–1.37	0.07–1.23
Metconazole	0.02–3.89	0.06–1.72
Tebuconazole	0.02–1.28	0.08–0.97
Prothioconazole	–	0.02–0.13

Source: Koga *et al.* (2010).

7.5 Management of Fungicide Resistance

Even without a clear answer as to the reasons for the reduced performance of triazoles observed 7 years after the introduction of *P. pachyrhizi* in Brazil, it reached a level at which rust control with straight azoles became unsatisfactory. Since 2007, the use of straight azoles has been decreasing and the use of triazole–strobilurin mixtures was intensified as the major strategy to reduce the risk of resistance.

FRAC recommendations for *P. pachyrhizi* are based on data presented at the annual meetings of Bayer CropScience and Syngenta. The results of the sensitive monitoring data presented showed that the performance of DMIs alone was reduced under extreme disease pressure situations, especially under curative application timings and/or extended spray intervals. Sensitivity shifts have been observed and variations were also found related to sampling timing and regional differences (FRAC, 2010).

The recommendations made by FRAC for the management of resistance are the general recommendations for SBIs. According to FRAC, in addition to ensuring an efficient disease control, it is also essential to apply fungicides preventively or as early as possible in the disease cycle, and also to apply DMI fungicides always in mixtures with effective non-cross-resistant fungicides. The products should always be applied considering the intervals recommended by the manufacturers and adjusted to disease epidemics (FRAC, 2010).

In Brazil, since ASR is one of the major diseases on soybean, with potential losses of up to 80%, the preferable use of triazole–strobilurin mixtures has been recommended since 2010 as the main strategy to avoid resistant selection.

Fungicides represent only one of the strategies to control ASR in Brazil. Cultural practices such as the free host period, early sowing with early maturity cultivars, monitoring fields from the beginning of crop development and planting resistant varieties, when available, will optimize the use of fungicides and reduce the selection pressure toward the development of fungicide resistance. The integration of disease resistance and fungicide use can prolong not only the effectiveness of fungicides but also the lifespan of resistant cultivars. There are no other reports in the literature of the loss of ASR control efficiency by triazoles in other regions where the disease occurs.

References

Alves, S.A.M., Furtado, G.Q. and Bergamin Filho, A. (2006) Influência das condições climáticas sobre a ferrugem da soja. In: Zambolim, L. (ed.) *Ferrugem asiática da soja*. Suprema Gráfica e Editora Ltda, Viçosa, MG/Brazil, pp. 37–59.

Bonde, M.R., Nester, S.E., Austin, C.N., Stone, C.L., Frederick, R.D., Hartman, G.L., *et al.* (2006) Evaluation of virulence of *Phakopsora pachyrhizi* and *P. meibomiae* isolates. *Plant Disease* 90, 708–716.

Brent, K.J. and Hollomon, D.W. (2007) *Fungicide Resistance: The Assessment of Risk*, 2nd edn. FRAC Monograph No. 2. Aimprint, UK.

Bromfield, K.R. (1984) *Soybean Rust. Monograph 11*. American Phytopathological Society, St Paul, Minnesota.

Bromfield, K.R. and Hartwig, E.E. (1980) Resistance to soybean rust and mode of inheritance. *Crop Science* 20, 254–255.

Calegaro, P., Geraldes, J., Kemper, K., Pereira, R., Santos, C. and Singer, P. (2009) Resultados do monitoramento de resistência de *Phakopsora pachyrhizi* a fungicidas em soja. In: Godoy, C.V., Seixas, C.D.S. and Soares, R.M. (eds) *Reunião do Consórcio Antiferrugem Safra 2008-09*. Londrina, PR/Brazil, pp. 55–57.

Conab (2010) Séries históricas (http://www.conab.gov.br/, accessed 9 November 2010).

Del Ponte, E.M., Godoy, C.V., Li, X. and Yang, X.B. (2006) Predicting severity of Asian soybean rust epidemics with empirical rainfall models. *Phytopathology* 96, 797–803.

Embrapa (1997) Recomendações técnicas para a cultura da soja na cultura da soja na região central do Brasil. Embrapa Soja, Londrina, 171pp.

Embrapa (2008) Tecnologias de produção de soja – região central do Brasil 2009 e 2010. Embrapa Soja, Embrapa Cerrados, Embrapa Agropecuária Oeste, Londrina, PR/Brazil, 262pp.

Embrapa (2010) Tecnologias de produção de soja – região central do Brasil 2011. Embrapa Soja, Embrapa Cerrados, Embrapa Agropecuária Oeste, Londrina, PR/Brazil, 255pp.

Erickson, E.O. and Wilcox, W.F. (1997) Distributions of sensitivities to three sterol demethylation inhibitor. *Phytopathology* 87, 784–791.

Fehr, W.R., Caviness, C.E., Burmood, D.T. and Pennington, J.S. (1971) Stage of development descriptions for soybeans, *Glycine max* (L.) Merrill. *Crop Science* 11, 929–931.

FRAC (2010) General recommendations for the use of SBI fungicides (http://www.frac.info/frac/, accessed 9 November 2010).

Garcia, A., Calvo, E., Souza Kiihl, R., Harada, A., Hiromoto, D. and Vieira, L. (2008) Molecular mapping of soybean rust (*Phakopsora pachyrhizi*) resistance genes: discovery of a novel locus and alleles. *Theoretical and Applied Genetics* 117, 545–553.

Gisi, U., Chin, K.M., Knapova, G., Kung Farber, R., Mohr, U., Parisi, S., *et al.* (2000) Recent developments in elucidating models of resistance to phenylamide, DMI and strobilurin fungicides. *Crop Protection* 19, 863–872.

Godoy, C.V. (2005a) Resultados da rede de ensaios para controle químico de doenças na cultura da soja. Safra 2003/04. Série Documentos 251. Embrapa, Londrina, PR/Brazil.

Godoy, C.V. (2005b) Ensaios em rede para controle de doenças na cultura da soja. Safra 2004/05. Série Documentos 266. Embrapa, Londrina, PR/Brazil.

Godoy, C.V., Pimenta, C.B., Miguel-Wruck, D.S., Ramos Junior, E.U., Siqueri, F.V., Feksa, H.R. *et al.* (2007) Eficiência de fungicidas para controle da ferrugem asiática da soja, *Phakopsora pachyrhizi*, na safra 2006/07. Resultados sumarizados dos ensaios em rede. Circular Técnica 42. Embrapa, Londrina, PR/Brazil.

Godoy, C.V., Silva, L.H.C.P., Utiamada, C.M., Siqueri, F.V., Lopes, I.O.N., Roese, A.D. *et al.* (2009) Eficiência de fungicidas para controle da ferrugem asiática da soja, *Phakopsora pachyrhizi*, na safra 2008/09. Resultados sumarizados dos ensaios cooperativos 2009. Circular Técnica 69. Embrapa, Londrina, PR/Brazil.

Godoy, C.V., Utiamada, C.M., Silva, L.H.C.P., Siqueri, F.V., Henning, A.A., Roese, A.D. *et al.* (2010) Eficiência de fungicidas para o controle da ferrugem asiática da soja, *Phakopsora pachyrhizi*, na safra 2009/10: resultados sumarizados dos ensaios cooperativos 2010. Circular Técnica 80. Embrapa, Londrina, PR/ Brazil.

Grasso, V., Palermo, S., Sierotzki, H., Garibaldi, A. and Gisi, U. (2006) Cytochrome *b* gene structure and consequences for resistance to Qo inhibitor fungicides in plant pathogens. *Pest Management Science* 62, 465–472.

Hartman, G.L., Wang, T.C. and Tschanz, A.T. (1991) Soybean rust development and the quantitative relationship between rust severity and soybean yield. *Plant Disease* 75, 596–600.

Hartwig, E.E. (1986) Identification of a fourth major gene conferring to rust in soybeans. *Crop Science* 26, 1135–1136.

Henning, A.A. and Godoy, C.V. (2006) Situação da ferrugem da soja no Brasil e no mundo. In: Zambolim, L. (ed.) *Ferrugem Asiática da Soja.* UFV, Viçosa, MG/Brazil, pp. 1–14.

Ivancovich, A. (2005) Soybean rust in Argentina. *Plant Disease* 89, 667–668.

Koga, L.J., Lopes, I.O.N. and Godoy, C.V. (2010) Sensitivity monitoring of *Phakopsora pachyrhizi* population to triazoles in Brazil. In: *Programmes and Abstracts of 16th International Reinhardsbrunn Symposium on Modern Fungicides and Antifungal Compounds.* Friedrichroda, Germany, 168 pp.

Koller, W. and Scheinpflug, H. (1987) Fungal resistance to sterol biosynthesis inhibitors: a new challenge. *Plant Disease* 71, 1066–1074.

Li, X., Esker, P.D., Pan, Z., Dias, A.P., Xue, L. and Yang, X.B. (2010) The uniqueness of soybean rust pathosysthem. An improved understanding of the risk in different regions of the word. *Plant Disease* 94, 796–806.

McLean, R.J. and Byth, D.E. (1980) Inheritance of resistance to rust (*Phakopsora pachyrhizi*) in soybeans. *Australian Journal Agriculture Research* 31, 951–956.

Marchetti, M.A., Melching, J.S. and Bromfield, K.R. (1976) The effects of temperature and dew period on germination and infection by urediospores of *Phakopsora pachyrhizi. Phytopathology* 66, 461–463.

Mehl, A. (2009) *Phakopsora pachyrhizi*: sensitivity monitoring and resistance management strategies for DMI and QoI fungicides. In: *V Congresso Brasileiro de Soja, Mercosoja.* Embrapa Soja, Goiânia, GO/ Brazil, CD-Rom.

Melching, J.S., Dowler, W.M., Koogle, D.L. and Royer, M.H. (1989) Effects of duration, frequency, and temperature of leaf wetness periods on soybean rust. *Plant Disease* 73, 117–122.

Miles, M.R., Hartman, G.L., Levy, C. and Morel, W. (2003) Current status of soybean rust control by fungicides. *Pesticide Outlook* 14, 197–200.

Miles, M.R., Frederick, R.D. and Hartman, G.L. (2006) Evaluation of soybean germplasm for resistance to *Phakopsora pachyrhizi*. Online. *Plant Health Progress* doi 10.1094/PHP-2006-0104-01-RS.

Miles, M.R., Levy, C., Morel, W., Mueller, T., Steinlage, T., Rij, N., *et al.* (2007) International fungicide efficacy trials for the management of soybean rust. *Plant Disease* 91, 1450–1458.

Ono, Y., Buritica, P. and Hennen, J.F. (1992) Delimitation of *Phakopsora, Physopella* and *Cerotelium* and their species on Leguminosae. *Mycological Research* 96, 825–850.

Pasche, J.S., Wharam, C.M. and Gudmestad, N.C. (2004) Shift in sensitivity of *Alternaria solani* in response to QoI fungicides. *Plant Disease* 88, 181–187.

Patil, P.V. and Anahosur, K.H. (1998) Control of soybean rust by fungicides. *Indian Phytopathology* 51, 265–268.

Pivonia, S. and Yang, X.B. (2004) Assessment of the potential year-round establishment of soybean rust throughout the world. *Plant Disease* 88, 523–529.

Pontzen, R. and Scheinpflug, H. (1989) Effects of triazole fungicide on sterol biosynthesis during spore germination of *Botrytis cinerea, Venturia inaequalis* and *Puccinia graminis* f. sp. *tritici. Netherland Journal of Plant Pathology* 95, 152–160.

Rossi, R.L. (2003) First report of *Phakopsora pachyrhizi*, the causal organism of soybean rust in the Province of Misiones, Argentina. *Plant Disease* 87, 102.

Russell, P.E. (2005) *Sensitivity Baselines in Fungicide Resistance Research and Management.* FRAC Monograph No. 3. Aimprint, UK.

Rytter, J.L., Dowler, W.M. and Bromfield, K.R. (1984) Additional alternative hosts of *Phakopsora pachyrhizi*, causal agent of soybean rust. *Plant Disease* 68, 818–819.

Scherm, H., Christiano, R.S.C., Esker, P.D., Del Ponte, E.M. and Godoy, C.V. (2009) Quantitative review of fungicide efficacy trials for managing soybean rust in Brazil. *Crop Protection* 28, 774–782.

Schnabel, G., Bryson, P.K., Bridges, W.C. and Brannen, P. (2004) Reduced sensitivity in *Monilinia fructicola* to propiconazole in Georgia and implications for disease management. *Plant Disease* 88, 1000–1004.

Schneider, R.W., Holier, C.A., Whitam, H.K., Palm, M.E., McKemy, J.M., Hernandez, J.R., *et al.* (2005) First report of soybean rust caused by *Phakopsora pachyrhizi* in the continental United States. *Plant Disease* 89, 774.

Seixas, C.D.S. and Godoy, C.V. (2007) Vazio sanitário: panorama nacional e medidas de monitoramento. In: *Anais do simpósio brasileiro de ferrugem asiática da soja*. Embrapa, Londrina, PR/Brazil, pp. 23–33.

Sinclair, J.B. and Hartman, G.L. (eds) (1996) *Soybean Rust Workshop, 9–11 August 1995*. College of Agricultural, Consumer, and Environmental Sciences, National Soybean Research Laboratory Publication Number 1, Urbana, Illinois.

Sinclair, J.B. and Hartman, G.L. (1999) Soybean rust. In: Hartman, G.L., Sinclair, J.B. and Rupe, J.C. (eds) *Compendium of Soybean Diseases*, 4th edn. APS Press, Saint Paul, Minnesota, pp. 25–26.

Singer, P., Calegaro, P., Geraldes, J., Pereira, R. and Santos, C. (2008) Fungo monitorado. *Cultivar* 11, 16–18.

Siqueri, F.V. (2005) Ocorrência da ferrugem asiática (*Phakopsora pachyrhizi*) no estado do Mato Grosso. Safra 2004/2005. In: Juliatti, F.C., Polizel, A.C. and Hamawaki, O.T. (eds) *I Workshop Brasileiro sobre a Ferrugem Asiática*. EDUFU, Uberlândia, MG/Brazil, pp. 93–99.

Slaminko, T.L., Miles, M.R., Frederick, R.D., Bonde, M.R. and Hartman, G.L. (2008) New legume hosts of *Phakopsora pachyrhizi* based on greenhouse evaluations. *Plant Disease* 92, 767–771.

Soares, R.M. (2007) Balanço da ferrugem asiática na safra 2006–07 com base nos números do sistema de alerta. In: *Anais do simpósio brasileiro de ferrugem asiática da soja*. Embrapa, Londrina, PR/Brazil, pp. 11–16.

Staub, T. (1991) Fungicide resistance: practical experience with antiresistance strategies and the role of integrated use. *Annual Review of Phytopathology* 29, 421–442.

Yang, X.B., Tschanz, A.T., Dowler, W.M. and Wang, T.C. (1991) Development of yield loss models in relation to reductions of components of soybean infected with *Phakopsora pachyrhizi*. *Phytopathology* 81, 1420–1426.

Yorinori, J.T., Paiva, W.M., Frederick, R.D. and Fernandez, P.F.T. (2002) Ferrugem da soja (*Phakopsora pachyrhizi*) no Brasil e no Paraguai, nas safras 2000/01 e 2001/02. In: *Anais do Congresso Brasileiro de Soja*. Foz do Iguaçu, PR/Brazil, p. 94.

Yorinori, J.T., Paiva, W.M., Frederick, R.D., Costamilan, L.M., Bertagnolli, P.F., Hartman, G.E., *et al.* (2005) Epidemics of soybean rust (*Phakopsora pachyrhizi*) in Brazil and Paraguay from 2001 to 2003. *Plant Disease* 89, 675–677.

8 Resistance to Carboxylic Acid Amide (CAA) Fungicides and Anti-Resistance Strategies

Ulrich Gisi

Syngenta Crop Protection, Biology Research Centre, Stein, Switzerland

8.1 Introduction

The carboxylic acid amide (CAA) fungicides were officially announced by FRAC (Fungicide Resistance Action Committee, www.frac.info) in 2005 as group number 40 in the FRAC code list, including the three subclasses, cinnamic acid amides (dimethomorph, flumorph, pyrimorph), valinamide carbamates (benthiavalicarb, iprovalicarb, valiphenalate) and mandelic acid amides (mandipropamid) (Fig. 8.1; Gisi *et al.*, 2011). The reason for this classification was a common cross-resistance pattern among all members for the vast majority of the tested isolates of *Plasmopara viticola*. Other common features are the specific and rather narrow spectrum of activity against Oomycetes including pathogens of the Peronosporales such as *Bremia lactucae* on lettuce, *Peronospora* spp. on tobacco, pea, onion and other vegetables, *Pseudoperonospora* spp. on cucurbits and hops, *P. viticola* in grapes and several *Phytophthora* spp. on many crops such as potato, tomato, pepper and pineapple. However, the entire genus *Pythium* is insensitive, as are all other pathogens outside the Oomycetes. Dimethomorph (Fig. 8.1, 1) was the first in the class to be introduced in 1988 (Albert *et al.*, 1988), followed by iprovalicarb (Fig. 8.1, 4) in 1998 (Stenzel *et al.*, 1998), flumorph (Fig. 8.1, 2) in 2000 (Liu *et al.*, 2000), benthiavalicarb (Fig. 8.1, 5) in 2003 (Miyake

et al., 2003), mandipropamid (Fig. 8.1, 7) in 2005 (Huggenberger *et al.*, 2005; Lamberth *et al.*, 2006) and valifenalate (Fig. 8.1, 6) in 2008 (Gonzalez-Rodriguez *et al.*, 2011). Pyrimorph (Fig. 8.1, 3) is expected to be launched in the near future (Yan *et al.*, 2010).

8.2 Biological Activity of CAA Fungicides

CAAs inhibit the germination of cystospores and sporangia (but not zoospore release and motility); they affect growth of germ tubes and mycelium, thus preventing infection of the host tissue (Cohen and Gisi, 2007). After foliar application, CAAs provide strong preventive activity and some curative and antisporulant effects, depending on the active ingredient, the quantity taken up into the leaf and its distribution based on translaminar movement. Dimethomorph has good preventive and pronounced curative and antisporulant activities. Iprovalicarb is reported to be more systemic than dimethomorph (Cohen *et al.*, 1995; Stenzel *et al.*, 1998; Gisi, 2002) and mandipropamid (Hermann *et al.*, 2005) when used as a soil drench or foliar application. Iprovalicarb is translocated in the apoplast and also protects untreated leaves against infection, especially in grapes (Stübler *et al.*, 1999).

96

©CAB International 2012. *Fungicide Resistance in Crop Protection: Risk and Management* (ed. T.S. Thind)

Cinnamic acid amides:

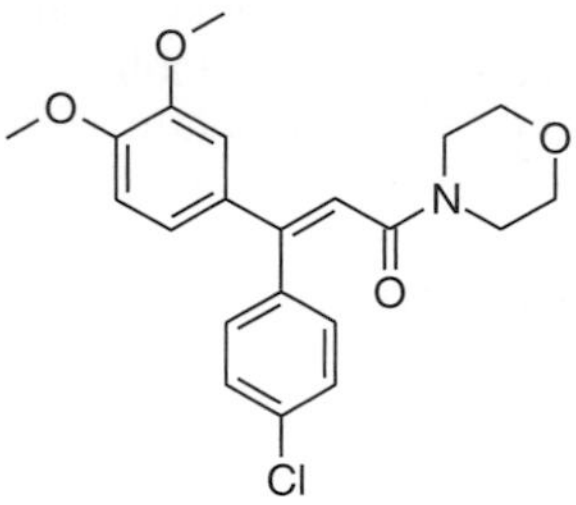

| 1 | 2 | 3 |
| Dimethomorph | Flumorph | Pyrimorph |

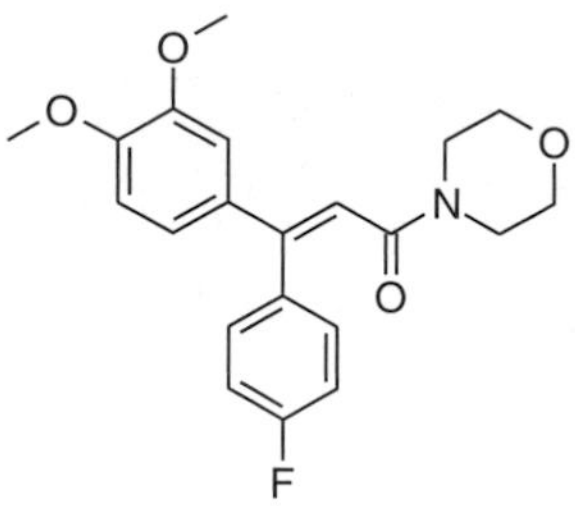

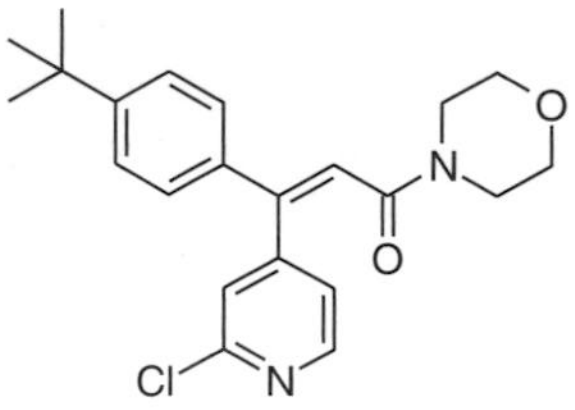

Valinamides:

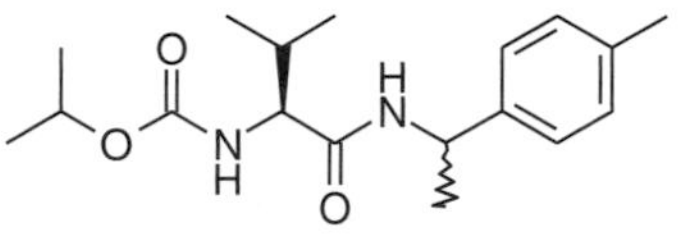

| 4 | 5 |
| Iprovalicarb | Benthiavalicarb |

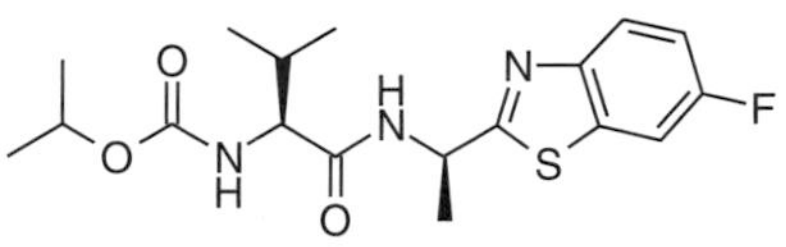

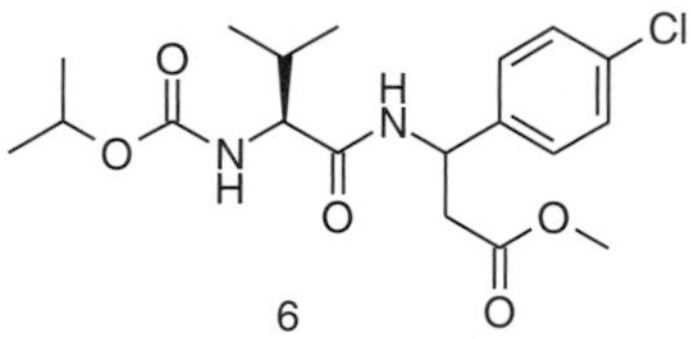

6
Valifenalate

Mandelic acid amides:

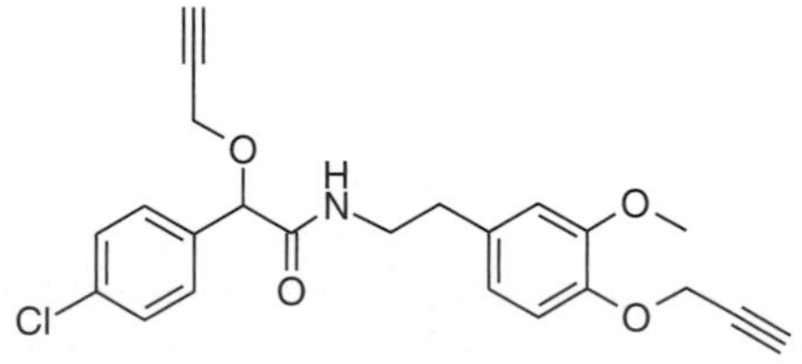

7
Mandipropamid

Fig. 8.1. Carboxylic acid amide (CAA) fungicides (Gisi *et al.*, 2011).

Benthiavalicarb delivers long-lasting preventive and curative (Miyake *et al.*, 2003; Hofman *et al.*, 2003; Reuveni, 2003) and some loco-systemic but low translaminar activity in grape leaves (Reuveni, 2003). Mandipropamid binds rapidly and tightly to the wax layer of the leaf surface, providing a rainfast and long-lasting barrier against infections (Hermann *et al.*, 2005). It delivers strong preventive and translaminar activity and provides robust control of both *P. infestans* and *P. viticola* also under severe disease pressures (Huggenberger *et al.*, 2005). In situations where resistance problems influence the effectiveness of CAAs (e.g. in some European vineyards), the curative and antisporulant potential may not be fully exploitable and products should be used in a preventive manner.

8.3   Mode of Action of CAA Fungicides

Cytological studies have demonstrated that dimethomorph, iprovalicarb and benthiavalicarb inhibit processes involved in cell wall biosynthesis and assembly (Albert *et al.*, 1991; Kuhn *et al.*, 1991; Thomas *et al.*, 1992; Jende *et al.*, 2002; Mehl and Buchenauer, 2002). The process of cell wall synthesis in Oomycetes is rather complex and still not well investigated. The altered architecture of the cell wall after CAA treatment was associated with effects on cytoskeletal elements or membrane-bound components (e.g. receptors, enzymes) responsible for transport of cell wall precursors. Enzymes associated with cell wall biosynthesis, such as glucanases and synthases of β-1,3 glucans were not inhibited (Jende *et al.*, 2002; Mehl and Buchenauer, 2002). CAAs were postulated to inhibit the three-dimensional arrangement and cross-linkage of the complex glucan structure necessary for germ tube and hyphal growth (Kuhn *et al.*, 1991). In *P. melonis*, a disruption of F-actin and microfilament organization was described as a consequence of CAA activity (Zhu *et al.*, 2007a). Alterations in phospholipid biosynthesis were also proposed, with an inhibition of phosphatidylcholine (lecithin) biosynthesis as the main target (Griffiths *et al.*, 2003). However, sequence analyses of two genes coding for choline-phosphotransferase (CPT) in wild-type and laboratory mutants of *Phytophthora capsici* selected with pyrimorph on amended agar did not show any amino acid differences in the two potential target genes (Sun *et al.*, 2010), suggesting an alternative site of action for CAA fungicides. Recent studies performed with mandipropamid contributed to elucidating the mode of action for CAA fungicides in *P. infestans*. The

incorporation of ^{14}C-labelled glucose into the β-1,4 glucan (cellulose) fraction of cell walls of germinating cystospores was inhibited in the presence of mandipropamid (Blum *et al.*, 2010a). Gene sequencing of artificially generated mutants of *P. infestans* which were resistant to mandipropamid revealed an amino acid substitution in the cellulose synthase (*CesA3*) gene at position 1105 from glycine to serine, G1105S. In addition, the transformation and expression of a mutated *CesA3* allele in a sensitive *P. infestans* isolate resulted in a CAA-resistant phenotype (Blum *et al.*, 2010a). Thus, cellulose synthase can be postulated as the primary target enzyme for CAA activity.

8.4   History of Resistance to CAA Fungicides

Resistance to CAAs in *P. infestans* and *B. lactucae* has never been detected in field populations, even though dimethomorph has been in use for more than 15 years. The natural variation of sensitivity in field populations is about 80-fold and is characterized by a continuous, unimodal log-normal distribution (Cohen *et al.*, 2007) without any (resistant) isolates being outside this distribution. The lack of resistant isolates in nature encouraged several researchers to produce artificial mutants of *P. infestans in vitro* (Chabane *et al.*, 1993; Bagirova *et al.*, 2001; Stein and Kirk, 2004; Yuan *et al.*, 2006; Ziogas *et al.*, 2006; Rubin *et al.*, 2008). Mutants resistant to dimethomorph, flumorph or mandipropamid were produced but were found to show reduced growth rates, reduced frequency of infections on leaves and tubers and lower fitness or survival over several generations compared with wild-type isolates. However, resistance to CAAs in *P. viticola* had already been reported in 1994, shortly after the introduction of dimethomorph in France (Chabane *et al.*, 1996). As a consequence, intensive sensitivity monitoring was carried out across European vineyards by several companies, and resistant isolates were detected repeatedly, mainly in some of the grape growing regions in France, Italy, Switzerland and Germany. Resistance has increased gradually in these countries over the past 3 years (since 2008). Also, resistance in *Pseudoperonospora cubensis* was reported from several trial sites in the USA, Israel and China (CAA FRAC Working Group, www.frac.info; Zhu *et al.*, 2007b). Cross-resistance was found among all CAAs for the vast majority of isolates (Gisi *et al.*, 2007). As to be expected, no cross-resistance was found between CAAs and other modes of action such as phenylamides, quinone outside inhibitor (QoI) fungicides and zoxamide. Because the group members of the CAA fungicides express different intrinsic activities, resistance factors (difference in sensitivity between wild-type and resistant isolates) can vary significantly.

8.5 Mechanism of Resistance to CAA Fungicides

To obtain more information on CAA resistance, the segregation pattern was investigated in sexual crosses made between CAA-sensitive and -resistant single sporangiophore isolates

of *P. viticola* (Gisi *et al.*, 2007). All F1 progeny isolates were sensitive to CAA fungicides reflecting a segregation pattern of s:r = 1:0. When F1 progeny isolates (siblings) were crossed, the segregation in the F2 progeny was about 9:1 (s:r) (Gisi *et al.*, 2007). This segregation pattern suggests that resistance to CAAs may be controlled by recessive nuclear genes. In the F2 progeny, resistance co-segregated for all tested CAAs (mandipropamid, dimethomorph, iprovalicarb). Sequencing of the four cellulose synthase genes, *CesA1*, *CesA2*, *CesA3* and *CesA4*, in a CAA-sensitive and -resistant single sporangiophore isolate of *P. viticola* revealed five single nucleotide polymorphisms (SNPs) affecting the amino acid structure of the protein (Blum *et al.*, 2010b). SNP inheritance in F1, F2 and F3 progeny isolates confirmed resistance to be correlated with one specific SNP located in the *CesA3* gene. If present in both alleles, this SNP led to the substitution of glycine for serine at position 1105 (G1105S), thus conferring CAA resistance (Blum *et al.*, 2010b). The G1105S mutation co-segregated in all F2 and F3 progeny isolates with the resistant phenotype (Fig. 8.2).

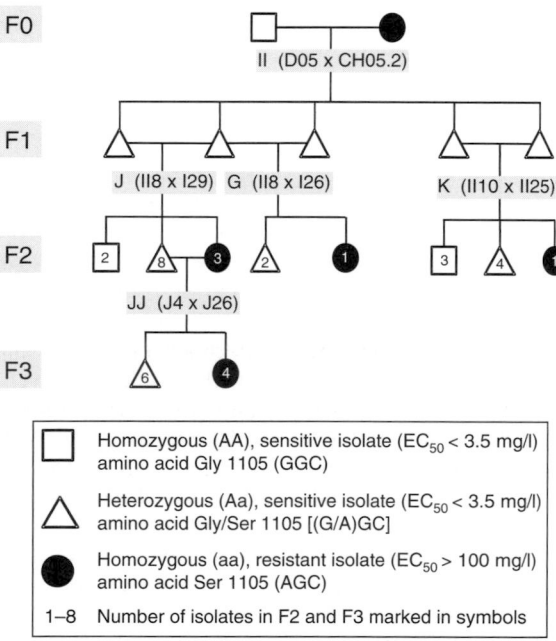

Fig. 8.2. Co-segregation of CAA resistance and G1105A in F1, F2 and F3 progeny of *Plasmopara viticola* (after Blum *et al.*, 2010b).

The results demonstrate that one recessive mutation in the *CesA3* gene causes inheritable resistance to CAAs and confirms cellulose synthase to be the target enzyme for CAA activity.

8.6   Resistance Risk and Implementation of Anti-Resistance Strategies for CAA Fungicides

Several elements contribute to estimating the inherent resistance risk (pathogen and fungicide risks), such as sensitivity structure of wild-type (baseline) and treated (selected) populations (log-normal or discontinuous bimodal distribution), cross- and multiple resistance, stability of resistance (with or without selection), biology of pathogen (especially number per season and duration of generations, frequency of sexual recombination, fitness and overseasoning of resistant individuals), persistence of fungicide activity (selection pressure), emergence of mutants upon forced selection and artificial mutagenesis, location of mutations (target site), biochemical mode of action (single-, multi-site), molecular mechanism of resistance (target gene, coding mutations) and inheritance of resistance (genetics) (Gisi and Staehle-Csech, 1988; Brent and Hollomon, 2007; Gisi and Sierotzki, 2008a). This estimation has to be done separately for the interaction of each fungicide/pathogen species combination and should also include the local agronomic conditions (agronomic risk). As a consequence, adapted use strategies can be designed to delay resistance development (management risk).

For CAA fungicides and *P. viticola*, the following factors contribute to an increased resistance risk: presence of CAA-resistant isolates in recent field populations; stability of CAA-resistant isolates (after transfer on untreated leaves); inheritable resistance (phenotype) after sexual crossing; and presence of an SNP (single nucleotide polymorphism, point mutation), G1105S, in the target gene (*CesA3*) of resistant isolates (Gisi *et al.*, 2007; Blum *et al.*, 2010b). However, other factors contribute to a decreased resistance risk, such as rather slow spread of CAA resistance in field populations (limited to some European countries; CAA FRAC Working Group, www.frac.info), decline of CAA resistance in absence of product use (Gisi and Sierotzki, 2008b), recessive nature of CAA resistance and resistance and G1105S expressed in homozygous isolates only (both alleles mutated; Blum *et al.*, 2010b). Therefore, the risk and extent of resistance for the entire CAA group of fungicides were classified by FRAC as moderate for *P. viticola*. As a consequence, CAA fungicides are recommended only in mixtures with multi-site fungicides or other effective non-cross-resistant partner fungicides. In Europe, a maximum of four treatments during one season may contain a CAA fungicide which should be used in a preventative manner. A moderate resistance risk can be attributed also to *P. cubensis*. In fact, resistant isolates have been detected in field populations in several countries (USA, Israel, China; CAA FRAC Working Group reports, www.frac.info).

In contrast to the evolution of resistance in *P. viticola*, no resistant isolates have been detected in field populations of *P. infestans*, although some CAAs have been in use for more than 15 years. Artificial mutants (produced artificially, then transferred and crossed) did not express stable resistance or were less fit, and upon resistant individuals were detected upon enforced selection (Cohen *et al.*, 2007; Rubin *et al.*, 2008). No heterozygous isolates (at position 1005 in the *CesA3* gene) have been detected so far in field populations (Blum *et al.*, 2011). Therefore, the resistance risk for CAAs was classified as low for *P. infestans*. In fact, fungicide resistance in *P. infestans* has evolved 'only' for one class of fungicides, the phenylamides, whereas for *P. viticola* all major fungicide groups active against Oomycetes are affected (to different degrees) by resistance problems (e.g. phenylamides, CAAs, QoIs, cymoxanil). The reasons for this differential behaviour of *P. infestans* are largely unknown, but may be related to rather rare sexual recombination, chromosomal abnormalities, heterokaryosis (Catal *et al.*, 2010) and quite frequent occurrence of triploid (or trisomic) individuals in field populations (Syngenta internal data). As a consequence, CAAs (especially mandipropamid) can be recommended as solo products for the control of *P. infestans*. However, as a resistance management precaution, the number of CAA applications per season is limited to no more than 50% of the treatments against *P. infestans*. Alternation with products of a different mode of action group should be considered (CAA FRAC Working Group recommendations; www.frac.info).

8.7 Conclusions

Resistance to CAA fungicides has evolved recently in some European countries for *P. viticola* and at some sites in the USA, Israel and China for *P. cubensis*, whereas no resistance has been detected for *P. infestans* and *B. lactucae*. Inheritance of resistance to CAA fungicides is based on one recessive nuclear gene. Resistance is based mainly on the G1105S mutation in the *CesA3* gene; the phenotype and genotype always co-segregate in progeny isolates. Resistance to CAA fungicides is expressed in homozygous isolates only; heterozygous isolates are sensitive. Therefore, resistance is recessive. The resistance risk is classified as low for *P. infestans* and moderate for *P. viticola* and *P. cubensis*. For the control of grape downy mildew, CAAs are recommended only in mixture with other effective, non-cross-resistant fungicides, and the number of applications per season is limited to a maximum of four.

References

Albert, G., Curtze, J. and Drandarevski, C.A. (1988) Dimethomorph (CME 151), a novel curative fungicide. In: *Brighton Crop Protection Conference*, Volume 1. British Crop Protection Council, Thornton Heath, Surrey, UK, pp. 17–24.

Albert, G., Thomas, A. and Gühne, M. (1991) Fungicidal activity of dimethomorph on different stages in the life cycle of *Phytophthora infestans* and *Plasmopara viticola*. 3rd International Conference on Plant Diseases. ANPP, Paris, pp. 887–894.

Bagirova, S.F., Li, A.Z., Dolgova, A.V., Elansky, S.N., Shaw, D.S. and Dyakov, Y.T. (2001) Mutants of *Phytophthora infestans* resistant to dimethomorph fungicide. *Journal of Russian Phytopathological Society* 2, 19–24.

Blum, M., Boehler, M., Randall, E., Young, V., Csukai, M., Kraus, S., *et al.* (2010a) Mandipropamid targets the cellulose synthase like PiCesA3 to inhibit cell wall biosynthesis in the oomycete plant pathogen, *Phytophthora infestans*. *Molecular Plant Pathology* 11, 227–243.

Blum M., Waldner, M. and Gisi, U. (2010b) A single point mutation in the novel *PvCesA3* gene confers resistance to the carboxylic acid amide fungicide mandipropamid in *Plasmopara viticola*. *Fungal Genetics and Biology* 47, 499–510.

Blum, M., Sierotzki, H. and Gisi, U. (2011) Comparison of cellulose synthase 3 (CesA3) gene structure in different Oomycetes. In: Dehne, H.W., Deising, H.B., Gisi, U., Kuck, K.H., Russell, P.E. and Lyr, H. (eds) *Modern Fungicides and Antifungal Compounds VI*. DPG Selbstverlag, Braunschweig, Germany, pp. 151–154.

Brent, K.J. and Hollomon, D.W. (2007) *Fungicide Resistance: The Assessment of Risk*. FRAC Monograph No. 2, 2nd revised edn. FRAC, Brussels.

Catal, M., King, L., Tumbalam, P., Wiriyajitsomboon, P., Kirk, W.W. and Adams, G.C. (2010) Heterokaryotic nuclear conditions and a heterogeneous nuclear population are observed by flow cytometry in *Phytophthora infestans*. *Cytometry Part A*, 77A, 769–775.

Chabane, K., Leroux, P. and Bompeix, G. (1993) Selection and characterization of *Phytophthora parasitica* mutants with ultraviolet-induced resistance to dimethomorph or metalaxyl. *Pesticide Science* 39, 325–329.

Chabane, K., Leroux, P., Maia, N. and Bompeix, G. (1996) Resistance to dimethomorph in laboratory mutants of *Phytophthora parasitica*. In: Lyr, H., Russell, P.E. and Sisler, H.D. (eds) *Modern Fungicides and Antifungal Compounds*. Intercept, Andover, UK, pp. 387–391.

Cohen, Y. and Gisi, U. (2007) Differential activity of carboxylic acid amide fungicides against various developmental stages of *Phytophthora infestans*. *Phytopathology* 97, 1274–1283.

Cohen, Y., Baider, A. and Cohen, B. (1995) Dimethomorph activity against oomycete fungal plant pathogens. *Phytopathology* 85, 1500–1506.

Cohen, Y., Rubin, E., Hadad, T., Gotlieb, D., Sierotzki, H. and Gisi, U. (2007) Sensitivity of *Phytophthora infestans* to mandipropamid and the effect of enforced selection pressure in the field. *Plant Pathology* 56, 836–842.

FRAC (www.frac.info).

Gisi, U. (2002) Chemical control of downy mildews. In: Spencer, P.T.N., Gisi, U. and Lebeda, A. (eds) *Advances in Downy Mildew Research*. Kluwer, Dordrecht, The Netherlands, pp. 119–159.

Gisi, U. and Sierotzki, H. (2008a) Molecular and genetic aspects of fungicide resistance in plant pathogens. In: Dehne, H.W., Deising, H.B., Gisi, U., Kuck, K.H., Russell, P.E. and Lyr, H. (eds) *Modern Fungicides and Antifungal Compounds V*. DPG Selbstverlag, Braunschweig, Germany, pp. 53–61.

Gisi, U. and Sierotzki, H. (2008b) Fungicide modes of action and resistance in downy mildews. *European Journal of Plant Pathology* 122, 157–167.

Gisi, U. and Staehle-Csech, U. (1988) Resistance risk evaluation of phenylamide and EBI fungicides. *Proceedings Brighton Crop Protection Conference*, Volume 1. British Crop Protection Council, Thornton Heath, Surrey, UK, pp. 359–366.

Gisi, U., Waldner, M., Kraus, N., Dubuis, P.H. and Sierotzki, H. (2007) Inheritance of resistance to carboxylic acid amide (CAA) fungicides in *Plasmopara viticola*. *Plant Pathology* 56, 199–208.

Gisi, U., Lamberth, C., Mehl, A. and Seitz, T. (2011) Carboxylic acid amide (CAA) fungicides. In: Krämer, W. and Schirmer, U. (eds) *Modern Crop Protection Compounds*, 2nd revised edn. Wiley-VCH, Weinheim, Germany (in press).

Gonzalez-Rodriguez, R.M., Cancho-Grande, B. and Simal-Gandara, J. (2011) Decay of fungicide residues during vinification of white grapes harvested after the application of some new active substances against downy mildew. *Food Chemistry* 125, 549–560.

Griffiths, R.G., Dancer, J., O'Neill, E. and Harwood, J.L. (2003) A mandelamide pesticide alters lipid metabolism in *Phytophthora infestans*. *New Phytologist* 158, 345–353.

Hermann, D., Bartlett, D.W., Fischer, W. and Kempf, H.J. (2005) The behaviour of mandipropamid on and in plants. *Proceedings British Crop Protection Conference, International Congress*, Volume 1. British Crop Protection Council, Thornton Heath, Surrey, UK, pp. 93–98.

Hofman, T.W., Boon, S.M., Coster, G., van Oudheusden, Z., Ploss, H. and Nagayama, K. (2003) New fungicide benthiavalicarb-isopropyl + mancozeb for foliar use in potatoes in Europe. *Brighton Crop Protection Conference*, Volume 2. British Crop Protection Council, Thornton Heath, Surrey, UK, pp. 413–418.

Huggenberger, F., Lamberth, C., Iwanzik, W. and Knauf-Beiter, G. (2005) Mandipropamid, a new fungicide against oomycete pathogens. *Proceedings British Crop Protection Conference, International Congress*, Volume 1. British Crop Protection Council, Thornton Heath, Surrey, UK, pp. 87–92.

Jende, G., Steiner, U. and Dehne, H.W. (2002) Microscopical characterization of fungicidal effects on infection structures and cell wall formation of *Phytophthora infestans*. In: Dehne, H.W., Gisi, U., Kuck, K.H., Russell, P.E. and Lyr, H. (eds) *Modern Fungicides and Antifungal Compounds III*. AgroConcept, Bonn, Germany, pp. 83–90.

Kuhn, P.J., Pitt, D., Lee, S.A., Wakley, G. and Sheppard, A.N. (1991) Effects of dimethomorph on the morphology and ultrastructure of *Phytophthora*. *Mycological Research* 95, 333–340.

Lamberth, C., Cederbaum, F., Jeanguenat, A., Kempf, H.J., Zeller, M. and Zeun, R. (2006) Synthesis and fungicidal activity of *N*-2-(3-methoxy-4-propargyloxy) phenethyl amides. Part II: Anti-oomycetic mandelamides. *Pest Management Science* 62, 446–451.

Liu, W.C., Li, Z.L., Zhang, Z.J. and Liu, C.L. (2000) Antifungal activity and prospect of flumorph and its mixtures. *Zhejiang Chemical Industry* 31, 87–88.

Mehl, A. and Buchenauer, H. (2002) Investigations of the biochemical mode of action of iprovalicarb. In: Dehne, H.W., Gisi, U., Kuck, K.H., Russell, P.E. and Lyr, H. (eds) *Modern Fungicides and Antifungal Compounds III*. AgroConcept, Bonn, Germany, pp. 75–82.

Miyake, Y., Sakai, J., Miura, I. and Nagayama, K. (2003) Effects of a novel fungicide benthiavalicarb-isopropyl against oomycete fungal diseases. *Brighton Crop Protection Conference*, Volume 1. British Crop Protection Council, Thornton Heath, Surrey, UK, pp. 105–112.

Reuveni, M. (2003) Activity of the new fungicide benthiavalicarb against *Plasmopara viticola* and its efficacy in controlling downy mildew in grape vines. *European Journal of Plant Pathology* 109, 243–251.

Rubin, E., Gotlieb, D., Gisi, U. and Cohen, Y. (2008) Mutagenesis of *Phytophthora infestans* for resistance against carboxylic acid amide and phenylamide fungicides. *Plant Disease* 90, 741–749.

Stein, J.M. and Kirk, W.W. (2004) The generation and quantification of resistance to dimethomorph in *Phytopthora infestans*. *Plant Disease* 87, 1283–1289.

Stenzel, K., Pontzen, R., Seitz, T., Tiemann, R. and Witzenberger, A. (1998) SZX 0722: a novel systemic oomycete fungicide. *Brighton Crop Protection Conference*, Volume 2. British Crop Protection Council, Thornton Heath, Surrey, UK, pp. 367–374.

Stübler, D., Reckmann, U. and Noga, G. (1999) Systemic action of iprovalicarb (SZX 0722) in grapevines. *Pflanzenschutz-Nachrichten Bayer* 52(1), 33–48.

Sun, H., Wang, H., Stammler, G., Ma, J. and Zhou, M. (2010) Baseline sensitivity of populations of *Phytophthora capsici* from China to three carboxylic acid amide (CAA) fungicides and sequence analysis of cholinephosphotranferases from a CAA-sensitive isolate and CAA-resistant laboratory mutants. *Journal of Phytopathology* 158, 244–252.

Thomas, A., Albert, G. and Schlösser, E. (1992) Use of video-microscope systems to study the mode of action of dimethomorph on *Phytophthora infestans. Mededelingen Faculteit Landbouwwetenschappen Rijksuniversiteit Gent* 57, 189–197.

Yan, X., Qin, W., Sun, L., Qi, S., Yang, D., Qin, Z., *et al.* (2010) Study of inhibitory effects and action mechanism of the novel fungicide pyrimorph against *Phytophthora capsici. Journal of Agricultural and Food Chemistry* 58, 2720–2725.

Yuan, S.K., Liu, X.L., Si, N.G., Dong, J., Gu, B.G. and Jiang, H. (2006) Sensitivity of *Phytophthora infestans* to flumorph: *in vitro* determination of baseline sensitivity and the risk of resistance. *Plant Pathology* 55, 258–263.

Zhu, S.S., Liu, X.L., Liu, P.F., Li, J.Q., Wang, H.M., Yuan, S.K., *et al.* (2007a) Flumorph is a novel fungicide that disrupts microfilament organization in *Phytophthora melonis. Phytopathology* 97, 643–649.

Zhu, S.S., Liu, X.L., Wang, Y., Wu, X.H., Liu, P.F., Li, J.Q., *et al.* (2007b) Resistance of *Pseudoperonospora cubensis* to flumorph on cucumber in plastic houses. *Plant Pathology* 56, 967–975.

Ziogas, B.N., Markoglou, A.N., Theodosiou, D.I., Anagnostou, A. and Boutopoulou, S. (2006) A high multidrug resistance to chemically unrelated oomycete fungicides in *Phytophthora infestans. European Journal of Plant Pathology* 115, 283–292.

9 New Modes of Action Contribute to Resistance Management

Derek W. Hollomon
Orchard House, Chew Stoke, Bristol, UK

9.1 Introduction

A literature survey covering 275 fungal pathogens causing crop and postharvest losses identified resistance in 151 pathogens (Fig. 9.1), involving mainly benzimidazole (MBC, B1[1]), phenylamide (A1), dicarboxamide (E3) and quinone outside inhibitors (QoIs; strobilurin, C3) fungicides. In 37 pathogens, resistance was generated solely in the laboratory, but for the remaining 114, resistant isolates were recovered from field populations. However, in only 27 cases, especially involving powdery and downy mildews, grey mould, potato blight and banana Sigatoka diseases, was there evidence that fungicide performance was eroded to an extent that there was no longer useful disease control. Despite this, in all cases control was not lost, as fungicides with alternative modes of action were available. Loss of a fungicide was not without cost for the user; in addition to some crop loss, alternatives were generally more expensive and often less effective. For the manufacturer, the financial cost was more significant, since the inevitable sales loss reduced income, some of which was needed to support research and development of new modes of action.

Availability of different modes of action has been at the core of anti-resistance management programmes. Despite the diversity of the modes of action available (see section 9.2), the number registered for use on a particular crop may be limited and loss of one for whatever reason increases the risk of resistance to remaining fungicides developing. The resistance problem is particularly acute for many minor crop uses, especially in horticulture, where only a single mode of action may be available. Given the seemingly increasing reliance on fungicides with a single biochemical target site, which generally increases the risk that selection will generate resistance, a steady stream of new modes of action is essential if resistance is to be managed effectively.

Not only resistance but also legislative changes affecting registration, and competition from generic manufacturers, all impact on the resources available for the discovery and development of new modes of action. More efficient high throughput screening of 'natural products' rather than just 'chemical' libraries, coupled with miniaturized whole plant assays, have helped maintain active research programmes in this area. But new modes of action generally are still identified by chance during the discovery process, which is then used to steer fungicide chemistry in directions that will yield fungicides that meet registration requirements.

[1] Code assigned by the Fungicide Resistance Action Committee to a mode of action.

©CAB International 2012. *Fungicide Resistance in Crop Protection: Risk and Management* (ed. T.S. Thind)

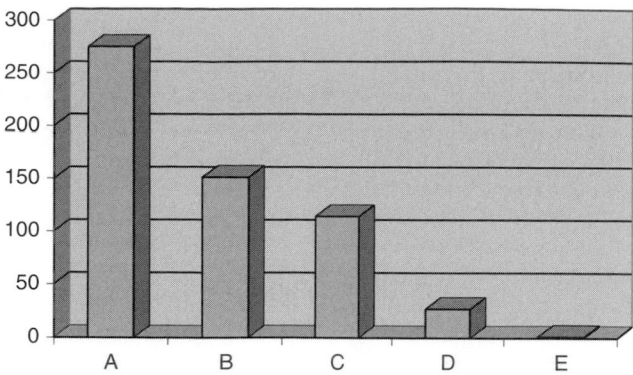

Fig. 9.1. Resistance in fungal pathogens causing crop and postharvest losses. A, total pathogens; B, number in which resistance was reported; C, resistance present in field isolates; D, number where disease control was eroded; E, disease control lost.

In this chapter, I review some recently introduced novel modes of action that have added much needed diversity to resistance management programmes. New modes of action have shifted away from single enzyme targets in biosynthetic pathways to regulatory steps in core developmental processes which impact on infection, growth and sporulation in many ways. After 8 years in a medical research environment, I am well aware of the wealth of new information about cell development and how it is regulated. Although much of this information is derived from animal and yeast systems, the processes are fundamental to all eukaryotes, including filamentous pathogenic fungi.

9.2 Exploiting Existing Modes of Action

Information presented by the Fungicide Resistance Action Committee (FRAC, 2009, www.frac.info/publications) identifies 28 modes of action targeted by different active ingredients, together with 9 proposed modes of action, 21 multi-site inhibitors and 6 compounds with unknown modes of action. This information emphasizes that understanding a mode of action has become more complex, and it is possible to exploit more than one site within a target protein. Benzimidazoles, which target a β-tubulin, are considered high-risk fungicides (Brent and Hollomon,

2007), and the amino acid changes causing resistance are well known (Ma and Michailides, 2005), yet this has not prevented exploiting other sites within β-tubulin. The benzamide, zoxamide (B3), interacts with β-tubulin and disrupts microtubules (Young and Slawecki, 2005), but shows no cross-resistance with MBC. As zoxamide is particularly active against oomycete diseases, it has proved a useful additional mode of action in anti-resistance strategies, especially in the control of early blight of potato (*Phytophthora infestans*). As more is learned about microtubules, new fungicides may be discovered that disrupt their function but which are not cross-resistant to other β-tubulin inhibitors.

Exploiting diversity within both sterol 14α demethylase, a very successful target in sterol biosynthesis (G1, DMIs), and azole chemistry has yielded recent fungicides such as prothioconazole, introduced in 2003 and which is important in controlling especially cereal diseases, where earlier DMIs failed because of resistance. Recent work has shown that prothioconazole generates a type 1 cytochrome P_{450} spectral change when it binds to the target demethylase, unlike other azoles which generate a type 2 spectrum (Parker *et al.*, 2011). This suggests that prothioconazole interacts with the target demethylase, and the haem in particular, differently from other azoles and may explain why this latest azole fungicide controls some cereal diseases which are resistant to older azoles (Leroux and Walker, 2011).

As crystal structures of target enzymes from pathogenic fungi and improved protein modelling become available, the impact of amino acid changes that cause resistance can be evaluated, allowing rational design of compounds that attack the same target but which avoid resistance (Frey *et al.*, 2010). Advising growers that durable resistance management programmes should avoid over-use of fungicides with the same mode of action needs revision. New fungicides that attack the same biochemical target but show no cross-resistance with existing products will be useful in anti-resistance strategies.

9.3   Targeting the Cytoskeleton

Microtubules are just one of several protein complexes that form the cystoskeleton frame-work within animal, plant and fungal cells. Together with actin protein filaments, micro-tubules provide 'rails' along which metabo-lite-carrying organelles are moved to the growing hyphal tip by component motor pro-teins, kinesins and dyneins. Another part of the cytoskeleton involves spectrins, integrins and their adapter ankyrin proteins. They help stabilize the otherwise fluid plasma mem-brane and provide a scaffold to anchor the many proteins involved in various membrane functions. The cytoskeleton is a dynamic cell component and the repeating polymeric pro-teins are constantly degraded and reassem-bled by energy-utilizing processes. Disrupting the cytoskeleton impacts strongly on growth and development.

Detailed characterization and under-standing of the function of spectrins are derived from animal studies, particularly using erythrocytes, and identical protein fila-ments appear absent from filamentous fungi. Nevertheless, smaller 'spectrin-like' proteins have been identified in fungi, including *P. infestans* (Cotado-Sampayo *et al.*, 2006; Toquin *et al.*, 2010), by a variety of immunological techniques using antibodies raised against chicken or human α-β spectrin. The smaller size may well result from proteolytic cleav-age, as spectrins are very susceptible to pro-tease digestion (Cotado-Sampayo *et al.*, 2006).

A bioinformatic approach failed to find evidence of spectrin genes in a number of fungi, but small 'spectrin-specific-repeat domains' were present in the genomes of *Phytophthora sojae*, *Neurospora crassa* and *Magnaporthe grisea* (Toquin *et al.*, 2010).

Fluopicolide (B5; Fig. 9.2) is a recently introduced benzamide (= acylpicolide) fungi-cide which is particularly active against oomycete pathogens (Toquin *et al.*, 2007). Treatment of *P. infestans* zoospores with a low dose (1 µg/ml) of fluopicolide stops swim-ming immediately, with swelling and lysis following within minutes. These rapid symp-toms are similar to those seen with QoIs, but fluopicolide shows no cross-resistance with QoIs, carboxylic acid amides (CAAs, F5) or phenylamides.

Using immunofluorescence and an antispectrin antibody, Toquin *et al.* (2010) showed that 'spectrin-like' proteins normally formed patches in the plasma membrane, but after treatment of *P. infestans* mycelium with fluopicolide, fluorescence quickly was redis-tributed randomly throughout the cyto-plasm. Preliminary experiments showed that expression of many genes associated with endoplasmic reticulum and Golgi function were upregulated, suggesting that fluopicol-ide might interfere with vesicle transport. A full understanding of the mode of action will require detailed biochemical analysis to unravel the function of these spectrin-like proteins, but there is already enough evi-dence to confirm that fluopicolide has a novel mode of action and exerts a rapid effect on the cytoskeleton, at least in *P. infestans*.

Fluopicolide adds to at least eight other modes of action available to control oomycete diseases. For many, there are already well-defined anti-resistance strategies involving mixtures and limited use of single products. Although resistance has emerged against sev-eral of these oomycete fungicides, especially in *Plasmopara viticola* (grape downy mildew), strategies generally have been durable and there has been no repeat of resistance prob-lems caused by extensive use of phenyla-mides following their introduction. Many of these modes of action are specific for oomyc-ete pathogens, no doubt reflecting that Oomycetes are related to brown algae and

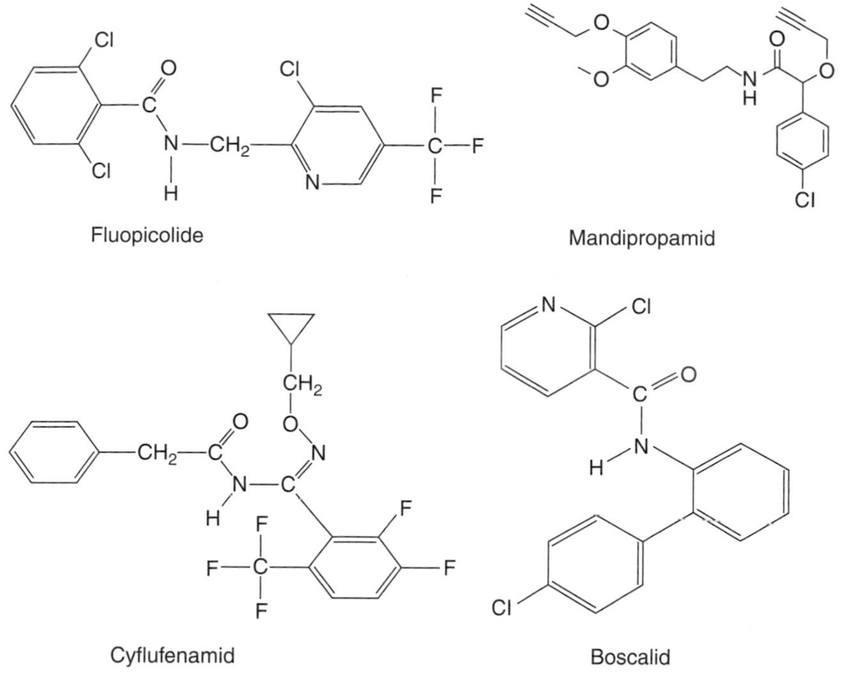

Fig. 9.2. Structure of four fungicides mentioned in the text.

have little similarity, at least at the genomic level, with true filamentous fungi (Rossman, 2006). Fluopicolide appears to target a core cellular process, which also occurs in true filamentous fungi, suggesting that a full understanding of its molecular action might identify targets in other fungal pathogens, broadening its disease control spectrum and adding another, much needed mode of action for the resistance management of Ascomycete, Basidiomycete and Deuteromycete (Fungi Imperfecti) pathogens.

9.4 Targeting Cell Walls

Despite chemical diversity, at least seven active compounds are grouped together as CAA fungicides, not because of a common mode of action but because of some degree of cross-resistance between them all. They are active primarily against Oomycetes. Although use has been confined largely to control of *Phytophthora* and *Plasmopara* spe-

cies, all cause rapid plasmolysis and leakage of cytoplasmic contents, suggesting that they may have at least a similar mode of action. Alteration in the composition of membrane lipids pointed towards inhibition of choline phosphotransferase (Griffiths *et al.*, 2003), and perhaps phospholipase D (Dubuis *et al.*, 2008). Other work with flumorph points to increased membrane permeability resulting from disruption of F-actin filaments affecting new cell wall formation (Zhu *et al.*, 2007). Recent evidence suggests that mandipropamid (Fig. 9.2) inhibits cellulose synthase, at least in *P. infestans*, and this target would explain its specificity since, unlike other fungi, oomycete cell walls contain 1–6 β-D glucans and cellulose with very little chitin (Blum *et al.*, 2010). Incorporation of D-(U-^{14}C)-glucose by germinating cysts into acid-washed crystalline cellulose was inhibited at concentrations that correlated with effects on germ tube growth. Resistance in a laboratory-generated mutant was linked to two point mutations in a gene (*CesA3*) coding for a protein known to be involved in plants in cellulose synthesis. Transformation of this mutant gene into wild-type *P. infestans* generated resistant transformants, confirming a role for this gene in mandipropamid resistance. Although strictly this explains a mechanism of resistance, the level of resistance in the original mutant is over 1000-fold, and such high levels of resistance are generally caused by a target-site change (Sierotzki and Gisi, 2003). Cellulose is synthesized in *P. infestans* in a complex organelle involving at least four proteins integrated into a membrane (Grenville-Briggs *et al.*, 2008) so that the mode of action of mandipropamid may involve a target closely related to the *CesA3* protein.

The significance of these findings is that it provides a framework to establish if other CAAs have the same mode of action and mechanism of resistance. In fact, mandipropamid-resistant mutants showed only a low level of resistance to dimethomorph (5- to 38-fold), indicating a different mechanism of resistance, and perhaps even target site, and that CAAs should not be treated as a group in anti-resistance strategies.

9.5   Targeting Signal Transduction

Successful pathogens not only need to enter and spread within their host but also, eventually, they must produce spores that can be dispersed in some way to infect a fresh host. These developmental processes must be closely coordinated and signals transferred to the nucleus in order to activate transcription of relevant genes. Signal transduction has been studied extensively in animals, yeast and some filamentous fungi (for a review see Li *et al.*, 2007) and often involves plasma membrane located G-protein coupled receptors (G-PCR). These receptors transfer signals to heterotrimeric G-proteins, which amplify the signal before it is passed through adenylcyclase and kinases to the nucleus (Fig. 9.3). G-proteins play an important role in detecting signals for nutrient acquisition, asexual sporulation, mating and pathogenesis. The complexity of signal

transduction, especially in recognizing changes in nutrient status governing the initiation and development of sporulation structures, arises from extensive 'crosstalk' between different signalling pathways. Regulation involves changes in GDP–GTP phosphorylation levels through the action of GDP–GTP exchange proteins and *ras* and *rho* GTPases.

The first evidence of a mode of action in some way involving G-protein signalling has been observed with the quinoline, quinoxyfen (E1), which only inhibits powdery mildews preventing germination and appressoria formation (Wheeler *et al.*, 2000). Barley powdery mildew (*Blumeria graminis* f. sp. *hordei*) mutants generated in the laboratory expressing changes in a *ras* GTP-ase and phosphokinase C were not only resistant to quinoxyfen but also conidiophores largely failed to differentiate conidia (Wheeler *et al.*, 2003). A recently introduced preventive and specific powdery mildew fungicide, the quinazolinone proquinazid (E1), also alters a *ras* GTP-ase, but in a way different from quinoxyfen (Gilbert *et al.*, 2009). The novel benzophenone, metrafenone (U8), shows similarities with quinoxyfen, interfering with appressoria formation but also affecting conidiophore differentiation, as seen in quinoxyfen-resistant mutants

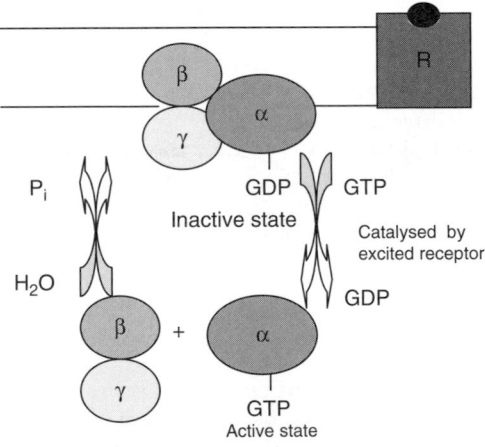

Fig. 9.3. Heteromeric G-protein cycle. R, receptor.

(Schmit *et al.*, 2006). Metrafenone disrupts the F-actin cap at the hyphal tip, perhaps through regulation of F-actin by *rho* and *ras* GTP-ases anchored to the cytoskeleton (Opalski *et al.*, 2006). Although there have been no mode of action studies, the phenylacetamide powdery mildew fungicide, cyflufenamid (U6), interferes with appressoria function and conidiation, suggesting that it also interferes with signal transduction.

Apart from providing an additional mode of action for anti-resistance management of high resistance risk powdery mildews, these four fungicides provide evidence that signal transduction pathways can be intercepted in several ways. Because of crosstalk between different pathways, resistance mechanisms may exert adverse pleiotropic effects on other development processes, which reduces resistance risk. Both metrafenone and cyflufenamid have useful activity against other pathogens, including for metrafenone the model non-pathogenic filamentous fungus, *Aspergillus nidulans*. This recognizes the importance of the core components of signal transaction pathways in filamentous fungi and that there is scope for novel chemistry showing broad spectrum disease control.

9.6 Targeting Respiration

The core respiration pathway involves four large protein complexes embedded in the inner mitochondrial membrane and denoted as complexes I, II, III and IV (Fig. 9.4). With the exception of the small, non-specific molecular inhibitors of complex IV (e.g. cyanide), all fungicides that block electron transport act at quinone binding sites involved in electron transfer from complexes I and II to complex III. One of these binding sites in complex III (Qo = outer site) oxidizes ubiquinol and passses one electron to the adjacent Rieske iron-sulfur protein, and the resulting semiquinone is recycled (Q cycle) to a second site in the cytochrome *b* subunit (Qi = inner site). The successful strobilurin and related QoI fungicides block electron flow through the Qo site.

Understanding how electrons flow through the Qo centre in the cytochrome *b* subunit of mitochondrial complex III explains why Qo-specific inhibitors (QoIs) differ in their cross-resistance patterns (Fisher and Meunier, 2001). Some natural QoIs, such as stigmatellin, bind within the 'distal' pocket of the Qo centre adjacent to the neighbouring Rieske iron-sulfur protein, whereas fungicides such as azoxystobin bind in the 'proximal pocket', blocking access of ubiquinol to

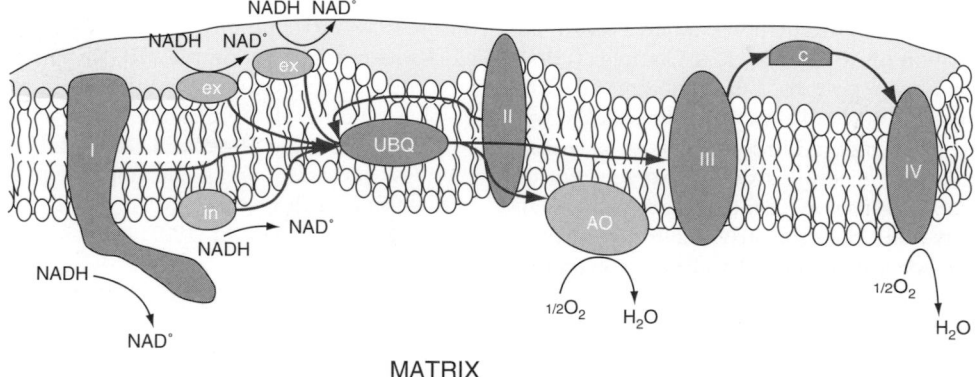

Inter-membrane space

MATRIX

Fig. 9.4. Fungal electron transport chain in the inner mitochondrial membrane. AO, alternative oxidase; C, cytochrome *c*; ex, external NAD(P)H dehydrogenase; in, internal NADH dehydrogenase; UBQ, ubiquinone. From Joseph-Horne *et al.* (2001).

the haem b_1. Stigmatellin shows no cross-resistance with azoxystrobin and related QoI fungicides where resistance is caused by the G143A amino acid change. Unfortunately, as yet, there are no fungicides that exploit these binding differences within the Qo centre.

Among the first successful systemic fungicides were the carboxamide, carboxin, and the phenyl benzamide, benodanil. They could be used as foliar sprays or seed treatments, but their disease control spectrum was limited just to rusts, smuts and Rhizoctonia (Basidiomycetes). Their mode of action quickly was shown to involve inhibition of succinate dehydrogenase (SDH; succinate-quinone reductase, SQR). Resistance never emerged as a major problem, but it could be generated easily in the laboratory, and the *Ustilago maydis* carboxin-resistance gene was developed as a selectable marker for use in fungal transformation systems (Keon *et al.*, 1991). This prompted a detailed analysis of the mechanism of resistance, which was caused by a single amino acid change.

Succinate dehydrogenase links the citric acid cycle (= TCA) with respiration, and couples succinate oxidation with reduction of ubiquinone. The enzyme is made up of four protein subunits (SDH, A–D), forming complex II of the electron transport chain, and which is half-buried in the inner mitochondrial membrane. The hydrophilic SDH-A oxidizes succinate to fumarate and passes the two electrons through an iron-sulfur centre to SDH-B. Subunits SDH-C and -D anchor the two surface proteins and contain a haem, the function of which is unclear. Complex II does not pump protons, but inhibition of SDH impacts on the TCA cycle and quickly will reduce energy (ATP) conservation. Inhibition will also generate reactive oxygen species (ROS), which may be the primary way SDHIs exert their effect. Clearly, inhibition of SDH will have a strong impact on growth and development.

High levels of resistance to boscalid were selected quickly in field populations of a number of pathogens, including *Botrytis cinerea* (Kim and Xiao, 2010), *Corynesporim cassiicola* (Corynespora leaf spot of cucurbits; Miyamoto *et al.*, 2009) and *Alternaria alternata* (alternaria late blight of pistachio; Avenot *et al.*,

2009). Sequence comparisons identified the target site mutations that caused this resistance and altered not only the boscalid binding site in SDH-B (P225L/F/T; H272Y/R) but also amino acid changes in SDH-C and -D, which were not involved directly in fungicide binding (Avenot and Michailides, 2010). Although cross-resistance extended to other SDH fungicides, different SDHIs seemed to reflect different levels of resistance. Indeed, the pyridinyl-ethyl-benzamide, fluopyram, showed no cross-resistance when highly boscalid- and penthiopyrad-resistant isolates of *A. alternata* (Avenot and Michailides, 2009), *C. cassiicola* and *Podosphaera xanthi* (cucurbit powdery mildew; Ishii *et al.*, 2011) were tested. This suggests that the complex II target site accommodates diversity of resistance mechanisms which can be linked to different SDHIs and which impact on the development of anti-resistance strategies. SDHIs may have a common target enzyme, but not precisely the same mode of action or the same resistance risk, and it will be interesting to see if other members of the new chemical class of pyridinyl-ethyl-benzamides, bixafen and penflufen (which will be available as a seed treatment), show the same cross-resistance patterns as fluopyram.

So far, only two fungicides have been commercialized that block the Qi site of complex III (cyazofamid and amisulbrom; C4), but unlike broad-spectrum QoIs, these are specific against Oomycetes and are used especially to reduce sporulation. A third compound, a triazolo-pyrimidylamine, ametoctradin, is also specific against Oomycetes and blocks respiration in complex III. Its precise molecular target is not known, but does not involve binding at the Qo site (Gold *et al.*, 2010). Targeting the 41-protein complex I has been less successful and so far only the pyrimidinamine, diflumetorim (C1), which was introduced in Japan in 1997, acts at this target.

Clearly, respiration inhibitors have been important components of disease control and anti-resistance strategies for many years, although resistance has reduced the effectiveness of some QoIs. The Qi site is equally important in the core respiration pathway and energy conservation and it is surprising that compounds targeting this site control only Oomycetes. At present, there is no cross-resistance between Qo and Qi site fungicides and so broad-spectrum Qi site fungicides would be a welcome addition for resistance management.

9.7 Future Needs

Resistance mechanisms that cause practical disease control problems are protein phenomena in which amino acid changes reduce fungicide binding yet maintain function. Attempts to generate resistant mutants in the laboratory during the development of a novel mode of action form part of resistant risk assessments (Brent and Hollomon, 2007). This is resource-intensive, especially for obligate pathogens, and it would be useful to incorporate synthesis strategies early in development that predict, and overcome, possible resistance mutations. Modern biochemical and molecular techniques speed up identification of novel modes of action,

and coupled with the gradual improvements in protein modelling and crystallography, make it possible to predict resistance mutations. No studies have been published using filamentous fungi, but the scope of this approach can be seen in recent work published involving novel propargyl-based antifolate inhibitors of dihydrofolate reductase in methicillin-resistant *Staphylococcus aureus* (MRSA; Frey *et al.*, 2010). The shape of the inhibitor binding site was determined using a protein engineering algorithm and confirmed by crystallography. This identified a double amino acid change (V31Y/F92I) which hardly altered function, yet reduced binding of the lead inhibitor (compound 1, Table 9.1), generating a resistance factor (RF) of 18. Simple changes to compound 1 increased inhibitor binding (K_i; Table 9.1) and reduced resistance without altering activity against the wild type. Although RFs are small by agricultural standards, improvements in the accuracy of *in silico* resistance mutations, and which require greater computing power,

Table 9.1. Inhibition assays (K_i nM) for dihydrofolate reductase and propargyl-linked antifolates.

		V31Y, F92 I	Wild type	Resistance factor
	1	180	10	18
	2	60	8.5	7.1
	3	65	9.0	7.2

Note: From Frey *et al.* (2010) *PNAS* 107, 13707–13712.

will allow a lead design strategy against targets susceptible to resistance mutations. Coupled with rapid molecular diagnostics to detect resistance mutations early, it should be possible to define which fungicide analogues are needed to maintain durable anti-resistance strategies.

Chemical synthesis has dominated the search for novel modes of action, but new technologies offer other approaches to disease control, and greater diversity of modes of action. Non-coding RNAs regulate developmental pathways in a number of ways (Hüttenhofer *et al.*, 2005), including gene silencing in a specific way at a post-transcriptional level. Silencing involves double-stranded RNA (dsRNA; hairpin RNA), which is trimmed into small (20–30 nucleotides) interfering RNA (siRNA) by an RNA-ase III (dicer enzyme). siRNAs are integrated into an RNA-induced silencing complex (RISC), which includes an endonuclease, that degrades the target mRNA (for a more detailed account of the pathway see Price and Gatehouse, 2008). This process of RNA interference (RNAi) is highly conserved in eukaryotes and has been a widely used tool in functional genomics.

RNAi is already used in clinical trials to control human cancers and viral disease, and the feasibility of controlling insects on plants has been demonstrated in laboratory experiments using key pests by silencing cytochrome P450 (Mao *et al.*, 2007; Bautista *et al.*, 2009), vacuola ATPase (Baum *et al.*, 2007) and a Rieske iron-sulfur protein (Gong *et al.*, 2011). Work with plant pathogens has been limited to *in vitro* studies, but RNAi has been used to 'knock down' expression of targeted genes in *Magnaporthe oryzae* (Kadotani *et al.*, 2003), *Venturia inaequalis* (Fitzgerald *et al.*, 2004) and *B. cinerea* (Patel *et al.*, 2008). Although transient silencing of the cellulose synthase gene family in *P. infestans* reduced appressoria formation, its impact on disease control was not reported (Grenville-Briggs *et al.*, 2008).

Despite its potential, there are significant challenges that must be overcome before RNAi technology is a commercial option in crop protection (Price and Gatehouse, 2008; Huvenne and Smagghe,

2010). Selecting the target gene is not always straightforward and silencing may occur *in vitro* but not *in planta* (Patel *et al.*, 2008). The length of dsRNA, or whether to use instead the shorter siRNA, is likely to affect the efficacy of disease control. Effects may be transient where a pathogen lacks sufficient RNA-dependent RNA polymerase activity to amplify the RNA and maintain it at an optimal level for silencing. Application of RNAi technology in crop protection is likely to rely on delivery by transgene-encoded dsRNA in plants, but a recent report of transient control of diamond back moth (*Plutella xylostella*) using siRNA synthesized with some modified nucleotides to increase its stability (Gong *et al.*, 2011) suggests that spraying may be possible. There are also resistance risks, since silencing depends on sequence specificity, and polymorphism in the target gene may allow selection of resistant pathogen populations. How the efficacy of RNAi technology will compare with conventional fungicides or resistant crops produced by plant breeding or genetic manipulation is, of course, unknown, but it offers the potential to expand greatly the modes of action available for anti-resistance strategies.

9.8   Conclusion

Fungicide resistance causes losses to both growers and manufacturers but, so far, no disease has suffered total loss of control. This has been due largely to the availability of several different modes of action, but further losses due to resistance and changes in registration requirements mean that new modes of action are needed urgently to help ensure sufficient food for a growing world population. Many existing fungicide targets expose the importance of core developmental processes that are common in all eukaryotes. There is a need to transfer fundamental advances in developmental biology, generated largely through medical research, to the biochemistry of fungal plant pathogens in order to identify relevant targets and define appro-priate screens to detect activity. Integrating novel modes of action generated through chemistry or exposed through new tech-nologies such as RNAi, with crop plant resistance produced by conventional plant breeding or transgenic methods, will produce the wide spectrum of approaches needed to maintain durable disease control strategies.

References

Avenot, H. and Michailides, T. (2009) Monitoring the sensitivity to boscalid of *Alternaria alternata* populations from California pistachio orchards (Abstr). *Phytopathology* 99 (6, Suppl), S6.

Avenot, H.F. and Michailides, T.J. (2010) Progress in understanding molecular mechanisms and evolution of resistance to succinate dehydrogenase inhibiting (SDHI) fungicides in phytopathogenic fungi. *Crop Protection* 29, 643–651.

Avenot, H., Sellam, A. and Michailides, T. (2009) Characterization of mutations in the membrane-anchored subunits AaSDHC and AaSDHD of succinate dehydrogenase from *Alternaria alternata* isolates confer-ring field resistance to the fungicide boscalid. *Plant Pathology* 58, 1134–1143.

Baum, J.A., Bogaert, T., Clinton, W., Heck, J.R., Feldmann, G.R., Ilagan, O., *et al.* (2007) Control of coleopteran insect pests through RNA interference. *Nature Biotechnology* 25, 1322–1326.

Bautista, M.A.M., Miyata, T., Miura, K. and Tanaka, T. (2009) RNA interference-mediated knockdown of cytochrome P450, *CYB6G1*, from the diamond back moth, *Plutella xylostella*, reduces larval resistance to permethrin. *Insect Biochemistry and Molecular Biology* 39, 38–46.

Blum, M., Boehler, M., Randall, E., Young, V., Csukai, M., Kraus, S., *et al.* (2010) Mandipropamid targets the cellulose synthase-like PiCesA3 to inhibit cell wall biosynthesis in the oomycete plant pathogen *Phytophthora infestans*. *Molecular Plant Pathology* 11, 227–243.

Brent, K.J. and Hollomon, D.W. (2007) *Fungicide Resistance: The Assessment of Risk*. FRAC Monograph 2, 2nd edn. FRAC, Brussels, 52 pp.

Cotado-Sampayo, M., Ojha, M., Ortega Pérez, R., Chappuis, M.-L. and Barja, F. (2006) Proteolytic cleavage of a spectrin-related protein by calcium-dependent protease in *Neurospora crassa*. *Current Microbiology* 53, 311–316.

Dubuis, P.-H., Waldner, M., Bochler, M., Fonné-Pfister, R., Sierotzki, H. and Gisi, U. (2008) Molecular approaches to evaluating resistance mechanisms in CAA fungicides. In: Dehne, H.-W., Deeising, H.B., Gisi, U., Kuck, K.H., Russell, P.E. and Lyr, H. (eds) *Modern Fungicides and Antifungal Compounds*. DPG Selbstverlag, Braunschweig, Germany, pp. 79–84.

Fisher, N. and Meunier, B. (2001) Effects of mutations in mitochondrial cytochrome b in yeast and man. *European Journal of Biochemistry* 268, 1155–1162.

Fitzgerald, A., van Kan, J.A.L. and Plummer, K.M. (2004) Simultaneous silencing of multiple genes in the apple scab fungus, *Venturia inaequalis*, by expression of RNA with chimeric reverted repeats. *Fungal Genetics and Biology* 41, 963–971.

Frey, K.M., Georgiev, I., Donald, B.R. and Anderson, A.C. (2010) Predicting resistance mutations using protein design algorithms. *Proceedings of National Academy of Sciences of the United States of America* 107, 13707–13712.

Gilbert, S.R., Cools, H.J., Fraaije, B.A., Bailey, A.M. and Lucas, J.A (2009) Impact of proquinazid on appres-sorial development of the barley powdery fungus *Blumeria graminis* f. sp. *hordei*. *Pestcides Biochemistry and Physiology* 95, 127–132.

Gold, R., Klappach, K., Speakman, J., Schlehuber, S. and Brix, H.-D. (2010) Initium® a new innovative fun-gicide for the control of Oomycetes. (Abstr). *16th International Reinhardsbrunn Symp – Modern Fungicides and Antifungal Compounds*.

Gong, L., Yang, X., Zhang, B., Zhong, G. and Hu, M.Y. (2011) Silencing of Rieske iron-sulfur protein using chemically synthesized siRNA as a potential bio-pesticide against *Plutella xylostella*. *Pest Management Science* 67, 514–520.

Grenville-Briggs, L.J., Anderson, V.L., Fugelstad, J., Avrova, A.O., Bouzenzana, J., William, A., *et al.* (2008) Cellulose synthesis in *Phytophthora infestans* is required for normal appressorium and successful infection of potato. *Plant Cell* 20, 720–738.

Griffiths, R.G., Dancer, J., O'Neill, E. and Harwood, J.L. (2003) A mandelamide pesticide alters lipid metab-olism in *Phytophthora infestans*. *New Phytologist* 158, 345–353.

Hüttenhofer, A., Schattener, P. and Polacek, N. (2005) Non-coding RNAs: hope or hype? *Trends in Genetics* 21, 289–297.

Huvenne, H. and Smagghe, G. (2010) Mechanisms of dsRNA uptake in insects and potential of RNAi for pest control: a review. *Journal Insect Physiology* 56, 227–235.

Ishii, H., Miyamoto, T., Ushio, S. and Kakishima, M. (2011) Lack of cross resistance to a novel succinate dehydrogenase inhibitor fluopyram in highly boscalid-resistant isolates of *Corynespora cassiicola* and *Podosphaera xanthii*. *Pest Management Science* 67, 474–482.

Joseph-Horne, T., Hollomon, D.W. and Wood, P.M. (2001) Fungal respiration: a fusion of standard and alternative components. *Biochimica et Biophysica Acta* 1504, 179–195.

Kadotani, N., Nakayashiki, H., Tosa, Y. and Mayama, S. (2003) RNA silencing in the phytopathogenic fungus *Magnaporthe oryzae*. *Molecular Plant-Microbe Interactions* 16, 769–776.

Keon, J.P.R., White, G.A. and Hargreaves, J.A. (1991) Isolation, characterisation and sequence of the gene conferring resistance to the systemic fungicide carboxin from the maize smut pathogen *Ustilago maydis*. *Current Genetics* 19, 475–481.

Kim, Y.K. and Xiao, C.L. (2010) Resistance to pyraclostrobin and boscalid in populations of *Botrytis cinerea* from stored apples in Washington State. *Plant Disease* 64, 604–612.

Leroux, P. and Walker, A.-S. (2011) Multiple mechanisms account for resistance to sterol 14α-demethylation inhibitors in field isolates of *Mycosphaerella graminicola*. *Pest Management Science* 67, 44–59.

Li, L., Wright, S.J., Krystofova, S., Park, G. and Borkovich, K.A. (2007) Heteromeric G protein signalling in filamentous fungi. *Annual Review of Microbiology* 61, 423–453.

Ma, Z. and Michailides, T.J. (2005) Understanding molecular mechanisms of fungicide resistance and molecular detection of resistant genotypes in phytopathogenic fungi. *Crop Protection* 24, 853–863.

Mao, Y.-B., Cai, W.-J., Wang, J.-W., Hong, G.-J., Tao, X.-Y., Wang, L.-J., *et al.* (2007) Silencing a cotton bollworm P450 monooxygenase gene by plant-mediated RNA impairs larval tolerance of gossypol. *Nature Biotechnology* 25, 1307–1313.

Miyamoto, T., Ishii, H., Seko, T., Kobori, S. and Tomita, Y. (2009) Occurrence of *Corynespora cassiicola* isolates resistant to boscalid on cucumber in Ibaraki Prefecture, Japan. *Plant Pathology* 58, 1144–1151.

Opalski, K.S., Tresch, S., Kogel, K.-H., Grossmann, H., Köhle, H. and Hückelhoven, R. (2006) Metrafenone: studies on the mode of action of a novel cereal powdery mildew fungicide. *Pest Management Science* 62, 393–401.

Parker, J.E., Warrilow, A.G.S., Cools, H.J., Martel, C.M., Nes, W.D., Fraaije, B.A., *et al.* (2011) Prothioconazole binds to *Mycosphaerella graminicola* CYP51 by a different mechanism compared to other azole antifungals. *Applied Environmental Microbiology* 77, 1460–1465.

Patel, R.M., Hennghan, M.W., van Kan, J.A.L., Bailey, A.M. and Foster, G.D. (2008) The pOT vector system: improving ease of transgene expressing in *Botrytis cinerea*. *Journal of General and Applied Microbiology* 54, 367–376.

Price, D.R.G. and Gatehouse, J.A. (2008) RNAi-mediated crop protection against insects. *Trends in Biotechnology* 26, 393–400.

Rossman, A.Y. (2006) Why are Phytophthora and other Oomycetes not true fungi? *Outlooks on Pest Management* 17, 217–219.

Schmitt, M.R., Carzaniga, R., Cotter, H. van T., O'Connell, R. and Hollomon, D.W. (2006) Microscopy reveals disease control through novel effects on fungal development: a case with an early generation benzophenone fungicide. *Pest Management Science* 62, 383–392.

Sierotzki, H. and Gisi, U. (2003) Molecular diagnostics for fungicide resistance in plant pathogens. In: Voss, G. and Ramos, G. (eds) *Chemistry of Crop Protection: Progress and Prospects in Science and Regulation.* Wiley-VCH Verlag GmbH, Weinheim, Germany, pp. 71–88.

Toquin, V., Barja, F., Sirven, C. and Beffa, R. (2007) Fluopicolide, a new anti-Oomycetes fungicide with a new mode of action inducing peturbation of s spectrin-like protein. In: Krämer, W. and Schirmer, U. (eds) *Modern Crop Protection Compounds*, Volume 2. Wiley-VCH Verlag GmbH, Weinheim, Germany, pp. 675–682.

Toquin, V., Barja, F., Sirven, C., Gamet, S., Mauprivez, L., Peret, P., *et al.* (2010) Novel tools to identify the mode of action of fungicides as exemplified with fluopicolide. In: Gisi, U., Gullino, M.L. and Chet, I. (eds) *Recent Developments in Disease Management.* Springer, Dordrecht, Germany, pp. 19–36.

Wheeler, I.E., Hollomon, D.W., Longhurst, C. and Green, E. (2000) Quinoxyfen signals a stop to infection in powdery mildews. *Proceedings of Brighton Crop Protection Conference – Pests and Diseases.* BCPC, Alton, UK, pp. 841–846.

Wheeler, I.E., Hollomon, D.W., Gustafson, G., Mitchel, J.C., Longhurst, C., Zhang, C., *et al.* (2003) Quinoxyfen perturbs signal transduction in barley powdery mildew (*Blumeria graminis* f. sp. *hordei*). *Molecular Plant Pathology* 4, 177–186.

Young, D.H. and Slawecki, R.A. (2005) Cross-resistance relationships between zoxamide, carbendazim and diethofencarb in field isolates of *Botrytis cinerea* and other fungi. In: Dehne, H.-W., Gisi, U., Kuck, K.H., Russell, P.E. and Lyr, H. (eds) *Modern Fungicides and Antifungal Compounds IV*. BCPC, Alton, UK, pp. 125–131.

Zhu, S.S., Liu, X.L., Liu, P.F., Li, Y., Li, J.Q., Wang, H.M., Yuan, S.K. and Si, N.G. (2007) Flumorph is a novel fungicide that disrupts microfilament organisation in *Phytophthora melonis*. *Phytopathology* 97, 643–649.

10 Field Kit- and Internet-Supported Fungicide Resistance Monitoring

Guido Schnabel,[1] Achour Amiri[2] and Phillip M. Brannen[3]
[1]*Department of Entomology, Soils, and Plant Sciences, Clemson University, Clemson, USA;* [2]*Plant Pathology, Gulf Coast REC, Wimauma, Florida, USA;* [3]*Plant Pathology Department, University of Georgia, Athens, Georgia, USA*

10.1 Introduction

Reduced risk fungicides have become the mainstay for plant disease management in modern agriculture. They are highly effective in managing both pre- and postharvest diseases and have replaced many of the older, more toxic chemicals. Fungicides classified as 'reduced risk' by EPA have low risk to human health, low toxicity toward non-target organisms, low potential for contamination of water or other environmental resources and support the implementation of integrated pest management (IPM) practices. Key reduced risk fungicides include the methyl benzimidazole carbamates (MBCs), the quinone outside inhibitors (QoIs), the succinate dehydrogenase inhibitors (SDHIs), anilinopyrimidines (APs), phenylpyrroles and hydroxyanilides. However, their high efficacy and comparably low toxicity towards the environment and humans come with a price. Reduced risk fungicides, as well as some important conventional fungicides such as demethylation inhibitor (DMI) fungicides, are prone to resistance development because of their site-specific mode of action.

Monitoring for resistance generally allows one to: (i) investigate suspected field resistance; (ii) predict the emergence of resistance problems; (iii) assess resistance management strategies; (iv) track the population dynamics of resistant populations; (v) help select fungicides on a local level; and (vi) advance our basic understanding of fungicide resistance (Brent, 1988). Monitoring for fungicide resistance in commercial fruit, vegetable and field crop production areas has been conducted almost exclusively by research scientists and, in most cases, it has involved short-term investigations into individual cases of suspected resistance. Shifts are determined with standard laboratory methods, such as mycelial growth inhibition, germination rates and the detection and quantification of point mutations using molecular methods in pathogen populations (Fraaije *et al.*, 2002; Russell, 2004). While such information is valuable, it does not allow for study of the evolution, spread and distribution of resistant populations, sensitivity shifts and molecular mechanisms of resistance on a scientific level; nor does such data address the needs of individual producers. Most of these assays take a week or longer to conduct before results are available, or they require high-tech equipment and molecular expertise. Furthermore, these methods are costly with regard to the labour and supplies needed, and thus using them as a service would require charging large fees. Fast, reliable and easy-to-use monitoring systems are needed to enable continued surveillance programmes that can be sustained over several years.

©CAB International 2012. *Fungicide Resistance in Crop Protection: Risk and Management* (ed. T.S. Thind)

Such programmes could be used to determine location-specific resistance profiles and provide for individual resistance management recommendations. In the best-case scenario, producers themselves would conduct the monitoring assay and make decisions based on scientific results when disease is observed in their specific locations. However, as long as there are no commercially available, user-friendly field kits, the expertise of professional farm advisors or extension specialists is required.

In any given year, brown rot, caused by *Monilinia fructicola* (G. Wint.) Honey, can cause substantial damage on peach and other stone fruit. For example, disease loss estimates from Georgia (USA) indicated that brown rot alone was responsible for US$9.8 million in production losses (due to direct disease losses and cost of fungicide applications) in 2003, which amounted to > 75% of all combined disease-related losses in the peach crop (Williams-Woodward, 2004). *M. fructicola* can infect the flowers, shoots and fruit of peaches, causing blossom blight, stem canker and brown rot, respectively (Ogawa and English, 1995). Of these symptom types, preharvest brown rot is the most economically important. Every commercial peach cultivar is susceptible to this disease; hence, host resistance is not a management option at this time.

Due to the humid and warm climate observed in south-eastern USA, which is conducive to fungal and bacterial disease development and epidemics, production of high-quality peach fruit is dependent on highly effective fungicides. Given the explosive nature of preharvest brown rot epidemics, management depends heavily on the application of two or three reduced risk fungicide applications in the final 2–3 weeks before harvest (Brannen et al., 2005). Many fungicides are registered for brown rot control (Table 10.1). The most effective and commonly used fungicides for brown rot control are the MBCs, DMIs, QoIs and SDHIs. Other chemical classes, although registered, provide little brown rot disease protection under southeastern USA growing conditions (Schnabel et al., 2009), or residue restrictions prevent their preharvest usage.

Using brown rot of peach as a model system, we developed a field kit-supported resistance monitoring programme. The results from the kit, when combined with expert recommendations generated through an online web application, provide producers with science-based information that can be used to improve pre- and postharvest brown rot disease management (Schnabel et al., 2010).

10.2 Existing and Emerging Resistance to Reduced Risk Fungicides in *Monilinia fructicola*

MBC fungicides were considered the silver bullet for brown rot control when they were introduced in the 1970s. Compared to what growers were spraying prior to their introduction (mainly sulfur and captan), the MBCs, specifically benomyl, were indeed sensationally effective. However, soon after introduction, the fungicide failed to control disease in many commercial areas due to resistance development (Whan, 1976; Sonoda et al., 1983; Ogawa et al., 1984; Michailides et al., 1987; Zehr et al., 1991; Yoshimura et al., 2004). The MBCs provide a classic example of single-step or qualitative resistance, resulting in essentially no control of *M. fructicola* within a short time period. After the initial resistance establishment, the MBCs were recommended to be used sparingly and only in combination with a protectant fungicide. Still, resistant populations have persisted in South Carolina and Georgia, USA (Zehr et al., 1991; Schnabel et al., 2010). On the other hand, DMI fungicides were applied for almost two decades before any significant signs of resistance emerged. Initially, resistance was first reported from an experimental orchard at the Musser Fruit Research Farm, Clemson University, and from some commercial orchards in the west central and northern parts of South Carolina in 1995 (Zehr et al., 1999), but it later resulted in a major brown rot outbreak in commercial orchards in Georgia during 2002 (Brannen and Schnabel, 2003; Schnabel et al., 2004). Since then, DMI resistance has been detected in many locations of South Carolina, Ohio, New Jersey and New York (Schnabel et al., 2004;

Table 10.1. Overview of chemistries used to control brown rot blossom blight and fruit rot in commercial peach orchards of the USA and their resistance risk.

Chemical class (abbreviation)	Examples of common active ingredients	FRAC code	Efficacy		First report of field resistance in the USA	Resistance risk
			Blossom blight	Fruit rot		
Reduced risk fungicides						
Anilinopyrimidines (APs)	Pyrimethanil, cyprodinil[a]	9	++	++	Not reported	Medium
Hydroxyanilides	Fenhexamid	17	++	++	Not reported	Medium
Methyl benzimidazole carbamates (MBC)	Thiophanate-methyl	1	+++	+++	Sonoda et al. (1983)	High
Quinone outside inhibitors (QoIs)	Azoxystrobin, pyraclostrobin	11	++++	++++	Amiri et al. (2010)	High
Succinate dehydrogenase inhibitors (SDHIs)	Boscalid	7	+++++	+++++	Amiri et al. (2010)	Medium
Conventional fungicides						
Chloronitriles	Chlorothalonil[a]	M5	+++	–	Not reported	Low
Demethylation inhibitors (DMIs)	Propiconazole, fenbuconazole	3	+++++	+++++	Zehr et al. (1999)	Medium
Dicarboxamides	Iprodione[a]	2	+++	–	Not reported	High
Inorganic	Sulfur	M2	+	+	Not reported	Low
Phthalimides	Captan	M4	++	++	Not reported	Low

Note: [a]Not registered after bloom.

Luo *et al.*, 2008; Burnett *et al.*, 2010). A recent study shows that *M. fructicola* populations with reduced sensitivity to QoI and SDHI fungicides are emerging in the southeast USA (Amiri *et al.*, 2010). The failure of common fungicides to control field-isolated resistant *M. fructicola* strains can be demonstrated in detached fruit assays, where the label rates of registered products are unable to prevent disease development (Table 10.2).

Based on existing MBC, DMI, QoI and SDHI resistance in south eastern USA peach orchards, fungicide resistance management in *M. fructicola* has taken on a new urgency. Resistance to DMI, QoI and MBC fungicides has been detected in the Upstate and Midlands of South Carolina, as well as the Midlands of Georgia (Table 10.3). The only area in the southeast that has not yet been impacted by resistance is South Georgia, where limited disease pressure allows producers to use conventional, less expensive fungicides. Because MBC fungicides have been used since the early 1980s and because MBC resistance is known to persist for years in peach production areas (Zehr *et al.*, 1991), preharvest brown rot spray programmes have relied primarily on rotations of QoI and DMI fungicides in recent years. So far, we have not encountered populations with dual resistance, despite laboratory evidence that selection for DMI resistance may increase selection for QoI resistance (Luo and Schnabel, 2008). This finding is of critical importance because it allows producers to select efficacious fungicides for brown rot control in the presence of resistance to another chemical class. A region-wide resistance management programme must be implemented to determine location-specific resistance profiles and to provide customized resistance management strategies.

Table 10.2. *Monilinia fructicola* isolates from commercial orchards causing disease on fungicide-treated, detached peaches.[a]

Isolate	Phenotype	Propiconazole (DMI)		Thiophanate-methyl (MBC)		Azoxystrobin (QoI)	
		DI (%)	DS (%)	DI (%)	DS (%)	DI (%)	DS (%)
Harvs9-1	DMI resistant	100	34.4	–	–	–	–
By1-2	DMI resistant	100	26.8	–	–	–	–
Rb1-1	MBC resistant	–	–	100	23	–	–
Cor.8-1	MBC resistant	–	–	100	44.2	–	–
Harv5.2	QoI resistant	–	–	–	–	100	15
Cor6-1	QoI resistant	–	–	–	–	100	34
By5-1	Sensitive	0	0	0	0	0	0
ByB1SS	Sensitive	0	0	0	0	0	0

Note: [a]Fungicides were applied prior to inoculations to detached fruit using commercial formulations at recommended label rates. DS, disease severity; DI, disease incidence.

Table 10.3. Occurrence of at least one fungicide-resistant *M. fructicola* strain per sample detected during the monitoring programme in commercial orchards from South Carolina and Georgia in 2008 and 2009.

	2008			2009			
	South Carolina		Georgia	South Carolina		Georgia	
Resistance	Upstate $n = 18$ (%)	Midlands $n = 7$ (%)	Midlands $n = 3$ (%)	Upstate $n = 16$ (%)	Midlands $n = 4$ (%)	Midlands $n = 16$ (%)	South $n = 15$ (%)
DMI	27.80	28.60	100.00	31.20	25.00	62.50	0.00
MBC	16.70	0.00	0.00	50.00	25.00	6.30	0.00
QoI	27.80	42.80	66.70	25.00	0.00	18.80	0.00

10.3   Distinguishing Sensitive and Resistant Isolates

Determination of discriminatory doses

Once the decision is made to start a resistance monitoring programme, a meaningful discriminatory dose of a fungicide aimed at distinguishing sensitive from resistant isolates must be determined. But what exactly is a 'resistant' isolate? In the literature, the term 'resistant' appears to be used rather loosely. When sampling isolates from the field, there usually is a wide range of sensitivities in baseline populations, and often a wider range of sensitivities in populations that have been pre-exposed to fungicides. For producers, and thus for the kit-supported monitoring programme, isolates with 'practical resistance' are the ones of concern. Organisms with practical resistance possess sufficient levels of resistance, frequency, pathogenicity and fitness to cause a decrease or loss of fungicide effectiveness in practice. In other words, the objective is not to distinguish sensitive isolates from isolates that *in vitro* are more resistant to fungicides and yet do not cause detectable practical problems. Consequently, the discriminatory dose for detection of field resistant isolates may not detect the ones that are considered resistant in the laboratory.

The first step for determining the optimal discriminatory dose is collection and isolation of the pathogen from 'problem' areas, that is areas where disease has developed at unacceptable levels despite appropriate fungicide applications, and to determine the *in vitro* sensitivity using germination rate and, more typically, mycelial growth tests using traditional Petri dish assays. *In vitro* sensitivity traditionally is determined by exposing isolates of the pathogen to different concentrations of the fungicide to determine the concentration at which a fungicide inhibits 50% of the mycelial growth (EC_{50} value). Based on *in vitro* results, isolates with low, medium and high EC_{50} values (representing the entire range of sensitivities) are selected for *in vivo* assays. For *M. fructicola*, we are using detached peach fruit sprayed with label rates of a fungicide and inoculated with

the pathogen (Schnabel *et al.*, 2004). It would be expected that isolates with high EC_{50} values (and maybe even those with medium EC_{50} values) are capable of developing disease on fungicide-treated detached fruit, as well as on the control fruit. A discriminatory dose can then be inferred based on *in vitro* and *in vivo* studies. Lastly, the kit-supported monitoring assay is conducted with the estimated discriminatory dose with the same reference strains used for the *in vivo* studies. If the distinction between sensitive and resistant isolates is not satisfactory, the discriminatory dose may have to be readjusted. The discriminatory dose will naturally vary from one chemical class to another, from one chemical within a chemical class to another and from one pathogen to another.

10.4   Development of a Rapid and Reliable Monitoring Assay

Considering the above-mentioned resistance issues with the most efficacious chemical classes, the need arises for location-specific resistance profiles. But how can regional resistance monitoring be conducted without the need for huge financial or high-tech resources? If county extension personnel are to be included in generating these profiles, many of the currently available, labour-intensive and expensive standard laboratory techniques would be unsuitable. Therefore, a field kit is needed in support of a monitoring programme that can be used either directly in the field or in county extension offices under semi-sterile conditions without the need for special equipment.

The centrepiece of an earlier version of the assay was a lip balm tube-based assay developed previously for quick and reliable assessments of resistance profiles in populations of *M. fructicola* from commercial peach orchards (Amiri *et al.*, 2008, 2009a). The idea of using a lip balm device was first described in 1984 to sample and assay conidia of *Venturia inaequalis* in the field by pressing a water agar surface of the core against a sporulating lesion (Lalancette *et al.*, 1984). Later, the tube was filled with Benlate-amended PDA medium for MBC resistance monitoring (Zehr *et al.*, 1991). For reasons explained later, we shifted to a 24-well plate assay that was prepared in the laboratory and sent out as a self-contained kit to participating county agents for location-specific resistance monitoring. The kit-based, location-specific resistance monitoring programme consists of the following system components.

Sample collection

Establishing a standard protocol for sample collection is important to ensure that a sample represents a specific location properly. Spores are collected from no less than 10 symptomatic fruit for each location. Each fruit must be from a different tree and at least two buffer trees must separate sampled

trees. The fruit is handled so that nothing touches the sporulating (diseased) area during collection and transport. Over time, we have transitioned to collecting spores with individually wrapped sterile cotton swabs, where the cotton tip is rubbed gently against the sporulating area of a fruit to capture as many spores as possible. In the process, the white cotton will turn from pure white to somewhat brown, indicating that sufficient spores have been collected. After the spores are collected from the fruit, the swab is returned to its wrapper, labelled, and is ready to be used for the actual sensitivity assay (Fig. 10.1). This technique eliminates the potential problem of fruit-to-fruit cross-contamination, avoids having to handle rotting fruit and allows for medium-term storage (months) of the samples in the refrigerator.

Fig. 10.1. Schematic representation of the practical steps involved in the kit-supported resistance monitoring system. Symptomatic peaches are chosen from a commercial orchard following a strict protocol (1,2); spores are collected using individually wrapped cotton swabs and returned to the wrapper (3); spores are then transferred from the swab to the centre of 24-well plates containing growth medium with or without fungicides of certain discriminatory doses (4,5); the lids of the plates are returned and plates are incubated for 72 h at room temperature (6); mycelial growth is assessed visually and entered into a web application (7); the web application calculates resistance factors and provides resistance management recommendations.

The lip balm tube assay

As mentioned above, this assay was our first attempt to develop a field kit for resistance monitoring. The assay was performed as described previously (Amiri *et al.*, 2008, 2009a), with slight modifications. Briefly, standard lip balm tubes (15 × 60 mm) were sterilized and filled with fungicide-amended or -unamended potato dextrose agar (PDA) plus 100 μg/ml streptomycin. Treatments were propiconazole, thiophanate-methyl and azoxystrobin at the rates indicated below. After the medium had cooled, tubes were capped and stored at 4°C. To slice off disks for the assay, the cap was removed and the medium uplifted by turning the bottom shaft of the lip balm tube. Two agar disks of ~3 mm thickness were cut from each treatment tube and placed in a quarter section of the Petri dish (each treatment had its own section). Conidia were transferred directly from the fruit or the cotton swab to the centre of the disks, going back and forth from the fruit/swab to each disk. The eight disks in one Petri dish (three fungicide treatments and one non-treated control treatment) were inoculated with spores from the same fruit. The entire procedure has been illustrated with colour photographs in a recent publication (Schnabel *et al.*, 2010).

To conduct the assay, agents and specialists were provided with a kit containing enough supplies to process five 10-fruit samples: 1 lighter, 5 data sheets, 1 scalpel handle and scalpel blades, 1 user manual, 5 pairs of latex gloves, 4 permanent markers of different colour with one colour for each treatment, 10 plastic containers with lids for fruit collection, 40 lip balm tubes filled with PDA amended with water (control; 10 tubes); propiconazole (Orbit 3.6EC, Syngenta Crop Protection, Greensboro, NC; at the discriminatory dose of 0.3 μg/ml; 10 tubes); thiophanate-methyl (Topsin-M 70W, Ceraxagri, King of Prussia, PA; at the discriminatory dose of 50 μg/ml; 10 tubes); or azoxystrobin (Abound 2.08F, Syngenta; at the discriminatory dose of 3 μg/ml) plus salicyl-hydroxamic acid (SHAM) at 100 μg/ml, 60 sterile toothpicks and Petri dishes and 75 parafilm strips.

Although the lip balm tube assay was a breakthrough in that it enabled agents and consultants to generate valuable data for their growers, the assay had certain weaknesses. Petri dishes had to be labelled; disks had to be sliced off and positioned in the Petri dish for inoculation; and dishes had to be wrapped with parafilm. The entire assay took 30–50 min for a 10-fruit sample, sometimes longer depending on experience. In addition, the assay allowed for only four treatments, due to space limitations in the 90 mm Petri dish, and PDA-filled tubes tended to become contaminated on the bottom shaft after 4 weeks of storage, due to PDA leakage.

Well plate assay

The next (and current) device is based on a standard laboratory sterile 24-well (6 × 4 wells) plate (12.5 × 8.5 × 2 cm). If held vertically, there are 4 wells per row, allowing for four samples to be processed, and 6 wells per column, allowing for six treatments (one control and five fungicide treatments; Fig. 10.1). Consequently, for a 10-fruit sample, three 24-well plates are required, with the choice of leaving two columns empty or processing 12 instead of 10 fruit per location. Similar to the lip balm tubes, the wells are prefilled with fungicide-amended or -unamended PDA (1.5 ml/well) plus 100 mg/l streptomycin. When all the wells are filled with PDA, the plates are sealed with removable plastic film for each column individually (Fig. 10.1). A kit shipped to agents may contain 15 prefilled 24-well plates ready for inoculation, 5 data sheets, 1 user manual, 10 individually wrapped anti-microbial hand wipes, 60 toothpicks and 60 individually wrapped cotton swabs; sufficient supplies to monitor five locations with 10–12 observations (fruit) per sample. Prior to inoculation, the seal from the first column is removed and spores are transferred to the centre of the PDA-filled wells. Spores from the same fruit are transferred directly from the fruit or from the cotton swab into each well of a single column using a wooden toothpick. In our experience, the toothpicks do not need to be autoclaved for the assay when using antibiotics in the PDA. Furthermore, we have used the same toothpick to inoculate the 6 wells of a single column for a single fruit sample and then switched toothpicks when going to the next single-fruit sample (column) without noticing increased contamination problems. When all four columns are inoculated, the plastic lid is returned and the plate is labelled and incubated at room temperature (20–25°C) for exactly 72 h. Each batch of plates should undergo a standard quality control check prior to shipment to make sure the plates of the new batch perform as they should using DMI, MBC and QoI fungicide-sensitive and -resistant isolates.

Assessment of results

The proper scientific way to assess the results would be to measure crosswise the diameter of colonies developing in each disk/well and to record the average number. However, for simplicity and time constraints, we ask agents to assess growth visually as follows: (–) for absence of growth, (+) for less than 20% growth, (++) for less than 50% but more than 20% growth and (+++) for more than 50% growth compared to the control disk/well. In other words, we assess the data visually and subjectively rather than measuring mycelial growth precisely. Such visual assessments do not require the use of additional tools such as calipers or rulers and thus facilitate the process of determining the absence or presence of resistance.

Development of a web application

In the early stages of development of the resistance monitoring programme, we had asked agents to record the results on data sheets, communicate the results to the specialist and wait for resistance management recommendations. Specifically, the following additional steps were involved before a recommendation reached the producer: (i) data sheets were faxed to the extension specialist; (ii) data were analysed for each fungicide class; (iii) an appropriate response was formulated; and (iv) the response was communicated to

the agent and then to the grower. These actions added days to the turnaround time, since the recommendations went through numerous intermediaries. For producers, delays in obtaining critical spray recommendations are unacceptable and methods are available to overcome these delays completely – Internet technology.

A web application devoted to the resistance monitoring programme was developed in an effort to accelerate communication transfer and thus further improve the service to producers. Web applications are extremely useful due to the ubiquity of web browsers and the ability to update and maintain web applications without distributing and installing software on potentially thousands of client computers. The ability to update information for a web application also increases the utility of such programmes and, in the case of resistance management, new fungicides or resistance conditions on the ground would require immediate changes to the recommendations. A web application gives maximum flexibility in restoring the integrity of the programme when changes occur – immediate flexibility.

With our web assay system, visual observations from the 24-well plate or the lip balm assays are entered into the web application together with information about the origin of the sample and spray history of the orchard (Fig. 10.2). The web application validates that all required information is entered and requests a final review of the data entered. Then a resistance factor (RF) is calculated separately for each chemical class (see details below). This computed RF factor is then compared to an RF key to determine the resistance category: $RF \leq x$ = no resistance; $RF > x \leq y$ = very low danger of resistance; $RF > y \leq z$ = moderate resistance; $RF > z$ = high resistance (Fig. 10.2). We are currently using 0.5, 4 and 32 for x, y and z, respectively, but the RF key may be adjusted according to the specialist's preference. The location-specific portion of the final report is created by assembling pre-drafted resistance management recommendations for each fungicide and corresponding resistance category (no, low, moderate and high).

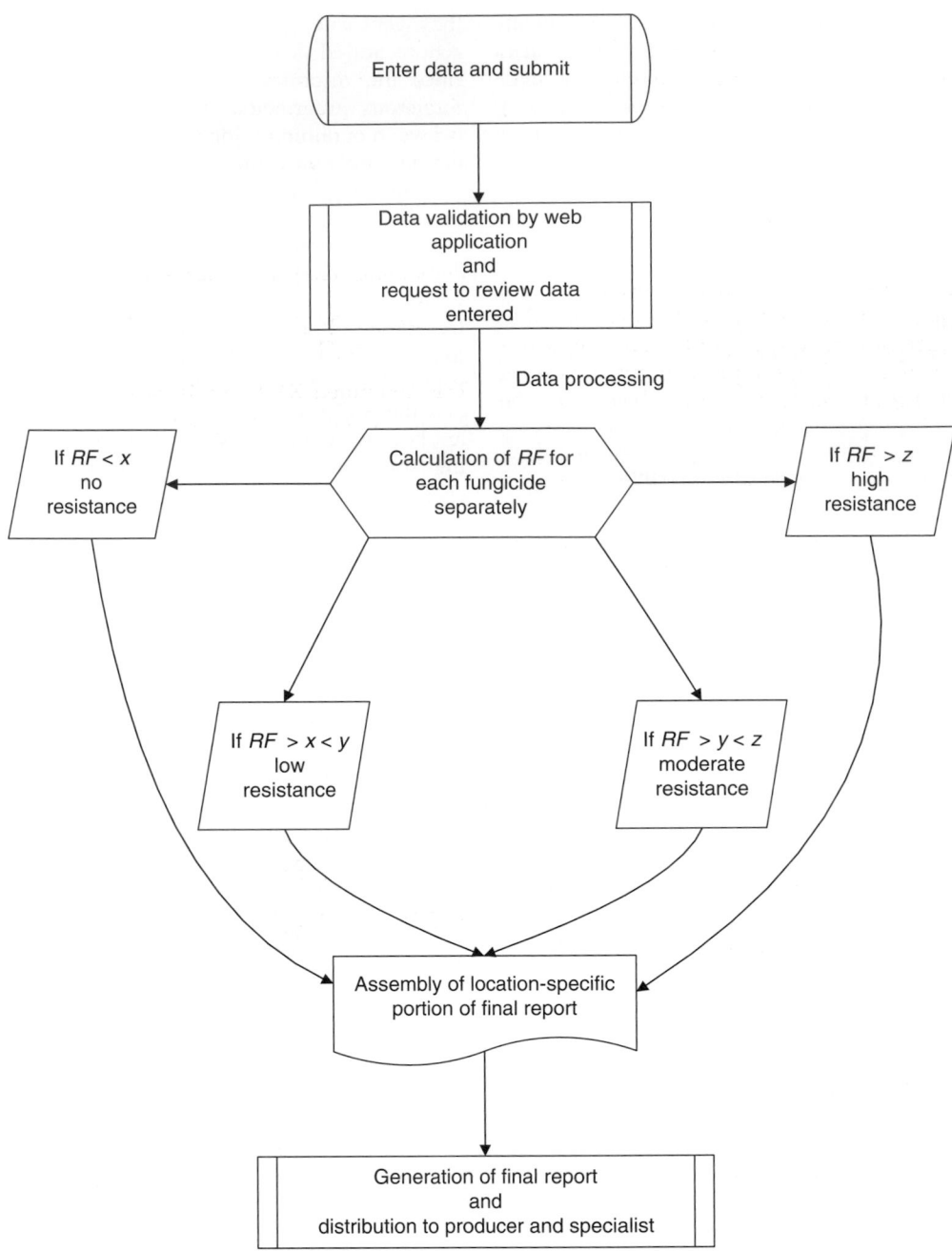

Fig. 10.2. Flow chart of web application in support of the resistance monitoring programme. Values for *x, y* and *z* may be equal to 0.5, 4 and 32, respectively.

On completion of the location-specific portion, the final report is generated, displayed and automatically e-mailed for review to the specialist in real time. The user of the web application can save the report in pdf format and share it instantly with the producer by e-mail attachment. To our knowledge, this is the first web application of its kind. The details on data entry, *RF* calculation and automated response service are provided below. The web application prototype for monitoring resistance in *M. fructicola* is currently located at www.peachdoc.com under 'Profile'.

Data entry

Users have to login with a username and password. On login, the user will be asked to enter information in regard to the sample origin, nature of the sample (fruit rot or blossom blight) and same-year spray history of the orchard from which the sample was taken. Similar to the original data recording sheet that county agents and specialists had been using before, users can enter – (dash), + (plus), ++ (double plus) and +++ (triple plus), indicating no growth, less than 20% growth, up to 50% growth and more than 50% growth, respectively, on disks/wells amended with a specific discriminatory dose of each fungicide. Any unusual observations can be entered as text directly next to each sample or into a separate text box. The program will accept a minimum of 8 and a maximum of 10 samples per location.

Calculation of the resistance factor (*RF*)

We developed a mathematical formula to calculate the *RF* for a sample of 8–10 fruit. The formula takes into account the disproportionate importance of a 'single plus' versus a 'double plus' versus a 'triple plus'. For example, if six out of ten observations are rated 'single plus' (less than 20% growth), with the remaining four being dashes (no growth), the entire resistance profile would indicate low levels of resistance. On the other hand, if only two of ten observations revealed 'triple pluses', this would indicate a major concern relative to resistance issues at that location.

Given *n* observations, where $8 \le n \le 10$, assign the value x_i to be 0, 1, 2 or 3 if the growth rating of observation *i* is a dash, single plus, double plus or triple plus, respectively. A resistance factor for observation *i*, denoted RF_i, is then computed as:

$$RF_i = x_i^5 \text{ for each } i = 1, \dots, n. \qquad (10.1)$$

These individual resistance factors for the observations are then averaged over the number *n* of observations to obtain the resistance factor *RF* as:

$$RF = \frac{1}{n} \sum_{i=1}^{n} (RF_i). \qquad (10.2)$$

Combining (10.1) and (10.2), we obtain:

$$RF = \frac{1}{n} \sum_{i=1}^{n} x_i^5. \qquad (10.3)$$

This computed *RF* factor is then compared to our *RF* key to determine the resistance category:

0–0.5 = no resistance; >0.5–4 = low resistance; >4–32 = moderate resistance; >32–243 = high resistance.

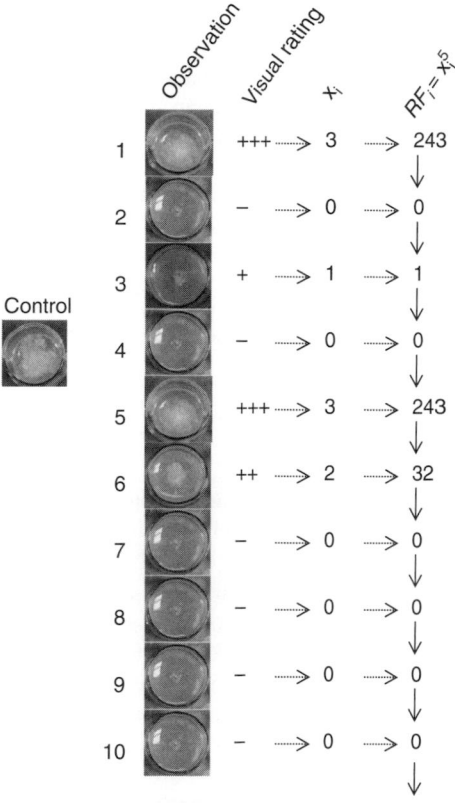

$$RF = \text{average } (RF_i1-10) = 51.9$$

Fig. 10.3. Illustration of resistance factor (*RF*) calculation from a 10-fruit sample.

For example, Fig. 10.3 shows $n = 10$ observations collected from one location. Each visual rating is assigned the appropriate x_i value of 0, 1, 2 or 3 and then each of these values is raised to the power 5 (e.g. a triple plus (+++) rating is assigned the number 3 and $3^5 = 243$) to obtain the resistance factor RF_i of (10.1) for observation i, as found in the last column. The resistance factor RF is then computed by averaging the ten resistance values, as in (10.2) with $n = 10$.

Disease control recommendation

After the data are entered by the user and processed by the web application, a report is prepared containing: (i) a summary of the data entered; (ii) the calculated RF factor for each chemical class; (iii) the resistance category for each chemical class; and (iv) the corresponding disease control recommendation. The latter (iv) consists of an introduction indicating general guidelines for brown rot management, profile-specific recommendations and finishing statements containing general considerations and a disclaimer. The disease control recommendations are managed and updated by the peach disease extension specialist in charge of the state in which participating producers reside. Below is an example of the location-specific portion of the final report. In this hypothetical case, resistance to propiconazole was detected. Units are kept in the US customary system to maintain authenticity.

The *RF* value for propiconazole indicates high resistance to DMI fungicides

Recommendation: Due to the absence of effective alternatives, DMI fungicides should still be used for preharvest brown rot control. In contrast to the benzimidazoles, however, DMI fungicides still work well against DMI-resistant populations if the rate is increased. Use Elite 45DF, Indar 2F, Indar 75WSP, or their generics at double the normal rate (highest allowable label rate). For Elite, that would be 8 oz/acre, for Indar 2F it would be 12 fl oz/acre and for Indar 75WSP it would be 4 oz/

acre in the USA. Do not use DMI fungicides more than once per season if possible (not during bloom and again, if possible, only once during the preharvest season). In particular, since resistance to DMI fungicides has developed, ensure Pristine (high rate) is used 14 days before the harvest window, and a DMI fungicide should not be used until 7 days before harvest.

The *RF* value for thiophanate-methyl indicates no resistance to MBC fungicides

Recommendation: Use a benzimidazole product (i.e. thiophanate-methyl) such as Topsin-M 70W or -WSB or Thiophanate Methyl 85WDG in combination with another material (such as captan) once a year either during bloom sprays or 3 weeks before harvest. Incorporating benzimidazoles, which are effective in the absence of resistance, into a disease management programme will maximize the efficacy and longevity of the other chemistries being used. Remember always to tank mix thiophanate-methyl with a fungicide having a different mechanism of action, such as captan. A combination of a benzimidazole with Vangard would also be a good option for blossom blight control, and a combination with Abound would be a good option for preharvest brown rot control. Do not cut label rates of the benzimidazole fungicide or its tank-mix partner; use both materials at labelled rates.

The *RF* value for azoxystrobin indicates no resistance to QoI fungicides

Recommendation: Use Pristine 38WG or Abound during the preharvest season as recommended in the most current *Southeastern Peach, Nectarine, and Plum Pest Management and Culture Guide* (http://www.ent.uga.edu/peach/Peach Guide.pdf). Use the high dose. Do not use respiration inhibitor fungicides (Pristine or Abound) more than once per season if possible (not during bloom and again if possible, only once during the preharvest season). Many years of field testing have shown that Pristine is best applied 14 days before harvest, followed by a DMI application at 7 days before harvest. Limit applications to one or two per season. Additional applications may be required at this point due to few alternatives, but use other effective materials as much as possible.

10.5 The Issue of Contamination

Whenever nutrient-rich media, such as PDA, are used there is the issue of contamination. Many fungi, yeast and bacteria can grow on such media and thus can cause false positive readings. As for bacteria, we inhibit the growth of about 99% of bacteria with the addition of streptomycin to the growth medium. Yeasts grow significantly slower than *M. fructicola* and are not a major fresh fruit contaminant, thus are not likely to cause a problem for the system we have developed. That leaves us with fungal contaminants, which even with great care cannot be avoided.

Most contaminants, just like the pathogen itself, are carried by wind and rain to the fruit. We observed that the percentage of contaminated disks/wells was low (less than 5%) early in the peach ripening season (June) and increased, as the season progressed, to about 15% in August (Amiri *et al.*, 2009a). This increase corresponds to inocula build-up throughout the season, releasing increasing amounts of spores into the air. For each crop, the composition of troublesome contaminants will likely vary. For peach fruit in south east USA, the main fungal contaminants have been *Alternaria* and *Cladosporium* species, followed by *Rhizopus stolonifer* later in the season. It takes some training to distinguish the target pathogen, in our case *M. fructicola*, from contaminants. For illustrations of fungal contaminants, download the user manual of the resistance monitoring kit under 'Profile' at www.peachdoc.com. The colony on the disk/well is likely a contaminant if (i) the colony develops from anywhere but the point of inoculation (often *Penicillium* spp.), (ii) the colony turns black in the centre after 3 days of incubation (*Cladosporium* spp. or *Alternaria* spp.), (iii) unusual amounts of aerial mycelium develop after 1–2 days of incubation (sign of *R. stolonifer* or *G. persicaria*) and (iv) the colony develops as a sporulating green or dark translucent colony (*Penicillium* or *Colletotrichum* spp., respectively).

An alternative to using complete media would be to use selective media. Selective media have been developed for many pathogens, including *M. fructicola* (Phillips and Harvey, 1975) and *Botrytis cinerea* (Kerssies, 1990; Edwards and Seddon, 2001), two of the fungi we consider most suitable for kit-supported resistance monitoring. However, selective media contain fungicides to control potential contaminants that may interfere with the activity of the fungicide of interest, either by additive or synergistic effects. For example, recently we developed APDA-F500, a medium suitable for isolating *Monilinia* species from fruit surfaces while inhibiting other fungi (Amiri *et al.*, 2009b). However, when adding propiconazole, for example, most *Monilinia* isolates no longer grew on APDA-F500 (G. Schnabel and A. Amiri, South Carolina, 2009, personal communication).

10.6 Involvement of Field Specialists in the Monitoring Process

The advantages of having non-academic field personnel, such as county extension agents, participating in this monitoring programme vastly outweigh the disadvantages. Extension agents are very much 'plugged in' when it comes to producer concerns. They deal with producers on a daily basis and build a very special relationship based on trust. Many agents have told us that the resistance monitoring programme has given them a great tool to provide science-based information that will make a direct impact on farm performance, be it in the form of avoidance of unnecessary sprays, improving fungicide performance, managing resistance or simply giving the producer the peace of mind that everything is all right. The programme further supports and strengthens agent/producer relationships. The only downside is the potential variability in growth assessments between different scorers conducting the tests. In order to alleviate any potential inaccuracies, all participants were and should be required to go through training before being given the opportunity to order kits. For those who cannot attend the training sessions, a video detailing the different steps and procedures to follow for the test completion has been developed and made available. Participants are trained in the proper collection of field samples, performing the assay, identification of potential contaminants, visual assessment of mycelial growth and data reporting.

Kits were shipped by overnight delivery to trained agents and specialists in South Carolina and Georgia, the main peach producing states in the region. The kit had to be shipped at 4-week intervals because of the relatively short shelf life of the lip balm tubes filled with fungicide-amended potato dextrose agar (Amiri _et al._, 2009a). Thus, agents and specialists had the kit available during the main southeastern USA peach production months of June, July and August. The shelf life of 24-well plates is 6–8 weeks (Fig. 10.4), thus only two shipments are required. Agents and specialists were asked to record the name of the producer, the origin of the sample, spray history of the orchard and the visual assessment of mycelial growth on detached disks on a data sheet (provided with the kit). During the first 2 years of the resistance monitoring programme, users

faxed the report to the specialist for analysis and recommendations. Starting with the third year, the web application (described above) was used for real-time data analysis and feedback.

10.7 Highlights of 3 Years of Monitoring

Resistance to all three chemical classes exists but management recommendations appear to work. As mentioned above, resistance to DMIs, QoIs and MBCs is present in South Carolina and Georgia (Table 10.1). DMI resistance was detected most frequently in 2008 (35.7%), less frequently in 2009 (31.4%) and appeared to be almost absent (3.7%) in 2010, despite an increase of sample number (Table 10.4). A similar trend was found for QoI resistance. MBC resistance was found less frequently in 2008 compared to DMI and QoI resistance. However, MBC resistance was stable throughout the 3 years of sampling. DMI and QoI fungicides have been, and still are, the mainstay products since 2001 for pre- and postharvest brown rot control. Prior to the establishment of the resistance monitoring programme, many

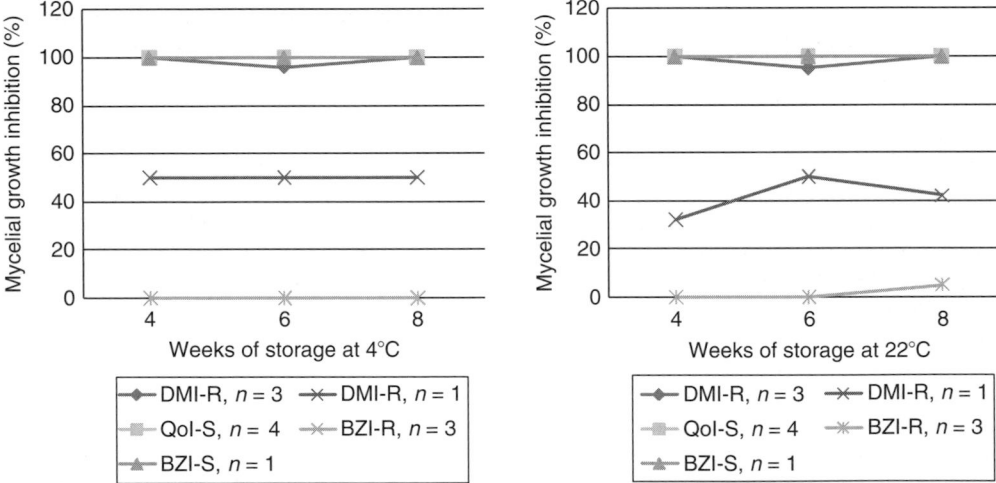

Fig. 10.4. Inhibition of mycelial growth of _Monilinia fructicola_ strains with different fungicide resistance phenotypes on 24-well plates filled with fungicide-amended or -unamended PDA after 4, 6 and 8 weeks of storage at 4 and 22°C.

Table 10.4. Occurrence of resistance in samples from Georgia and South Carolina collected between 2008 and 2010.

| Resistance to | 10-fruit samples (%) with at least one resistant isolate | | | |
	2008	2009	2010	
DMIs	35.7a[a]	31.4a	3.7b	$p < 0.001$
QoIs	35.7a	13.7b	3.7b	$p < 0.001$
MBCs	10.7a	19.6a	14.8a	$p = 0.64$
	$n = 28$	$n = 51$	$n = 54$	

Note: [a]Values with different letters within rows are significantly different according to the ANOVA pairwise multiple comparison procedure; mean separation by Holm–Sidak method.

growers tended to use exclusively DMIs or QoIs, which likely caused the observed build-up of resistance. We believe that when producers learned about the results of the monitoring programme, they became much more aware about the necessity of implementing anti-resistance management strategies. These included rotation of chemical classes (Schnabel *et al.*, 2004), increase of product dosage in respect to DMI fungicides (Brannen *et al.*, 2007) and skipping a chemical class altogether if resistance was severe. Some growers still use MBC fungicides at least once a year for the control of not only brown rot but also other diseases, such as peach scab caused by *Fusicladosporium carpophilum*. That may explain the stable occurrence of MBC-resistant isolates in the southeast.

We believe that the implementation of this monitoring programme in South Carolina and Georgia and the resulting recommendations for resistance management prevented disease epidemics in 2009, a year that was conducive for a disease epidemic. We estimate that losses due to pre- and postharvest brown rot would have been 10–15% in each state, which translates into US$4–8 million in revenue for each state, if key growers had not made adjustments to their spray programmes.

Between 2008 and 2009, there was no significant difference in the relationship between the spray history of the orchard and its resistance profile. Therefore, the data for the 2 years are shown combined (Fig. 10.5). Data for 2010 were not included because very little resistance was recorded, regardless of the spray history of the orchard. The data indicate that preharvest spray programmes consisting of DMI fungicides or DMI fungicides alternated with QoI fungicides selected significantly more for DMI resistance. Likewise, spray programmes based on QoI fungicides or QoIs applied in combination with DMI fungicides selected significantly more for QoI resistance. The data indicate that frequent applications of the same members of a specific chemical class will select for resistance more rapidly, which is consistent with our general understanding of how fungicide resistance develops in the field (Wade, 1988). The more unexpected result was that the rotation of DMI and QoI fungicides, which has been recommended as a general anti-resistance management strategy, did not appear to reduce the number of resistant isolates in the 10-fruit samples compared to programmes using either fungicide exclusively. However, outbreaks of brown rot were not observed, indicating that the QoI fungicides likely controlled the DMI-resistant portion of the population and the DMI fungicides likely controlled the QoI-resistant portion of the population. To date, we have not observed a population or even a single isolate with resistance to both chemical classes.

10.8 Other Potential Applications for Location-Specific Resistance Monitoring

This first prototype of a kit-supported resistance monitoring programme was developed

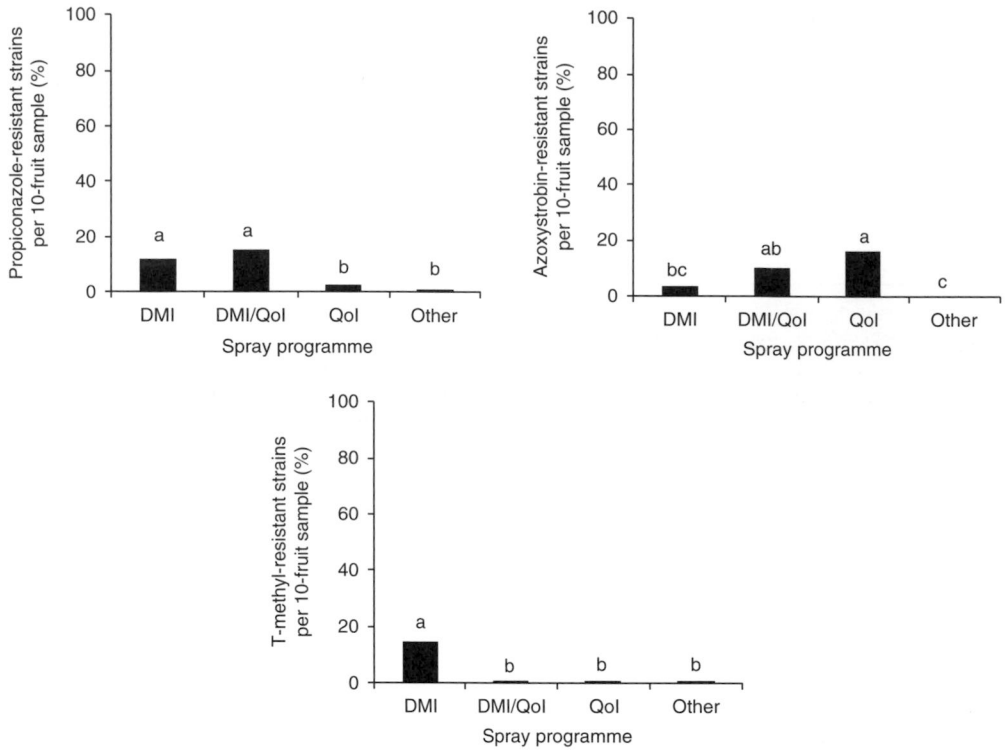

Fig. 10.5. Effect of four spray programmes on the occurrence of propiconazole (DMI), azoxystrobin (QoI) and thiophanate-methyl (MBC) resistant isolates per location. Shown is the average number of fungicide-resistant strains per 10-fruit sample. A fungal strain was considered resistant when assessed as double or triple plus. Combined data set of 2 years (2008 and 2009) were submitted to an ANOVA followed by Fisher's LSD test.

specifically for the brown rot disease of peach due to documented cases of field resistance to multiple chemical classes of fungicides in *M. fructicola*. Developing a monitoring programme for other pathogens is possible, but the pathogen would have to fulfil some important criteria to be considered suitable for this monitoring system. First, the pathogen has to sporulate abundantly on the agricultural crop so that spores can be collected easily and used for inoculation; second, the pathogen needs to be able to grow on an artificial medium (no obligate fungi); third, the fungus should grow relatively fast on artificial medium (so that most potential contaminants can be outgrown and results are obtained in a reasonable time). Pathogens that come to mind and fulfil these criteria are all *Monilinia* spp. affecting stone fruit; *Botrytis* spp. causing grey mould of fruit and vegetables; *Penicillium*

spp. causing rots in citrus and pome fruit; and *Alternaria* spp. causing disease on multiple crops.

10.9 Conclusions

Reduced risk fungicides and DMIs have become an important part of IPM, but their vulnerability towards resistance development requires the implementation of resistance monitoring programmes. Since resistance profiles may vary from one farm to another due to differences in spray history, location-specific testing is warranted for precision brown rot management. We developed a novel kit- and web-supported resistance monitoring system that provides county agents and consultants with a tool for timely determination of location-specific resistance profiles.

Acknowledgements

This material is based on work supported by the CSREES/USDA, under project number SC-1000642 and USDA CSREES S-RIPM grants 2006-34103-17007, 2007-41530-04350 and 2008-34103-19098. We would like to thank all participating county agents and extension specialists from South Carolina and Georgia, with special thanks to Clemson University Extension Agents A. Rollins and W.G. Henderson for exemplary support of this programme. We also thank William Bridges at Clemson University for statistical advice.

References

Amiri, A., Scherm, H., Brannen, P.M. and Schnabel, G. (2008) Laboratory evaluation of three rapid agar-based assays to assess fungicide sensitivity in *Monilinia fructicola*. *Plant Disease* 92, 415–420.

Amiri, A., Brannen, P.M. and Schnabel, G. (2009a) Validation of the lipbalm tube assay for evaluation of fungicide sensitivity in field isolates of *Monilinia fructicola*. Online. *Plant Health Progress* DOI: 10.1094/PHP-2009-1118-01-RS.

Amiri, A., Holb, I.J. and Schnabel, G. (2009b) A new selective medium for the recovery and enumeration of *Monilinia fructicola*, *M. fructigena* and *M. laxa* from stone fruit. *Phytopathology* 99, 1199–1208.

Amiri, A., Brannen, P.M. and Schnabel, G. (2010) Reduced sensitivity of *Monilinia fructicola* field isolates from South Carolina and Georgia to respiration inhibitor fungicides. *Plant Disease* 94, 737–743.

Brannen, P.M. and Schnabel, G. (2003) *Monilinia fructicola* (brown rot) resistance to propiconazole in Georgia – evidence and implications for disease management (http://newsletters.caes.uga.edu/srpn/3-5/SoutheastRegionalPeachNewsletterV3-5.pdf, accessed 25 October 2010).

Brannen, P., Horton, D., Bellinger, B. and Ritchie, D. (2005) 2005 Southeastern peach, nectarine and plum pest management and culture guide. Extension Bulletin 1171. University of Georgia, Athens, Georgia.

Brannen, P., Hotchkiss, M., Reilly, C.C. and Amiri, A. (2007) Evaluation of fungicide programs to manage DMI-resistant *Monilinia fructicola* in a late-ripening peach, 2006. *Plant Disease Management Reports*. Online. 1, STF003.

Brent, K.J. (1988) Monitoring for fungicide resistance. In: Delp, C.J. (ed.) *Fungicide Resistance in North America*. APS Press, St Paul, Minnesota, pp. 9–11.

Burnett, A., Lalancette, N. and McFarland, K. (2010) First report of the peach brown rot fungus *Monilinia fructicola* resistant to demethylation inhibitor fungicides in New Jersey. *Plant Disease* 94, 126.

Edwards, S.G. and Seddon, B. (2001) Selective media for the specific isolation and enumeration of *Botrytis cinerea* conidia. *Letter of Applied Microbiology* 32, 63–66.

Fraaije, B.A., Butters, J.A., Coelho, J.M., Jones, D.R. and Hollomon, D. (2002) Following the dynamics of strobilurin resistance in *Blumeria graminis* f. sp. *tritici* using quantitative allele specific real-time PCR measurements with the fluorescent dye SYBR green I. *Plant Pathology* 51, 45–54.

Kerssies, A. (1990) A selective medium for *Botrytis cinerea* to be used in a spore-trap. *Netherlands Journal of Plant Pathology* 96, 247–250.

Lalancette, N., Russo, J.M. and Hickey, K.D. (1984) A simple device for sampling spores to monitor fungicide resistance in the field. *Phytopathology* 74, 1423–1425.

Luo, C.X. and Schnabel, G. (2008) Adaptation to fungicides in *Monilinia fructicola* isolates with different fungicide resistance phenotypes. *Phytopathology* 98, 230–238.

Luo, C.X., Cox, D.K., Amiri, A. and Schnabel, G. (2008) Occurrence and detection of the DMI resistance-associated genetic element 'Mona' in *Monilinia fructicola*. *Plant Disease* 92, 1099–1103.

Michailides, T.J., Ogawa, J.M. and Opgenorth, D.C. (1987) Shift of *Monilinia* spp. and distribution of isolates sensitive and resistant to benomyl in California prune and apricot orchards. *Plant Disease* 71, 893–896.

Ogawa, J.M. and English, H. (1995) Brown rot. In: Ogawa, J.M., Zehr, E.I., Bird, G.W., Ritchie, D.F., Uriu, K. and Uyemoto, J.K. (eds) *Compendium of Stone Fruit Diseases*. APS Press, St Paul, Minnesota, pp. 7–10.

Ogawa, J.M., Manji, B.T., Bostock, R.M., Canez, V.M. and Bose, E.A. (1984) Detection and characterization of benomyl-resistant *Monilinia laxa* on apricots. *Plant Disease* 68, 29–31.

Phillips, D.J. and Harvey, J.M. (1975) Selective medium for detection of inoculum of *Monilinia* spp. on stone fruits. *Phytopathology* 65, 1233–1236.

Russell, P.E. (2004) Sensitivity baselines in fungicide resistance research and management (http://www.frac.info/frac/publication/anhang/monograph3.pdf, accessed 25 October 2010). CropLife International, Brussels.

Schnabel, G., Bryson, P.K., Bridges, W. and Brannen, P.M. (2004) Reduced sensitivity in *Monilinia fructicola* to propiconazole in Georgia and implications for disease management. *Plant Disease* 88, 1000–1004.

Schnabel, G., Hudson, S. and Bridges, W. (2009) Preharvest fungicide programs for pre- and postharvest brown rot control in nectarine and peach. *Plant Disease Management Reports.* Online. 3, STF002.

Schnabel, G., Amiri, A. and Brannen, P. (2010) Sustainable brown rot management of peaches in the southeastern United States. *Outlooks on Pest Management* 21, 208–211.

Sonoda, R.M., Ogawa, J.M., Manji, B.T., Shabi, E. and Rough, D. (1983) Factors affecting control of blossom blight in a peach orchard with low level benomyl-resistant *Monilinia fructicola*. *Plant Disease* 67, 681–684.

Wade, M. (1988) Strategies for preventing or delaying the onset of resistance to fungicides and for managing resistance occurrences. In: Delp, C.J. (ed.) *Fungicide Resistance in North America*. APS Press, St Paul, Minnesota, pp. 14–15.

Whan, J.H. (1976) Tolerance of *Sclerotinia fructicola* to benomyl. *Plant Disease Management Reporter* 60, 200–201.

Williams-Woodward, J.L. (2004) 2003 Georgia plant disease loss estimates. Bulletin 41. University of Georgia, Athens, Georgia.

Yoshimura, M.A., Luo, Y., Ma, Z. and Michailides, T.J. (2004) Sensitivity of *Monilinia fructicola* from stone fruit to thiophanate-methyl, iprodione, and tebuconazole. *Plant Disease* 8, 373–378.

Zehr, E.I., Toler, J.E. and Luszcz, L.A. (1991) Spread and persistence of benomyl-resistant *Monilinia fructicola* in South Carolina peach orchards. *Plant Disease* 75, 590–593.

Zehr, E., Luszcz, L.A., Olien, W.C., Newall, W.C. and Toler, J.E. (1999) Reduced sensitivity in *Monilinia fructicola* to propiconazole following prolonged exposure in peach orchards. *Plant Disease* 83, 913–916.

11 Fungicide Resistance in Oomycetes with Special Reference to *Phytophthora infestans* and Phenylamides

Dietrich Hermann and Ulrich Gisi

Syngenta Crop Protection, Biology Research Centre, Stein, Switzerland

11.1 Introduction

The control of diseases caused by plant pathogens of the Oomycetes has been a major target since the beginning of modern chemical crop protection. A wide range of fungicides is available and new products are being introduced to the market at regular intervals. Older products comprise several multi-site fungicides (e.g. mancozeb, folpet, copper formulations) and single-site inhibitors such as cyanoacetamide-oximes (like cymoxanil), carbamates (like propamocarb), phosphonates (like fosetyl-Al), phenylamides (e.g. metalaxyl-M), dinitroanilines (like fluazinam), quinone outside inhibitors (QoIs), strobilurins (e.g. azoxystrobin, famoxadone, fenamidone), toluamides (like zoxamid) and the first carboxylic acid amides (CAAs; like dimethomorph). In the latter fungicide group, new representatives have been introduced in recent years, such as iprovalicarb, benthiavalicarb, mandipropamid and flumorph. In the past few years, new modes of action have become available, including quinone inside inhibitors (QiIs; e.g. cyazofamid, amisulbrom), benzamides like fluopicolide and a new respiration inhibitor, ametoctradin, with unknown mode of action. In terms of the fungicide market for the control of Oomycetes, the most important foliar pathogen species are *Phytophthora infestans* (potato and tomato late blight), *Plasmopara viticola* (grape downy mildew), *Pseudoperonospora cubensis* (cucumber downy mildew), *Bremia lactucae* (lettuce downy mildew), *Peronospora* spp. (downy mildews on different crops such as peas, brassicas, tobacco) and soilborne pathogens of the genus *Pythium* (e.g. in turf, carrots, ornamentals).

The phenylamide fungicides include compounds such as metalaxyl, metalaxyl-M (mefenoxam), furalaxyl, benalaxyl, benalaxyl-M (kiralaxyl), ofurace and oxadixyl. Metalaxyl was introduced into the market in 1977 by Ciba-Geigy (now Syngenta) and became the most important compound of its class in this market segment. The phenylamides are a highly active class of fungicides, specifically controlling plant pathogenic Oomycetes (Peronosporales and Sclerosporales, as well as most members of the Pythiales) (Gisi, 2002). They penetrate the plant tissue rapidly, are translocated mainly in the apoplast within the plant and inhibit RNA synthesis in oomycete pathogens as primary target. The unique properties of phenylamide fungicides, such as the control of all members of the Peronosporales and most Pythiales, the long-lasting preventive and curative activity, the high systemicity and the excellent safety profile have been reviewed by Gisi and Ziegler (2002) and Müller and Gisi (2012). In 1996, the commercial introduction of the more active enantiomer of metalaxyl, metalaxyl-M (mefenoxam), was announced. By introducing a pure enantio-mer, replacing the racemic metalaxyl, a new chapter in both the control of Oomycetes by phenylamide fungicides and the use of chiral crop protection agents in general was opened. In all applications, such as foliar, soil or seed treatment, the outstanding level of control by metalaxyl-M is achieved at up to half the rate of its predecessor, metalaxyl (Nuninger *et al.*, 1996). Metalaxyl-M is used mostly in mixture with multi-site fungicides (e.g. mancozeb) to enlarge the spectrum of activity to non-oomycete diseases and to cope with the evolution of resistance (Gisi, 2002). The concept of using chiral phenylamides was also taken up by Isagro, who announced benalaxyl-M (kira-laxyl) in 2007 (Anon., 2009).

11.2   Mode of Action and Mechanism of Resistance for Phenylamide Fungicides

The phenylamide fungicides inhibit ribo-somal RNA synthesis, specifically RNA polymerization (polymerases). In mycelium of *Phytophthora megasperma*, metalaxyl affected the polymerase I complex of rRNA synthesis, which was considered as the primary site of action (Davidse, 1995). Endogenous RNA polymerase activity of isolated nuclei of *P. megasperma* and *P. infestans* was highly sen-sitive to metalaxyl, unless the isolates were from resistant isolates, suggesting that a mutation in the target site was responsible for resistance (Davidse, 1988). This hypothesis was further supported by the observation that (^{3}H)-metalaxyl bound to cell-free mycelial extracts of metalaxyl-sensitive but not-resistant isolates (Davidse, 1988). A secondary site of action was postulated for benalaxyl and the chemically related herbicides propachlor and metolachlor by Gozzo *et al.* (1984) and Davidse (1995), and again for kiralaxyl in 2009 (Anon., 2009). An increasing leakage of amino acids from zoospores was observed in the presence of kiralaxyl. However, this effect on membrane function occurred at much higher fungicide concentrations than those needed for RNA inhibition, suggesting that it might be an unspecific secondary rather than a primary

activity. Phenylamide fungicides especially affect hyphal growth and the formation of haustoria and sporangia in Oomycetes (Schwinn and Staub, 1995). Shortly after the commercial use of metalaxyl, resistant isolates were detected in *P. cubensis*, *P. infestans*, *Peronospora tabacina* and *P. viticola* (Gisi and Cohen, 1996). In most cases, this was coupled to a decline in disease control. As a conse-quence, strict recommendations for the use of phenylamides were designed and enforced by the PA-WG of FRAC (Phenylamide Working Group within the Fungicide Resistance Action Committee) to prevent and further delay resistance evolution (Urech and Staub, 1985). These recommendations have been imple-mented successfully and products containing phenylamides remain important fungicides, offering specific advantages for the control of diseases caused by Oomycetes, although resistant isolates can be found in all regions of the world and on many crops.

Phenylamide resistance has been described as monogenic. The majority of the F1 progeny produced from metalaxyl-resistant (r) and metalaxyl-sensitive (s) parental iso-lates of *P. infestans* had an intermediate sen-sitivity to metalaxyl. Crosses between two F1 isolates with intermediate sensitivity yielded a 1s:2i:1r ratio of progeny in the F2 generation (Shattock, 1986). This Mendelian segregation pattern reflects a single-gene (monogenic) resistance (Shattock, 1988) based on an incompletely dominant gene (Shaw and Shattock, 1991). Resistance to metalaxyl was also reported as being con-trolled by a single incompletely dominant gene in *Phytophthora capsici* (Lucas *et al.*, 1990), *Phytophthora sojae* (Bhat *et al.*, 1993) and *B. lactucae* (Crute and Harrison, 1988). However, a continuous sensitivity segrega-tion pattern was observed in the F1 genera-tion received from r × s crosses of European and Mexican *P. infestans* parents, suggesting that one semi-dominant locus, together with several minor loci, might be involved in resistance (Fabritius *et al.*, 1997). Resistance in these isolates was associated with two loci, *MEX1* and *MEX2*, the second locus mapping to the same linkage group as *MEX1* but to a distinct site (Judelson and Roberts, 1999). Although many investigations on the mode of action and mechanism of resistance to phenylamide fungicides have been undertaken over the past 30 years, the responsible resistance gene(s) and the site of mutation(s) in the genome have not yet been elucidated. Resistance has evolved in all economically important plant pathogens of the Oomycetes and is widespread (Table 11.1). Although phenylamides are considered to bear a high intrinsic resistance risk (Gisi and Staehle-Csech, 1988), they have failed to eliminate fully the sensitive subpopulations from nature, even after 30 years of intensive use (Gisi, 2002). The proportion of resistant isolates in *P. infestans* and *P. viticola* fluctuates from year to year and also within seasons (Gisi and Ohl, 1994). It increases within a season, more rapidly in fields treated with phenylamides than in untreated fields, starts to decline at the end of the season and is significantly lower at the

beginning of the next season compared with the proportion at the end of the previous year (Gisi and Cohen, 1996).

11.3 FRAC Use Recommendations for Phenylamide Fungicides

The Phenylamide (PA) Working Group was formed in 1982 and held annual meetings until 1997. In 1999, the FRAC Steering Committee agreed to transform the PA Working Group to a PA Expert Forum, organizing meetings only on a request-and-need basis expressed by more than one member company, but maintaining an 'information desk' for phenylamide-related matters handled by the chairman of the forum. Information on product performance and sensitivity of field populations would be made available through the FRAC website. The use strategies

Table 11.1. Comparison of resistance mechanism for QoI, phenylamide (PA) and CAA fungicides.

	QoIs	PAs	CAAs
Biochemical mode of action	Single-site, respiration complex III, cytochrome *bc*1, Qo site	Single-site, r-RNS biosynthesis, polymerase I	Single-site, cellulose biosynthesis, cellulose synthase 3 (CesA3)
Resistance gene	Monogenic, cyt *b* gene	Monogenic, MEX I/II loci	Monogenic, *cesA3* gene
Mutations	G143A, F129L	Site unknown	G1105S, G1105V
Inheritance of resistance (s × r)	F1: sensitive or resistant Population: s:r ~ 1:1 mitochondrial, maternal	F1: intermediate F2: s:i:r ~ 1:2:1 semi-dominant	F1: sensitive F2: s:r ~ 3:1 recessive
Resistance in *P. infestans*	None	Widespread	None
Resistance in *P. viticola*	Frequent, widespread, persistent, evolution rapid, declining in absence of selection	Frequent, widespread, persistent, evolution rapid, fluctuating over season and years	Variable, regional, evolution slow, declining in absence of selection
Resistance risk	Moderate for *P. infestans*, high for *P. viticola*	High for all species	Low for *P. infestans*, moderate for *P. viticola*
Resistance management and product use	Restriction of application number, only in mixture	Restriction of application number, only in mixture	Restriction of application number, solo for *P. infestans*, but only in mixture for *P. viticola* control

for PAs have been well established and published (www.frac.info). These involve the
preventive use of pre-packed mixtures with
well-defined amounts of non-phenylamide
fungicides for foliar application, a limited
number of applications per crop and per
season and no soil use for the control of
airborne pathogens. Current sensitivity monitoring data are produced by only a few
research groups and the presence of resistant
subpopulations is well known in several plant
pathogens on a range of crops worldwide
(www.frac.info). Sensitive subpopulations
have not disappeared, even though
PA-containing products have been used over
the past 30 years. This strongly suggests that the
recommended anti-resistance strategies are
successful and that resistance is not fixed perma-
nently in the pathogen genome. Sampling and
sensitivity test methods have been published
through FRAC (Gisi, 1992).

11.4   Sensitivity and Genotype Structure of Field Populations

When the sensitivity of *P. infestans* isolates
from culture collections, including representatives from the 1970s, were tested in the early
1990s, PA-resistant individuals were detected
(Daggett *et al.*, 1993), suggesting that PA
resistance had existed already at low proportions in wild-type populations before PA fungicides were used commercially (1977/78).
Thus, recurrent mutations give rise to resistant individuals at different locations and time
periods (Gisi and Cohen, 1996). In fact, resistant isolates of *P. infestans*, *P. viticola* and
P. cubensis had been detected already in field
populations 2 years after introduction into
the market of solo PA products which obviously had been selected through the use of
PAs, increased in frequency, survived and
then later migrated to other places. They can
compete with sensitive isolates even in the
absence of PAs. Therefore, resistant isolates
can be detected in current populations in
fields treated or not treated with PAs. In most
cases, mixed populations can be controlled
adequately by PA-containing products if the
proportion of resistant isolates is not too high

and if the number of applications is limited
(see use recommendations).

Potato and tomato late blight (*Phytophthora infestans*)

The dynamics of resistance evolution are
nowadays driven only partly by selection
through PA fungicides (number of applications per season has declined to around one);
equally important are dissemination and
migration of isolates in infected plant tissue
(seed potato, tomato seedlings). In many
European countries, the mating type distribution has changed in recent years from mainly
A1 in the 1990s to a majority of A2 (Gisi *et al.*,
2011). For unknown reasons, the mating type
structure can change within a few years from
dominance of one type during several years
(e.g. A2) followed by years favouring the
opposite type (e.g. A1), as has been described
for Israel (Cohen, 2002). No resistant isolates
have been detected for fluazinam and CAA
fungicides such as mandipropamid, whereas
the proportion of isolates resistant to PA fungicides (e.g. mefenoxam) is nowadays high
(about 65–85%) in north-western Europe (e.g.
UK, France, Germany, Belgium, Netherlands
and Switzerland) and has increased over the
past few years (since 2003; Gisi *et al.*, 2011).
However, PA resistance is rather low (<30%)
in Northern Europe (Denmark, Sweden,
Norway and Finland; A. Hannukkala, Finland,
2010, personal communication), Israel (Cohen,
2002) and Japan (Akino *et al.*, 2009). There is
no genetic link between mating type and PA
resistance; the two traits do not co-segregate
(Gisi and Cohen, 1996). Aggressiveness of
P. infestans isolates is often slightly higher for
PA-resistant compared to -sensitive isolates
and for isolates collected later compared to
earlier in the same season (Gisi *et al.*, 2011).
It is about equally high for A1 and A2 types
and for isolates collected in the late 1990s and
in 2007 (French and British isolates). Six different SSR (simple sequence repeat) genotype
families can be distinguished in European
populations. In 1997, populations were dominated by genotype families I, III and IV, which
declined significantly in 2007. They were

largely displaced by genotype families II ('blue 13' type) and V, which are by coincidence mainly A2/PA resistant and A1/PA sensitive, respectively (Gisi *et al.*, 2011). The appearance and dominance of new *P. infestans* genotypes in recent European populations (e.g. 'blue13' isolates, A2/PA resistant; Lees *et al.*, 2009) and some parts of the USA (US-22 isolates, mostly A2/PA sensitive; Ristaino, 2010) may be a result of migration (import of infected seed tubers) rather than emergence from local sexual recombination and selection through fungicide use.

A much more balanced and diverse genotype distribution is present in Swedish *P. infestans* populations, where sexual recombination seems to be frequent (B. Andersson, Sweden, 2010, personal communication). On the other hand, a rather high proportion of triploid isolates (outcrosses of diploid and tetraploid parents) and a mainly clonal structure of populations have been found in northwestern Europe (Gisi *et al.*, 2011), suggesting that such isolates may not survive for long during the process of local evolution. In addition, self-fertilization through hormonal induction and parasexual recombination with nuclear fusion in hyphae and sporangia occur quite frequently in *P. infestans*, resulting in heterokaryosis (different nuclear types in one individual, e.g. 2n, 3n, 4n, 2 × [0.75–1.75] n), triploid offspring and chromosomal abnormalities (Carter *et al.*, 1999; Catal *et al.*, 2010). Such processes may also allow the pathogen to adapt to new environmental conditions

without the need for sexual recombination. As a consequence, and in contrast to *P. viticola*, isolates with intermediate sensitivity to PA fungicides are quite rare in *P. infestans* populations. Another peculiar feature of *P. infestans* is the complete lack of fungicide resistance outside the phenylamides (e.g. QoIs, CAAs, cymoxanil; Table 11.2), whereas in *P. viticola*, resistance has evolved against almost all single-site fungicides (Kuck and Russell, 2006). In addition, *P. infestans* also can attack, outside potato, other solanaceous host plants like tomato and hairy nightshade, offering additional survival and inoculum sources. Although populations from tomato and potato can be separated from each other phenotypically and genotypically, there is a certain proportion of overlap in field populations (Knapova and Gisi, 2002). The aggressiveness of isolates is highest on the host of their origin and can be significantly lower for isolates collected from potato when tested on tomato.

Grape downy mildew (*Plasmopara viticola*)

The proportion of PA-sensitive *P. viticola* isolates has remained important and more or less stable for many years, especially in Italy and France. Since *P. viticola* undergoes sexual recombination every winter, the genetic diversity of the primary inoculum is very high and resistance is inherited according to Mendelian rules, i.e. all F1 progeny isolates

Table 11.2. Biological and molecular properties influencing evolution of fungicide resistance in pathogen species of Oomycetes.

Property	*P. infestans*	*P. viticola*	*P. cubensis*
Dynamics of epidemics	+++	+++	+++
Genetic diversity in populations (SSR genotypes)	+Europe; ++: SE, NL, MX	+++	++(?)
Frequency of sexual recombination	+	+++	+/−(?)
SNPs/diversity in genes (*CesA3*)	+/−	+++	++(?)
Dissemination of spores in air	++	+	+++
Spreading of inoculum (in plants)	+++	−	+
Cultivar resistance to pathogen	++	+	+++
Pathogen resistance to fungicides	−; (+++: PAs)	+++	+++

Note: Country codes: MX, Mexico; SE, Sweden; NL, The Netherlands. Symbols describe the intensity of property on resistance evolution from strong (+++) to none −; (?) = unknown.

are intermediate (i) in sensitivity; the proportion of sensitive, intermediate and PA-resistant isolates in F2 progeny is theoretically 1:2:1 (in sexual crosses 1:3:2; Gisi *et al.*, 2007). Therefore, the proportion of intermediates may be high in field populations. The sensitivity of populations fluctuates from year to year and within a season (Gisi, 2002). Sensitive, intermediate and resistant isolates can be detected in fields that have or have not been treated with PAs and are in a 'dynamic equilibrium' with each other. The dynamics of resistance evolution are driven not only by selection through PA fungicides, but also equally important are the Mendelian type of inheritance and the genetic background of resistance, as well as fitness and migration of isolates.

In contrast to the Mendelian inheritance of PA resistance, QoI resistance is encoded by the mitochondrial cyt *b* gene, resulting in 1:1 (s:r) segregation in unselected populations (Table 11.1). In fact, QoI resistance is rather high in many European *P. viticola* populations. Resistance to CAA fungicides is inherited by one recessive nuclear gene, resulting in a completely sensitive segregation in F1 and a 3:1 distribution of s:r progeny in F2 (Table 11.1), suggesting a much slower resistance evolution in field populations.

Other downy mildews

The presence of PA-resistant isolates in field populations has been confirmed in several other oomycete pathogens including *P. cubensis* (e.g. Israel, USA, Australia), *P. tabacina* (e.g. USA), *Phytophthora pisi* (e.g. New Zealand), *B. lactucae* (e.g. USA, UK, Italy), *Pythium* spp. (turfgrass in USA) and other pathogens on a range of crops in several countries (Gisi, 2002; list of resistant plant pathogenic organisms on www.frac.info). Resistance levels are not uniform and do not necessarily correlate with performance problems. Resistance to QoIs and CAAs has also been reported for *P. cubensis* in several countries. It is not known why fungicide resistance can evolve so quickly in this pathogen, although sexual recombination seems to be rare (Table 11.2).

Resistance risk

The relative resistance risk for different fungicides active against *P. viticola* and *P. infestans* can be estimated as follows:

Plasmopara viticola

Phenylamides ~ QoIs > CAAs ~ cymoxanil ~ fluopicolide >> zoxamide ~ fosetyl-Al > multi-sites (e.g. folpet, mancozeb).

Phytophthora infestans

Phenylamides >> fluopicolide > QoIs ~ QiIs > CAAs ~ cymoxanil ~ fluazinam ~ propamocarb ~ zoxamide > multi-sites (e.g. mancozeb).

Underlined are fungicide groups against which resistant field isolates are known. If no resistant field isolates exist, resistance risk is estimated based on mechanistic arguments for the respective mode of action. The estimated relative pathogen risk for resistance evolution in different Oomycetes is likely to decline in the following sequence of species:

> *P. viticola* > *P. cubensis* > *B. lactucae* ~
> *Peronospora* spp. > *Phytophthora* spp. with
> balanced sex (e.g. *P. capsici*) >
> *Phytophthora* spp. with unbalanced sex
> (e.g. *P. infestans*) >> *Pythium* spp.

Underlined are pathogen species with frequent sexual recombination. From the above list, it becomes clear that the highest overall resistance risk is represented by PA and QoI fungicides in *P. viticola*. Fungicide resistance in *P. infestans* seems to be a rather rare event, except for PAs (Table 11.2). Therefore, FRAC has regrouped *P. infestans* from formerly high to now medium risk in terms of fungicide evolution, although epidemics develop as dynamically in *P. infestans* as in *P. viticola*, which is classified as a high-risk pathogen.

11.5   Conclusions

Phenylamide (PA) fungicides with metalaxyl-M (mefenoxam) as the major representative of the FRAC group A1 (No. 4) compounds inhibit polymerase I in rRNA biosynthesis, but their molecular mode of action is not elucidated, although they have been in commercial use for more than 30 years. Widespread resistance has evolved in all major pathogen species of the Oomycetes, including *P. infestans*, shortly after PAs were introduced into the market. Although resistance risk for PAs is estimated as high (monogenic resistance, single-step selection), sensitive individuals have not been eliminated in populations by the use of PAs, partly because sound anti-resistance strategies have been established including restriction of application number to two per season (nowadays), preventive use and availability in preformulated mixtures only (e.g. with mancozeb). The dynamics of resistance in recent *P. infestans* populations is driven only partly by the selection process through fungicide use; more important are probably the dissemination and import of infected plant material such as seed potato and tomato seedlings, overwintering on alternative hosts and, in some countries, on dumps and leftover tubers. In a few countries (e.g. Sweden, Netherlands, Mexico),

the pathogen population contains a more or less equilibrated proportion of both A1 and A2 mating types, allowing sexual recombination resulting in a rather diverse genotype distribution. However, in north-western Europe, the current populations are more or less clonal with a few frequent genotypes dominating the population (e.g. 'blue13' type) which are, by coincidence, A2 and PA resistant. This distribution can change in a few years, as has occurred in the past in Israel and Japan with dominance of A1/PA-sensitive isolates or in some parts of the USA with new A2/PA-sensitive isolates. The pathogen can also undergo self-fertilization and parasexual recombination with nuclear fusion, resulting in heterokaryosis (different nuclear types in one individual), which may allow adaptation to new environmental conditions without the need for sexual recombination. However, resistance in *P. infestans* has evolved in 'only' one out of ten fungicide groups (PAs), in contrast to the behaviour of other Oomycetes such as *P. viticola*.

References

Akino, S., Kato, M. and Kondo, N. (2009) Recent genotypic changes in *Phytophthora infestans* in Japan (1997–2007). In: Schepers, H.T.A.M. (ed.) *Proceedings 11th Workshop European Network for Development of an Integrated Control Strategy of Potato Late Blight*. PPO Special Report Volume 13. Hamar, Norway, pp. 45–52.

Anon. (2009) Kiralaxyl. *Technical Bulletin*. Isagro, 25 pp.

Bhat, R.G., McBlain, B.A. and Schmitthenner, A.F. (1993) The inheritance of resistance to metalaxyl and to fluorophenylalanine in matings of homothallic *Phytophthora sojae*. *Mycological Research* 97, 865–870.

Carter, D.A., Buck, K.W., Archer, S.A., Van der Lee, T., Shattock, R.C. and Shaw, D.S. (1999) The detection of non-hybrid, trisomic and triploid offspring in sexual progeny of a mating of *Phytophthora infestans*. *Fungal Genetics and Biology* 26, 198–208.

Catal, M., King, L., Tumbalam, P., Wiriyajitsomboon, P., Kirk, W.W. and Adams, G.C. (2010) Heterokaryotic nuclear conditions and a heterogeneous nuclear population are observed by flow cytometry in *Phytophthora infestans*. *Cytometry Part A* 77A, 769–775.

Cohen, Y. (2002) Populations of *Phytophthora infestans* in Israel underwent three major genetic changes during 1983 to 2000. *Phytopathology* 92, 300–307.

Crute, I.R. and Harrison, J.M. (1988) Studies on the inheritance of resistance to metalaxyl in *Bremia lactucae* and on the stability and fitness of field isolates. *Plant Pathology* 37, 231–250.

Daggett, S.S., Götz, E. and Therrien, C.D. (1993) Phenotypic changes in populations of *Phytophthora infestans* from eastern Germany. *Phytopathology* 83, 319–323.

Davidse, L.C. (1988) Phenylamide fungicides: mechanism of action and resistance. In: Delp, C.J. (ed.) *Fungicide Resistance in North America*. APS Press, St Paul, Minnesota, pp. 63–65.

Davidse, L.C. (1995) Phenylamide fungicides: biochemical action and resistance. In: Lyr, H. (ed.) *Modern Selective Fungicides*, 2nd edn. Gustav Fischer, Jena, Germany, pp. 347–354.

Fabritius, A.L., Shattock, R.C. and Judelson, H.S. (1997) Genetic analysis of metalaxyl insensitivity loci in *Phytophthora infestans* using linked DNA markers. *Phytopathology* 87, 1034–1040.

FRAC www.frac.info.

Gisi, U. (1992) FRAC methods for monitoring the sensitivity of fungal pathogens to phenylamide fungicides. PA-FRAC of GIFAP. *EPPO Bulletin* 22, 297–322.

Gisi, U. (2002) Chemical control of downy mildews. In: Spencer, P.T.N., Gisi, U. and Lebeda, A. (eds) *Advances in Downy Mildew Research.* Kluwer, Dordrecht, The Netherlands, pp. 119–159.

Gisi, U. and Cohen, Y. (1996) Resistance to phenylamide fungicides: a case study with *Phytophthora infestans* involving mating type and race structure. *Annual Review of Phytopathology* 34, 549–572.

Gisi, U. and Ohl, L. (1994) Dynamics of pathogen resistance and selection through phenylamide fungicides. In: Heaney, S., Slawson, D., Hollomon, D.W., Smith, M., Russell, P.E. and Parry, D.W. (eds) *Fungicide Resistance.* BCPC Monograph No. 60. BCPC, Farnham, UK, pp. 139–146.

Gisi, U. and Staehle-Csech, U. (1988) Resistance risk evaluation of phenylamide and EBI fungicides. *Proceedings Brighton Crop Protection Conference,* Volume 1. British Crop Protection Council, Thornton Heath, Surrey, UK, pp. 359–366.

Gisi, U. and Ziegler, H. (2002) Fungicides, phenylamides, acylalanines. In: Plimmer, J.R. (ed.) *Encyclopedia of Agrochemicals,* Volume 2. John Wiley, New York, pp. 609–616.

Gisi, U., Waldner, M., Kraus, N., Dubuis, P.H. and Sierotzki, H. (2007) Inheritance of resistance to carboxylic acid amide (CAA) fungicides in *Plasmopara viticola. Plant Pathology* 56, 199–208.

Gisi, U., Walder, F., Resheat Eini, Z., Edel, D. and Sierotzki, H. (2011) Changes of genotype, sensitivity and aggressiveness in *Phytophthora infestans* isolates collected in European countries in 1997, 2006 and 2007. *Journal of Phytopathology* 159, 223–232.

Gozzo, F., Garavaglia, C. and Zagni, A. (1984) Structure–activity relationships and mode of action of acylalanines and related structures. *Proceedings Brighton Crop Protection Conference,* Volume 3. British Crop Protection Council, Thornton Heath, Surrey, UK, pp. 923–928.

Judelson, H.S. and Roberts, S. (1999) Multiple loci determining insensitivity to phenylamide fungicides in *Phytophthora infestans. Phytopathology* 89, 754–760.

Knapova, G. and Gisi, U. (2002) Phenotypic and genotypic structure of *Phytophthora infestans* populations on potato and tomato in France and Switzerland. *Plant Pathology* 51, 641–653.

Kuck, K.H. and Russell, P.E. (2006) FRAC: combined resistance risk assessment. In: Bryson, R.J., Burnett, F.J., Foster, V., Fraaije, B.A. and Kennedy, R. (eds) *Fungicide Resistance: Are We Winning the Battle but Losing the War?* AAB Conference Proceedings, Edinburgh, UK. *Aspects of Applied Biology* 78, 3–10.

Lees, A.K., Cooke, D.E.L., Stewart, J.A., Sullivan, L., Williams, N.A. and Carnegie, S. (2009) *Phytophthora infestans* population changes: implications. In: Schepers, H.T.A.M. (ed.) *Proceedings 11th Workshop European Network for Development of an Integrated Control Strategy of Potato Late Blight.* PPO Special Report Volume 13. Hamar, Norway, pp. 55–60.

Lucas, J.A., Greer, G., Oudemans, P.V. and Coffey, M.D. (1990) Fungicide sensitivity in somatic hybrids of *Phytophthora capsici* by protoplast fusion. *Physiological and Molecular Plant Pathology* 36, 175–187.

Müller, U. and Gisi, U. (2012) Newest aspects of nucleic acid synthesis inhibitors – metalaxyl-M. In: Krämer, W., Jeschke, P. and Witschel, M. (eds) *Modern Crop Protection Compounds,* 2nd edn. Wiley-VCH, Weinheim, Germany (in press).

Nuninger, C., Watson, G., Leadbitter, N. and Ellgehausen, H. (1996) CGA 329351: introduction of the enantiomeric form of the fungicide metalaxyl. *Proceedings Brighton Crop Protection Conference,* Volume 1. British Crop Protection Council, Thornton Heath, Surrey, UK, pp. 41–46.

Ristaino, J.B. (2010) The 2009 potato and tomato late blight epidemics: genealogical history, multiple sources and migration events. Contribution 5-S (special session), APS 2010 Annual Meeting, Charlotte, USA (abstract).

Schwinn, F. and Staub, T. (1995) Oomycetes fungicides: phenylamides and other fungicides against Oomycetes. In: Lyr, H. (ed.) *Modern Selective Fungicides,* 2nd edn. Gustav Fischer, Jena, Germany, pp. 323–346.

Shattock, R.C. (1986) Inheritance of metalaxyl resistance in the potato late blight fungus. *Proceedings Brighton Crop Protection Conference,* Volume 2. British Crop Protection Council, Thornton Heath, Surrey, UK, pp. 555–560.

Shattock, R.C. (1988) Studies on the inheritance of resistance to metalaxyl in *Phytophthora infestans. Plant Pathology* 37, 4–11.

Shaw, D.S. and Shattock, R.C. (1991) Genetics of *Phytophthora infestans*: the Mendelian approach. In: Lucas, J.A., Shattock, R.C., Shaw, D.S. and Cooke, L.R. (eds) *Phytophthora.* Cambridge University Press, Cambridge, UK, pp. 218–230.

Urech, P.A. and Staub, T. (1985) The resistance strategy for acylalanine fungicides. *EPPO Bulletin* 15, 539–543.

12 Resistance Risk to QoI Fungicides and Anti-Resistance Strategies

Andy Leadbeater

Syngenta Crop Protection AG, CH-4058 Basel, Switzerland

12.1 Introduction

The QoI (quinone outside inhibitor) class of fungicides represents one of the most widely used groups of fungicides for the control of important agricultural fungal pathogens. In addition to the strobilurins, the group includes famoxadone, fenamidone and pyribencarb, all chemically distinct from the strobilurins but with the same mode of action and contained within the same cross-resistance group. Indeed, eight chemical classes of Qo inhibitors have been identified to date (Table 12.1). The strobilurins were first commercialized in 1996 with the launch of azoxystrobin and kresoxim-methyl, and since this date there has been a steady stream of new QoI fungicides launched, with several more identified in company pipelines. There was a rapid uptake in the market of the strobilurins and within only a few years they were registered and sold in over 80 crops in 70 different countries (Bartlett *et al.*, 2002). Sales of strobilurins and other QoI fungicides amounted to approximately US$2.7 billion in 2008, representing 20–25% of the total global sales of fungicides. Sales of azoxystrobin alone in 2008 were approximately US$900 million, making it the world's biggest selling fungicide.

The discovery of the strobilurins was inspired by a group of natural fungicidal derivatives of β-methoxy-acrylic acid, the simplest of which were strobilurin A, oudemansin A and myxothiazole A. These derivatives are produced by the Basidiomycetes *Strobilurus*, *Mycena* and *Oudemansiella*, but were found unsuitable for commercial production as fungicides because of their photochemical instability and volatility. However, the knowledge of their structures and physical properties provided the starting point for many researchers to begin to optimize the chemistry towards commercially viable alternatives. Chemical synthesis programmes in the research and development based chemical companies, most notably Syngenta and BASF, resulted in the discovery of the first synthetic strobilurins azoxystrobin and kresoxim-methyl (Fig. 12.1), which subsequently were commercialized. A few years later, metominostrobin (Hayase *et al.*, 1995; Clough and Godfrey, 1998), trifloxystrobin (Margot *et al.*, 1998), picoxystrobin (Godwin *et al.*, 2000) and pyraclostrobin (Ammermann *et al.*, 2000) were announced. During the 1990s, famoxadone (Joshi and Sternberg, 1996) and fenamidone (Mercer *et al.*, 1998) were also discovered. More recently, pyribencarb (Kataoka *et al.*, 2010) has been announced and described as having the QoI mode of action. In addition, the fungicides coumoxystrobin, dicloaminostrobin, enoxastrobin, pyrametostrobin and pyraoxystrobin and triclopyricarb, all expected to be Qo inhibitors, have been reported by the

Table 12.1. The QoI fungicides developed by different agrochemical companies.

Class	Fungicide	Company	First sales
Methoxyacrylates	Azoxystrobin	Syngenta	1996
	Picoxystrobin	Syngenta	2002
	Enoxastrobin	SYRICI[a]	Ongoing
	Pyraoxystrobin	SYRICI[a]	Ongoing
	Coumoxystrobin	SYRICI[a]	Ongoing
Methoxycarbamates	Pyraclostrobin	BASF	2002
	Pyrametostrobin	SYRICI[a]	Ongoing
	Triclopyricarb	SYRICI[a]	Ongoing
Oximinoacetates	Kresoxim-methyl	BASF	1996
	Trifloxystrobin	Bayer Crop Science	1999
Oximinoacetamides	Metominostrobin	Shionogi	1999
	Dimoxystrobin	BASF	2006
	Dicloaminostrobin	SYRICI[a]	Ongoing
	Orysastrobin	BASF	2009
Oxazalidinediones	Famoxadone	Du Pont	1997
Dihydro-dioxazines	Fluoxastrobin	Bayer Crop Science	2005
Imidazolinones	Fenamidone	Bayer Crop Science	2001
Benzyl-carbamates	Pyribencarb	K-I Chemical	Ongoing

Note: [a]Shenyang Research Institute of Chemical Industry, China.

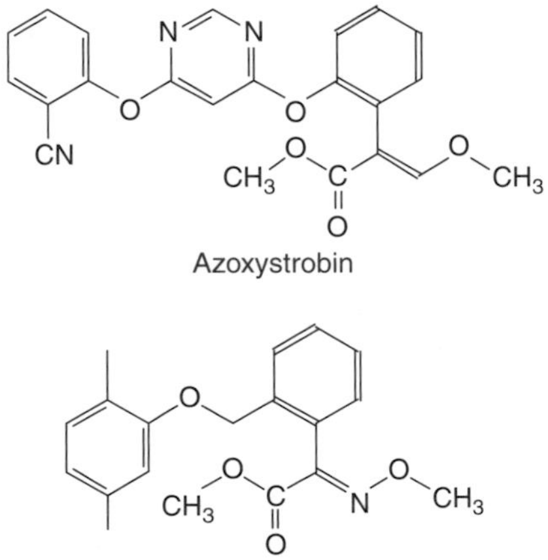

Azoxystrobin

Kresoxim-methyl

Fig. 12.1. Structures of the strobilurin fungicides azoxystrobin and kresoxim-methyl.

Shenyang Research Institute of Chemical Industry in China. The various QoI fungicides have very different physico-chemical properties, which confer different behaviours in the plant. For example, picoxystrobin is the most rapidly absorbed into plant tissue and the most xylem-systemic (Godwin et al., 2000). Azoxystrobin and metominostrobin are also xylem systemic. In contrast, kresoxim-methyl (Ammermann et al., 1992), trifloxystrobin and pyraclostrobin are all non-systemic.

12.2 Biochemical Mode of Action of the QoIs

The QoI fungicides are respiration inhibitors – their fungicidal activity comes from their ability to inhibit mitochondrial respiration by binding at the so-called Qo site (the outer quinol-oxidation site or ubiquinol site) of the cytochrome bc_1 enzyme complex (complex III), located in the inner mitochondrial membrane of fungi and other eukaryotes (Fig. 12.2). This inhibition blocks electron transfer between cytochrome b and cytochrome c_1, which, in turn, leads to an energy deficiency in fungal cells that halts reduced nicotinamide adenine dinucleotide (NADH) oxidation and adenosine triphosphate (ATP) synthesis, thereby preventing the production of ATP. This leads to the stopping of energy production and the fungus will eventually die. All the QoI fungicides share this common mode of action. The toxophore is similar in all compounds and always carries a carbonyl oxygen moiety, thought to be responsible for binding.

The fungicidal effect of the QoIs is particularly strong during the first period of the life cycle of fungi, i.e. at the spore germination stage. Spore germination and germ tube elongation of fungi are more susceptible to QoIs

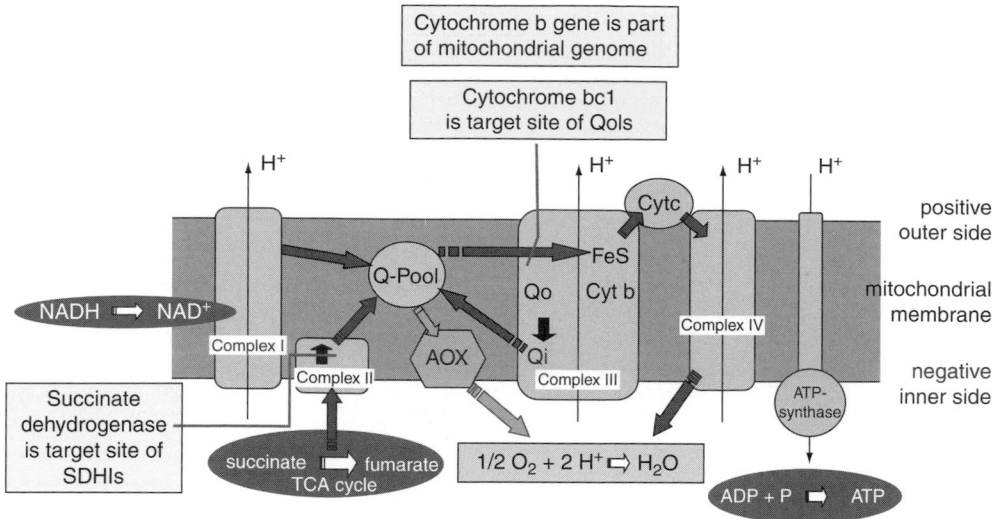

Fig. 12.2. The respiration chain and important inhibitors.
Source: U. Gisi.

than growth of fungal mycelium; therefore, they are highly effective against spore germination and early penetration of the host. Once the fungus is growing inside the leaf tissue, QoIs have only limited effect. For these reasons, the QoI fungicides are recommended to be used protectively rather than curatively or eradicatively, to maximize their effectiveness by targeting the optimum time of the pathogen life cycle as well as for resistance management purposes by gaining effective control of a smaller pathogen population. The spectrum of activity is very broad, including plant pathogen species of all major systematic groups such as Ascomycetes, Deuteromycetes, Basidiomycetes and Oomycetes.

12.3 Resistance Risk to the QoIs

With the mode of action of the QoI fungicides clearly being site specific, the question of fungicide resistance emerging in target diseases arose early in the development of these products. In order to assess the risk of practical resistance to a fungicide in the target pest(s), it is necessary to evaluate the different factors contributing to the risk, i.e. those inherent in the compound and its effect on the pest and those that might result from a particular use

pattern. The inherent resistance risk of a fungicide depends on various factors, some of which are associated with the product and others with the pest. Factors associated with the fungicide that have to be considered when assessing resistance risk include mode of action (single site, multi-site), type of resistance (monogenic or polygenic), persistence of activity and potency. Characteristics of the pathogen that have to be considered include genetic variability (high or low), generation time (short or long), inheritance of resistance and fitness of resistant individuals. The analysis of resistance risk to fungicides has been studied in detail by FRAC (Fungicide Resistance Action Committee) and other experts and is published comprehensively (Brent and Hollomon, 2007).

Resistance to strobilurin A, myxothiazol and stigmatellin, based on several point mutations in the cytochrome *b* gene, has been known in yeast since 1989 (di Rago *et al.*, 1989) and later reported in other species such as algae, bacteria, protozoa and mice (Bennoun *et al.*, 1991; Geier *et al.*, 1992; Degli Esposti *et al.*, 1993; Gennis *et al.*, 1993; Brasseur *et al.*, 1996). When the first synthetic strobilurins were developed, the significance of these findings for plant pathogens was probably underestimated. Production of stable mutations in yeast was unsuccessful, and several muta-tions were detected in many different posi-tions in the cytochrome *b* gene (including the relevant 129 and 143 positions), making it dif-ficult to predict which mutations might hap-pen in plant pathogens in nature. All mutations for several species exhibited a reduction in respiratory efficiency and were therefore all significantly less fit than the wild type. However, resistance factors were high-est for the mutation at position 143, G143A, in the cytochrome *b* gene.

The scientific findings showed clearly that it was possible to produce mutations conferring laboratory resistance to the Qo inhibitors in model systems and that there-fore it might be feasible for resistance to agri-cultural fungicides to occur in the field. However, there was no information on the real risk of significant mutations occurring in nature, of the likely fitness of any such resist-ant strains or of the potential for mutations conferring resistance to be fixed in field pop-ulations of plant pathogens. *In vivo* baseline sensitivity monitoring studies were carried out to establish the existing sensitivity levels within natural field populations of *Mycosphaerella graminicola* to QoIs using azoxystrobin (Godwin *et al.*, 1999). In baseline sensitivity monitoring, a normal distribution of LC_{50} and LC_{95} values was found with a rela-tively narrow range of sensitivity. In mutation studies, it was difficult to isolate stable mutants of *M. graminicola* to azoxystrobin, and the resistance factors of these mutants were low ($\times 5$ to $\times 20$). Target site resistance was considered to be the most likely route for the evolution of resistance to the QoIs, although other mechanisms for resistance were not discounted, such as detoxification, efflux of active ingredient and alternative mechanisms for respiration. Indeed, an alter-native oxidase mutant of *M. graminicola* with reduced sensitivity to azoxystrobin was iso-lated *in vitro*, but was found to be more sensi-tive *in vivo* to azoxystrobin than the wild types (Ziogas *et al.*, 1997). The mitochondrial site of action of QoIs originally was consid-ered less prone to resistance development than targets encoded by nuclear genes, with an expectation that resistance development should involve several genetic changes and

therefore resistance should evolve gradually. On the basis of the scientific evidence to hand at the time of market introduction, therefore, the likelihood of *M. graminicola* populations developing practical resistance to the QoI fungicides was judged to be moderate. This was based on the inherent properties of the fungicides, in combination with the medium-risk characteristics of these pathogens.

It should be noted that since the charac-teristics and mechanism of action of the stro-bilurins meant that clearly they were best used in protectant situations and tended to lack curative activity, they were recom-mended at an early stage of commercializa-tion to be used in programmes and mixtures with other fungicides with different charac-teristics and mode of action, such as the DMIs (triazoles). However, it was also very clear that these new fungicides were soon found to be extremely effective and therefore prone to being used by growers at rates significantly lower than those recommended by manufac-turers, with use rates of 50% or lower fre-quently reported.

To support the introduction of this important new class of fungicides into the market, the manufacturers (BASF, Syngenta and others later) initiated a worldwide moni-toring programme in key crops and diseases to provide data on the sensitivity status of fungal populations to the QoIs. This pro-gramme has continued to be carried out by the industry and by other collaborators ever since, resulting in a vast database of resist-ance knowledge.

12.4   History of Resistance to the QoI Fungicides

As already described earlier in this chapter, the strobilurin fungicides were introduced in 1996 and very rapidly became widely used in many crops around the world. The reason for this was their revolutionary levels of disease control and financial returns, and also their positive physiological effects in many crops, resulting in greener crops and increased yields. Possibly due to this rapid widespread use, resistance issues soon arose in some key diseases. In Europe, monitoring of the air spora of *Blumeria graminis* f. sp. *tritici* (wheat powdery mildew) in 1996 and 1997, conducted independently by members of the manufacturing industry, did not reveal the presence of any QoI-resistant strains anywhere across the continent. Further monitoring of the air spora throughout Europe during 1998, however, revealed the presence for the first time of resistant spores in northern Germany. Air spora sampled in southern Germany, France and the UK did not contain any resistant spores. It was reported at the time that the presence of such strains in northern Germany influenced the performance of strobilurins against this disease in a negative way, and should therefore be regarded as field resistance. Monitoring of other cereal pathogens in Europe in 1998, including *B. graminis hordei*, *M. graminicola* (= *Septoria tritici*), *Puccinia recondita* and *Pyrenophora teres* from air spora and/or leaf samples from trial sites did not reveal the presence of any strobilurin-resistant strains.

The development of the resistance situation in wheat powdery mildew in northern Germany was thought at the time to reflect a unique combination of factors leading to a significantly higher selection pressure for resistance.

During 1998, there were reports of resistance in cucurbit powdery mildew in Asia, a small number of resistant isolates were found in *Plasmopara viticola* at a trial site in South America and there were other reports of 'suspicious isolates' in *Venturia inaequalis* in the USA and *M. fijiensis* in Costa Rica. It was clear that action by industry and officials was required quickly and use recommendations were changed in recognition of the need both to manage resistance and to ensure lack of product failures in crops. A new working group was formed under the auspices of FRAC, the so-called STAR fungicides group. This was an acronym for STrobilurins And Related compounds. Cross-resistance studies clearly showed the existence of cross-resistance among all the strobilurins, and also with famoxadone and fenamidone (Chin *et al.*, 2000; Sierotzki *et al.*, 2000a). There was, however, no cross-resistance with other fungicides tested (DMI fungicides, the amines, cyprodinil or quinoxyfen). Later, the FRAC group

was renamed to reflect the mode of action, and is now called simply the QoI Working Group. This group considered the available data on resistance occurrence, cross-resistance patterns between QoIs, methodologies and product use (and abuse) history of the strobilurins. As a result, consolidated recommendations were produced for the 1999 season to cover all QoI fungicides. At the same time, the molecular mechanism of strobilurin resistance was elucidated for the first time in a plant pathogen, *B. graminis*, by Sierotzki *et al.* (2000b) (see below).

The FRAC QoI Working Group has worked with independent experts in the field of resistance and resistance management to monitor the spread and dynamics of resistance globally, and also to define and recommend resistance management strategies and tactics. The principles behind the resistance management recommendations have been to delay or prevent the appearance of resistance and also to ensure that when resistance does arise, the consequences in terms of field disease control are minimized. The success of this approach can be recognized by the fact that today the QoI fungicides continue to be the most successful fungicides in the market worldwide.

Over the past few years, the resistance management recommendations made by the FRAC Working Group have been adjusted based on field experience and advances in technology and the understanding of resistance. The basic recommendations valid today can be summarized as follows (FRAC, 2010):

* Fungicide programmes must deliver effective disease management. Apply QoI fungicide-based products at effective rates and intervals according to manufacturers' recommendations. Effective disease management is a critical component in delaying the build-up of resistant pathogen populations.

* The number of applications of QoI fungicide-based products in a total disease management programme must be limited, whether applied solo or in mixtures with other fungicides. This limitation is inclusive of all QoI fungicides. Limitation of QoI fungicides in a spray programme

provides time and space when the pathogen population is not influenced by QoI fungicide selection pressure.

- A consequence of the limitation of QoI fungicide-based products is the need to alternate them with effective fungicides from different cross-resistance groups.
- QoI fungicides containing only the solo product should be used in single or block applications in alternation with fungicides from a different cross-resistance group. Specific recommendations on the size of blocks are given for specific crops.
- QoI fungicides applied as tank mix or as a co-formulated mixture with an effective mixture partner should be used in single or block applications in alternation with fungicides from a different cross-resistance group. Specific recommendations on size of blocks are given for specific crops.
- Mixture partners for QoI fungicides should be chosen carefully to contribute to effective control of the targeted pathogen(s). The mixture partner must have a different mode of action and, in addition, it may increase the spectrum of activity or provide needed curative activity. Use of mixtures containing only QoI fungicides must not be considered as an anti-resistance measure. Where local regulations do not allow mixtures, then strict alternations with non-cross-resistant fungicides (no block applications) are necessary.
- An effective partner for a QoI fungicide is one that provides satisfactory disease control when used alone on the target disease.
- QoI fungicides are very effective at preventing spore germination and should therefore be used at the early stages of disease development (preventive treatment).

Following the confirmation of resistance to the QoI fungicides in *B. graminis* and some indications in other pathogens, an intensification in monitoring programmes for the key economic plant diseases was implemented, particularly in Europe. The first QoI-resistant

isolates of *M. graminicola* (= *S. tritici*) in wheat were detected retrospectively in the UK in 2001 at low frequency in QoI-treated research plots (Fraaije *et al.*, 2005) and subsequently in 2002 in five European countries (Gisi *et al.*, 2005). During 2003 and 2004, the frequency of resistant isolates increased rapidly in Northern Europe. This rapid increase of resistance was of great concern since the disease was considered the most economically damaging of cereal diseases in north-western Europe. The FRAC recommendations outlined above have been reconfirmed, strengthened and communicated so that QoIs are now recommended for use only in mixtures in cereals and with a limitation in spray numbers. Resistance in *M. graminicola* is now widespread across north-western Europe; as reported by frequent FRAC surveys, it exists at high levels across the entire UK, whereas in France and Germany resistance levels are higher in the north than in the south. The north–south gradient in resistance distribution is thought to be due to differences in the intensity of QoI use and lower disease pressure in southern regions.

Over time, resistance to QoI fungicides has been confirmed in many economically important pathogen species across the world. These occurrences are monitored by researchers, officials and by member companies of FRAC and reported (Table 12.2). It is interesting to note, however, that the occurrence, evolution, spread and significance of QoI resistance varies greatly with the pathogen and the geography involved. It has been shown that, in *M. graminicola*, resistance to the QoI fungicides has emerged simultaneously and independently at least four times in European populations rather than, as was first suspected, emerging in a single location and then spreading by migration to other countries (Torriani *et al.*, 2009). Similar findings have been reported for QoI resistance in *P. viticola*, showing that for this pathogen selected substitution conferring resistance to a fungicide has occurred independently several times (Chen *et al.*, 2007). It is known from intensive monitoring data of FRAC companies in Europe for *V. inaequalis* that in regions where resistance to QoI fungicides is present, the levels of resistance found are

Table 12.2. List of pathogens with field resistance towards QoI fungicides (FRAC, 2009, www.frac.info).

Species name	Host		Geographical distribution	Type of resistance
Alternaria alternata, Alternaria tenussima, Alternaria arborescens	Alternaria blight	Pistachio	USA	G143A
Alternaria alternata	Early blight	Potato and tomato	EU	G143A
Alternaria mali	Alternaria blotch	Apple	USA	G143A
Alternaria solani	Early blight	Potato	USA	F129L
Blumeria graminis f. sp. *tritici* and *hordei*	Powdery mildew	Wheat and barley	EU	G143A
Botrytis cinerea	Grey mould	Strawberries	EU	G143A
Cladosporium carpophilum	Scab	Almonds	USA	?
Colletotrichum graminicola	Anthracnose	Turf grass	USA	G143A
Corynespora cassiicola	Leaf spot, target spot	Cucumber	Japan	G143A
Didymella bryoniae	Gummy stem blight	Cucurbit	USA	G143A
Erysiphe necator	Powdery mildew	Grapes	USA, EU	G143A
Glomerella cingulata (= *Colletotrichum gloeosporioides*)	Anthracnose	Strawberries	Japan	G143A
Microdochium nivale Microdochium majus	Snow mould, ear blight	Wheat	EU	G143A
Mycosphaerella fijiensis	Black Sigatoka	Banana	Central and South America, CAM[a], PHIL[b]	G143A
Mycosphaerella graminicola	Septoria leaf spot	Wheat	EU	G143A
Mycosphaerella musicola	Yellow Sigatoka	Banana	South America, Australia	G143A
Mycovellosiella nattrassii	Leaf mould	Aubergine	Japan	G143A
Passalora fulva	Leaf mould	Tomato	Japan	F129L
Plasmopara viticola	Downy mildew	Grape	EU	G143A and F129L
Pseudoperonospora cubensis	Downy mildew	Cucurbits	EU, Asia	G143A
Pyrenophora teres	Net blotch	Barley	EU	F129L
Pyrenophora tritici-repentis	Tan spot	Wheat	EU	G143A, F129L and G137R
Pyricularia grisea	Grey leaf spot	Turf grass	USA	G143A and F129L
Pythium aphanidermatum	Pythium blight	Turf grass	USA	F129L
Ramularia areola	Grey mildew	Cotton	Brazil	?
Ramularia collo-cygni	Ramulariose	Barley	EU	G143A
Rhynchosporium secalis	Scald	Barley	EU	G143A
Sphaerotheca fuliginea	Powdery mildew	Cucurbits	EU, Asia	G143A
Stemphylium vesicarium	Brown spot	Asparagus and pear	EU	G143A
Venturia inaequalis	Scab	Apple	EU	G143A
Venturia pirina	Scab	Pears	USA	G143A

Note: [a]CAM, Cameroon; [b]PHIL, Philippines.

very heterogeneous, with values ranging from zero to high, even between neighbouring orchards. As another example, it was not clear why grape powdery mildew (*Erysiphe necator*), a pathogen classified as high risk for resistance development, did not exhibit resistance to the QoIs until 2002 in the USA (Wilcox *et al.*, 2003) and was not found in Europe until 2006, despite almost 10 years of commercial use. Although resistance to QoI fungicides has been found in *Pyricularia grisae* in turf, no resistance to the same pathogen in rice has been reported so far in Asia. And there remain several pathogens and many countries where resistance has not been found.

Important factors affecting the occurrence of fungicide resistance include disease pressure, product usage and fitness of resistant phenotypes. There is clear evidence that lower disease pressure, and therefore less intensive product use, has delayed the evolution of resistance to the QoI fungicides – an example was given previously with the north–south sensitivity distribution for *M. graminicola* in Germany and France. The fitness of pathogen isolates resistant to fungicides is significant in disease management. If fitness costs are associated with fungicide resistance, the frequency of resistant isolates will decline in the absence of fungicide. Fitness penalties have been observed in QoI-resistant populations of the oomycete *P. viticola* (Sierotzki *et al.*, 2008) and the rice pathogen *P. grisea* carrying the G143A substitution (Avila-Adame and Köller, 2003). Field populations of *P. viticola* collected between 2001 and 2003 in France and Italy and maintained on untreated plants for several generations reverted to full sensitivity over time (Genet *et al.*, 2006), which suggested that the resistant phenotypes were less fit than sensitive individuals. In contrast, fitness penalties are not apparent in the case of *B. graminis* (Heaney *et al.*, 2000), providing evidence that in this pathogen's QoI, resistance is stable in the absence of selection over several generations, and that resistant isolates appear to be equal in fitness to sensitive isolates found on untreated plants. Gisi *et al.* (2000, 2002), Ma and Michailides (2005) and Sierotzki *et al.* (2005) reported that the G143A mutation did not affect enzyme activity and did not compromise the fitness of a pathogen. It has

also been widely assumed that fitness costs are fixed. However, the cost of resistance to QoI fungicides seems to vary with environmental conditions such as temperature, being more costly under conditions that are suboptimal for the pathogen, as shown for *B. graminis* and *M. graminicola* (Brown *et al.*, 2006). The factors clearly act together to influence the occurrence and significance of resistance to the QoI fungicides in practical situations and help to explain why, despite high levels of resistance occurring in many situations, the QoI fungicides remain very important for disease control across the globe.

12.5 Mutations Associated with QoI Resistance

The molecular mode of resistance to QoI fungicides is now understood to an advanced level and at least 15 different point mutations have been described in the cytochrome *b* gene leading to resistance (Brasseur *et al.*, 1996). In field isolates of different plant pathogen species, the major basis of the resistance phenotype to QoI fungicides is a single amino acid substitution of glycine with alanine at position 143 (G143A) of the cytochrome b protein (Gisi *et al.*, 2002; Fig. 12.3). This amino acid substitution confers high levels of resistance.

Details of the nucleotide and amino acid sequences of the cyt *b* gene from a sensitive and a resistant isolate of *B. graminis* f. sp. *tritici* are shown in Table 12.3. The codon for the glycine or alanine at position 143 is underlined. The nucleotide and amino acid residues involved in the mutation G143A conferring resistance to QoI fungicides are represented in bold.

To date, three amino acid substitutions have been detected in the cytochrome *b* gene in plant pathogens that govern resistance to Qo inhibitors. These substitutions are:

- change from glycine to alanine at position 143 (G143A)
- change from phenylalanine to leucine at position 129 (F129L)
- change from glycine to arginine at position 137 (G137R).

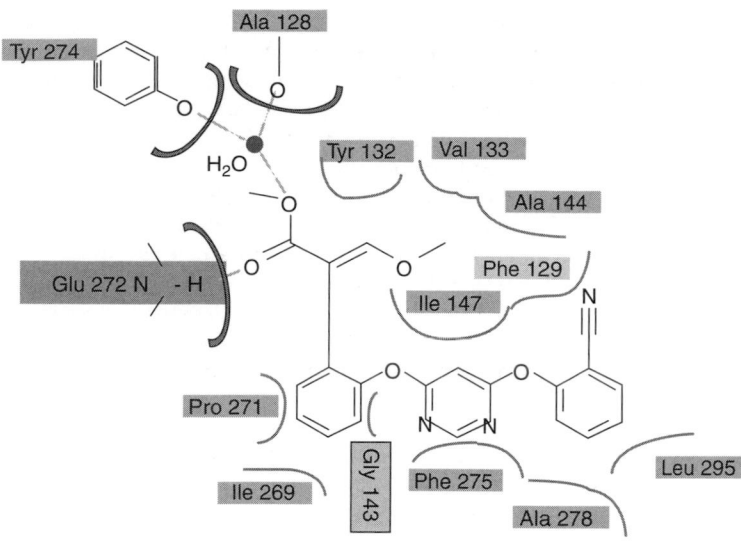

Fig. 12.3. Binding pocket with amino acids (yeast numbering) of the cytochrome bc_1 enzyme complex (of bovine heart) based on the co-crystal structure with the Qo inhibitor azoxystrobin (after Gisi *et al.*, 2002).

Table 12.3. Nucleotide and amino acid sequences of the cyt *b* gene from a sensitive and a resistant isolate of *Blumeria graminis* f. sp. *tritici*.

B. graminis f. sp. tritici isolates	Nucleotide sequence	Amino acid sequence
Sensitive	5' ... TGG <u>GGT</u> GCA ... 3'	...W**G**A...
Resistant	5' ... TGG <u>GCT</u> GCA ... 3'	...W**A**A...
	143	143

Notes: Underlining shows the codons at position 143; bold type shows the changes involved in the G143A mutation conferring resistance.

All mutations, G143A, G137R and F129L, are based on single nucleotide polymorphisms in the cytochrome *b* gene; the selection process is qualitative (single step). Based on current knowledge, resistance factors (RF = ED_{50} [resistant strain]/ED_{50} [sensitive wild-type strain]) associated with G143A, G137R and F129L are different. Resistance factors caused by F129L and G137R are usually low, ranging between 5 and 15, and in a very few cases up to 50, while resistance factors related to G143A are, in most cases, greater than 100 and usually greater than several hundreds. Isolates carrying G143A express high (complete) resistance. Isolates with F129L or G137R express moderate (partial) resistance. QoIs applied at the manufacturer's recommended rates may still provide effective control of diseases with the F129L or G137R mutation. In contrast, a severe loss in disease control is always seen in populations where G143A predominates and QoIs are used alone. G143A has been shown to be responsible for QoI resistance in more pathogen species than F129L (19 out of 25 plant pathogen species carry G143A). F129L has been detected in 3 out of 25 plant pathogens and 3 out of 25 pathogens possessing both mutations. G137R has been found only in *P. tritici-repentis*, at very low frequency. Details of the mutations found in the species where resistance has been confirmed are given in Table 12.2.

The dynamics of resistance evolution to the QoI fungicides has been shown to depend on the pathogen. In some pathogens, resistance caused by the G143A mutation arises fast and is strong. Examples are *B. graminis*, *P. viticola*, *M. fijiensis*, *M. graminicola*, *Alternaria alternata* and *P. tritici-repentis* (Sierotzki *et al.*, 2000a,b, 2007; Gisi *et al.*, 2002). In other cases, resistance caused by the G143A mutation has been slow to arise and spread but is also strong (*V. inaequalis, Uncinula necator*). Resistance caused by the F129L mutation tends to be only moderate in strength and significance but can also arise quickly (*A. solani*). And in some pathogens, notably the rusts *Monilinia fructicola* and *Phytophthora infestans*, resistance has not, until now, been confirmed.

12.6   The 'Intron Story'

As observed in the previous section, there are some pathogens in which resistance to the QoI fungicides has not been reported so far. There could be several reasons for this, including effective anti-resistance strategies, lack of exposure of the pathogen to QoI fungicides or strong fitness penalties of mutant strains to survive or be competitive. However, in the case of the rusts, belonging to the Basidiomycetes, treatment frequency with QoIs has been at least that experienced by the cereal powdery mildews, where resistance arose rapidly. The strobilurin-producing Basidiomycetes *Strobilurus tenacellus* and *Mycena galopoda* exhibit 'natural resistance' to QoIs, and the molecular mechanisms of this 'natural resistance' are known to be point mutations in the cyt *b* gene. This phenomenon was therefore investigated for *Puccinia* species (Grasso *et al.*, 2006).

In different *Puccinia* species, the presence of an intron has been observed directly after the triplet GGT that encodes for glycine at position 143. In all rust species included in the study, as well as in *A. solani* and *P. teres*, the codon GGT at position 143 is located exactly at the exon/intron boundary and is likely part of the signal sequences essential for the recognition of the intronic RNA to be excised. The authors predict that a nucleotide substitution in codon 143 (GGT → GCT), which is two nucleotides upstream from the exon/intron junction, will affect the splicing process strongly, leading to a deficient cytochrome *b*. The substitution of guanine to cytosine obviously does not allow a proper pairing of the exonic nucleotides with the intronic IGS sequence in the pre-mRNA molecule. Therefore, this substitution will be lethal and individuals carrying this mutation

will not survive. As a consequence, it is concluded that resistance to QoI fungicides based on the G143A mutation is not likely to evolve in species such as rusts (*Puccinia* spp., *Uromyces appendiculatus, Phakopsora pachyrhizi, Hemileia vastatrix*), *P. teres* and *A. solani*. However, species lacking an intron at position 143 in the cyt *b* gene can easily acquire resistance to QoIs. The presence of such an intron has also been reported in *Monilinia laxa, M. fructicola* (Luo *et al.*, 2010; Miessner and Stammler, 2010) and *Guignardia bidwellii* (Miessner *et al.*, 2011). In the fungal species investigated so far, the presence of an intron was conserved over all investigated isolates within a species, even after many years of high selection pressure by QoIs. There is only one exception, *Botrytis cinerea*, where two forms of the cytochrome *b* gene have been reported (Banno *et al.*, 2009). However, it cannot be excluded that mutations other than G143A conferring resistance may arise in upcoming populations selected by the use of QoI fungicides. For *A. solani* and *P. teres*, the mutations F129L and/or G137R have been reported (Sierotzki *et al.*, 2007; FRAC) as a mechanism for QoI tolerance. Both mutations are of minor importance, however, because generally they lead to lower resistance factors than the G143A mutation and it has been found that these two mutations have no, or only limited impact on the field efficacy of QoIs (Semar *et al.*, 2007). The results give some confidence around the continued sustainability of disease control with QoI fungicides in pathogens containing an intron after codon 143 in the cytochrome *b* gene, providing responsible resistance management practices are implemented. No resistant rust isolates have been detected in sensitivity monitoring programmes performed by several agrochemical companies, universities and private institutions over the past decade.

12.7   Monitoring for Resistance to QoI Fungicides

Conventional monitoring methods are based on sensitivity tests *in vitro* or on leaf discs (for obligate pathogens such as downy and powdery mildews). Bioassays provide essential information on sensitivity to fungicides. However, these methods can be slow and costly, especially for obligate pathogens that cannot be grown on artificial media. DNA-based methods targeted at specific SNPs (single nucleotide polymorphisms) offer rapid, cost-effective alternatives. Molecular methods can provide especially powerful tools to detect the early appearance of resistant isolates or to quantify resistance in populations. These detection methods are only of value if there is a very high correlation with the bioassays detecting resistant phenotypes qualitatively (which is the case for QoI resistance). To develop appropriate molecular methods for the detection of resistance, it is important to know the gene(s) coding for the altered target protein and to characterize the alterations.

Based on the characterization of the cytochrome *b* gene and identification of the point mutations identified as responsible for resistance to the QoI fungicides, specific primers and several Q-PCR (quantitative polymerase chain reaction) methods have been developed to monitor resistance in field populations (Wille *et al.*, 2002; McCartney *et al.*, 2003; Sierotzki and Gisi, 2003). In addition, the quantification of resistance due to a point mutation, such as G143A or F129L, can be done with novel and powerful molecular technologies. Such tests offer fast and simple opportunities to improve the evaluation of current field populations and the assessment of the risk of resistance development, to optimize resistance management and to support new product development.

A suite of conventional monitoring methods has been established by FRAC and its collaborators in academia and is published on the FRAC website (www.frac.info). It is recommended that these methods are used in order to provide a level of standardization in methodology and increase the meaningfulness of the results.

12.8 Conclusions

The QoI fungicides represent a case study in the occurrence, spread and impact of fungicide resistance in modern agriculture. Despite resistance in the field occurring within 3 years of commercial introduction, this class of fungicides remains the most economically important worldwide, with new products continuing to be developed. An understanding of the fundamental causes of resistance, the population dynamics and the inheritance of resistance has resulted in the ability to implement disease control strategies that have, on the most part, been successful. Of course, there are diseases and crops where resistance is at such levels where QoI fungicides are no longer effective – this is to be expected where there are such high resistance factors, no fitness penalties, high disease pressure and reliance on fungicides, as is the case for *B. graminis* in Europe. But there are still many situations where disease pressure is moderate or other factors have contributed to either slow down the evolution of resistance or prevent it completely. The QoI fungicides underline the issues that can occur when very effective novel products are introduced to the market, they are commercially successful and are used (or abused) widely. They become relied upon by growers, are sometimes used continually within a season and from season to season, often without respecting obvious use strategies to reduce resistance risk such as spray limitation and the use of mixtures and alternations with fungicides with alternative modes of action. It is in the interest of industry and growers alike to maintain the effectiveness of fungicides in the market and it is hoped that the experience of the QoI fungicides will help learning for the future novel classes of fungicides that will be introduced.

References

Ammermann, E., Lorenz, G., Schelberger, K., Wenderoth, B., Sauter, H. and Rentzea C. (1992) BAS 490F – a broad-spectrum fungicide with a new mode of action. *Proceedings Brighton Crop Protection Conference, Pests and Diseases.* British Crop Protection Council, Surrey, UK, pp. 403–410.

Ammermann, E., Lorenz, G., Schelberger, K., Müller, B., Kirgsten, R. and Sauter, H. (2000) BAS 500F: the new broad-spectrum strobilurine fungicide. *Proceedings Brighton Crop Protection Conference, Pests and Diseases.* British Crop Protection Council, Surrey, UK, pp. 541–548.

Avila-Adame, C. and Köller, W. (2003) Characterization of spontaneous mutants of *Magnaporthe grisea* expressing stable resistance to the Qo inhibiting fungicide azoxystrobin. *Current Genetics* 42, 332–338.

Banno, S., Yamashita, K., Fukumori, F., Okada, K., Uekusa, H., Takagaki, M., *et al.* (2009) Characterization of QoI resistance in *Botrytis cinerea* and identification of two types of mitochondrial cytochrome b gene. *Plant Pathology* 58,120–129.

Bartlett, D.W., Clough, J.M., Godwin, J.R., Hall, A.A., Hamer, M. and Parr-Dobrzanski, B. (2002) The strobilurin fungicides. *Pest Management Science* 58, 649–662.

Bennoun P., Delosme, M. and Kück, U. (1991) Mitochondrial genetics of *Chlamydomonas reinhardteii*: resistance mutations marking the cytochrome *b* gene. *Genetics* 127, 335–343.

Brasseur, G., Saribas, A.S. and Daldal, F. (1996) A compilation of mutations located in the cytochrome *b* subunit of the bacterial and mitochondrial bc1 complex. *Biochimica et Biophysica Acta* 1275, 61–69.

Brent, K.J. and Hollomon, D.W. (2007) Fungicide resistance: the assessment of risk. *FRAC Monograph No. 2*, second edn. Global Crop Protection Federation, Brussels, 53 pp.

Brown, J.K.M., Atkinson, M.A., Ridout, C.J., Sacristán, S., Tellier, A. and Wyand, R.A. (2006) Natural selection in plant–pathogen interactions: from models to laboratory to field. *Abstracts of the XIII Congreso de la Sociedad Española de Fitopatología*, Murcia, Spain, p. 15.

Chen, W.-J., Delmotte, F., Richard-Cervera, S., Douence, L., Greif, C. and Corio-Costet, M.-F. (2007) At least two origins of fungicide resistance in grapevine downy mildew populations. *Applied and Environmental Microbiology* 73(16), 5162–5172.

Chin, K.M., Küng-Färber, R. and Laird, D. (2000) Aspects of fungicide cross-resistance and implications for strobilurins. *Proceedings Brighton Crop Protection Conference, Pests and Diseases*. British Crop Protection Council, Surrey, UK, pp. 415–420.

Clough, J.M. and Godfrey, C.R.A. (1998) The strobilurin fungicides. Fungicidal activity, chemical and biological approaches to plant protection. Wiley *Series in Agrochemicals and Plant Protection*. John Wiley & Sons, Chichester, UK, pp. 109–148.

Degli Esposti, M., de Vries, S., Crimi, M., Ghelli, A., Patarnello, T. and Meyer A. (1993) Mitochondrial cytochrome *b*: evolution and structure of the protein. *Biochimica et Biophysica Acta* 1143, 243–271.

di Rago, JP., Coppee, J.Y. and Colsons, A.M. (1989) Molecular basis for resistance to myxothiazol, mucidin (strobilurin A), and stigmatellin. Cytochrome *b* inhibitors acting at the center of the mitochondrial ubiquinol-cytochrome *c* reductase in *Saccharomyces cerevisiae*. *Journal of Biological Chemistry* 264(24), 14543–14548.

Fraaije, B.A., Burnett, F.J., Clark, W.S., Motteram, J. and Lucas, J.A. (2005) Resistance development to QoI inhibitors in populations of *Mycosphaerella graminicola* in the UK. In: Dehne, H.W., Gisi, U., Kuck, K.H., Russell, P.E. and Lyr, H. (eds) *Modern Fungicides and Antifungal Compounds IV*. BCPC, Alton, Hants, UK, pp. 63–71.

FRAC (2010) www.FRAC.info.

Geier, B.M., Schägger, H., Brandt, U., Colson, A.-M. and von Jagow, G. (1992) Point mutation in cytochrome b of yeast ubihydroquinone: cytochrome-c oxidoreductase causing myxothiazol resistance and facilitated dissociation of the iron-sulfur subunit. *European Journal of Biochemistry* 208, 375–380.

Genet, J.-L., Jaworska, G. and Deparis, F. (2006) Effect of dose rate and mixtures of fungicides on selection for QoI resistance in populations of *Plasmopara viticola*. *Pest Management Science* 62,188–194.

Gennis, R.B., Barquera, B., Hacker, B., Van Doren, S.R., Arnaud, S., Crofts, A.R., Davidson, E., Gray, K.A. and Daldal, F. (1993) The bc 1 complexes of *Rhodobacter sphaeroides* and *Rhodobacter capsulatus*. *Journal of Bioenergetics and Biomembranes* 25(3), 195–209.

Gisi, U., Chin, K.M., Knapova, G., Färber, R.K., Mohr, U., Parisi, S., Sierotski, H. and Steinfeld, U. (2000) Recent developments in elucidating modes of resistance to phenylamide, DMI, and strobilurin fungicides. *Crop Protection* 19, 863–872.

Gisi, U., Sierotzki, H., Cook, A. and McCaffery, A. (2002) Mechanisms influencing the evolution to Qo inhibitor fungicides. *Pest Management Science* 58, 859–867.

Gisi, U., Pavic, L., Stanger, C., Hugelshofer, U. and Sierotzki, H. (2005) Dynamics of *Mycosphaerella graminicola* populations in response to selection by different fungicides. In: Dehne, H.W., Gisi, U., Kuck, K.-H., Russell, P.E. and Lyr, H. (eds) *Modern Fungicides and Antifungal Compounds IV*. BCPC, Alton, Hants, UK, pp. 73–80.

Godwin, J.R., Bartlett, D.W. and Heaney, S.P. (1999) Azoxystrobin: implications of biochemical mode of action. Pharmakokinetics and resistance management for spray programmes against Septoria diseases in wheat. In: Lucas, J.A., Bowyer, P. and Anderson, H.M. (eds) *Septoria on Cereals: A Study of Pathosystems*. CAB International, Wallingford, UK, pp. 299–315.

Godwin, J.R., Bartlett, D.W., Clough, J.M., Godfrey, C.R.A., Harrison, E.G. and Maund, S. (2000) Picoxystrobin: a new strobilurin fungicide for use on cereals. *Proceedings Brighton Crop Protection Conference, Pests and Diseases*. British Crop Protection Council, Surrey, UK, pp. 533–541.

Grasso, V., Palermo, S., Sierotzki, H., Garibaldi, A. and Gisi, U. (2006) Cytochrome *b* gene structure and consequences for resistance to Qo inhibitor fungicides in plant pathogens. *Pest Management Science* 62, 465–472.

Hayase, Y., Kataoka, T., Masuko, M., Niikawa, M., Ichinari, M., Takenaka, H., Takahashi, T., Hayashi, Y. and Takeda, R. (1995) Phenoxyphenyl alkoxyiminoacetamides. New broad-spectrum fungicides. *Synthesis and Chemistry of Agrochemicals IV*, ACS Symposium Series No 584. American Chemical Society, Washington, DC, pp. 343–353.

Heaney, S.P., Hall, A.A., Davies, S.A. and Olaya, G. (2000) Resistance to fungicides in the QoI-STAR cross-resistance group: current perspectives. *Proceedings Brighton Crop Protection Conference, Pests and Diseases*. British Crop Protection Council, Surrey, UK, pp. 755–762.

Joshi, M.M. and Sternberg, J.A. (1996) DPX-JE874: A broad-spectrum fungicide with a new mode of action. *Proceedings Brighton Crop Protection Conference, Pests and Diseases*. British Crop Protection Council, Surrey, UK, pp. 21–26.

Kataoka, S., Takagaki, M., Kaku, K. and Shimizu, T. (2010) Mechanism of action and selectivity of a novel fungicide, pyribencarb. *Journal of Pesticide Science* 35(2), 99–106.

Luo, C.X., Hu, M.J., Jin, X., Bryson, P.K. and Schnabel, G. (2010) Evidence for the unlikely development of the QoI fungicide resistance-related G143A mutation in the cyt *b* gene of *Monilinia fructicola*. *Pest Management Science* 66, 1308–1315.

Ma, Z. and Michailides, T.J. (2005) Advances in understanding molecular mechanisms of fungicide resistance and molecular detection of resistant genotypes in phytopathogenic fungi. *Crop Protection* 24, 853–863.

McCartney, H.A., Foster, S.J., Fraaije, B.A. and Ward, E. (2003) Molecular diagnostics for fungal plant pathogens. *Pest Management Science* 59, 129–142.

Margot, P., Huggenberger, F., Amrein, J. and Weiss, B. (1998) CGA 279202: a new broad-spectrum strobilurin fungicide. *Proceedings Brighton Crop Protection Conference, Pests and Diseases*. British Crop Protection Council, Surrey, UK, pp. 375–382.

Mercer, R.T., Lacroix, G., Gouot, J.M. and Latorse, M.P. (1998) RPA 407 213: a novel fungicide for the control of downy mildew, late blight and other diseases on a range of crops. *Proceedings Brighton Crop Protection Conference, Pests and Diseases*. British Crop Protection Council, Surrey, UK, pp. 319–328.

Miessner, S. and Stammler, G. (2010) *Monilinia laxa, M. fructigena* and *M. fructicola*: risk estimation of resistance to QoI fungicides and identification of species with cytochrome *b* gene sequences. *Journal of Plant Diseases and Protection* 117, 162–167.

Miessner, S., Mann, W. and Stammler, G. (2011) *Guignardia bidwellii*, the causal agent of black rot on grapevine has a low risk for QoI resistance. *Journal of Plant Diseases and Protection* 118(2), 51–53.

Semar, M., Strobel, D., Koch, A., Klappach, K. and Stammler, G. (2007) Field efficacy of pyraclostrobin against populations of *Pyrenophora teres* containing the F129L mutation in the cytochrome *b* gene. *Journal of Plant Diseases and Protection* 114, 117–119.

Sierotzki, H. and Gisi, U. (2003) Molecular diagnostics for fungicide resistance in plant pathogens. In: Voss, G. and Ramos, G. (eds) *Chemistry of Crop Protection*. Wiley-VCH, Weinheim, Germany, pp. 71–88.

Sierotzki, H., Parisi, S., Steinfeld, U., Tenzer, I., Poirey, S. and Gisi, U. (2000a) Mode of resistance to respiration inhibitors at the cytochrome *bc*1 complex of *Mycosphaerella fijiensis*. *Pest Management Science* 56, 833–841.

Sierotzki, H., Wullschleger, J. and Gisi, U. (2000b) Point-mutation in cytochrome *b* gene conferring resistance to strobilurin fungicides in *Erysiphe graminis* f. sp. *tritici* field isolates. *Pesticide Biochemistry and Physiology* 68, 107–112.

Sierotzki, H., Pavic, L., Hugelshofer, U., Stanger, C., Cleere, S., Windass, J. and Gisi, U. (2005) Population dynamics of *Mycosphaerella graminicola* in response to selection by different fungicides. In: Lyr, H., Russell, P.E., Dehne, H.-W., Gisi, U. and Kuck, K.-H. (eds) *Modern Fungicides and Antifungal Compounds II*. 14th International Reinhardsbrunn Symposium, AgroConcept, Bonn. Verlag Th. Mann, Gelsenkirchen, Germany, pp. 89–101.

Sierotzki, H., Frey, R., Wullschleger, J., Palermo, S., Karli, S., Godwin, J. and Gisi, U. (2007) Cytochrome *b* gene sequence and structure of *Pyrenophora teres* and *P. tritici-repentis* and implications for QoI resistance. *Pest Management Science* 63(3), 225–233.

Sierotzki, H., Kraus, N., Pepin, S., Fernandes, N. and Gisi, U. (2008) Dynamics of QoI resistance in *Plasmopara viticola*. In: Dehne, H.W., Deising, H.B., Gisi, U., Kuck, K.H., Russell, P.E. and Lyr, H. (eds)

Modern Fungicides and Antifungal Compounds V. Deutsche Phytomedizinische Gesellschaft, Braunschweig, Germany, pp. 151–158.

Torriani, S.F., Brunner, P.C., McDonald, B.A. and Sierotzki, H. (2009) QoI resistance emerged independently at least 4 times in European populations of *Mycosphaerella graminicola. Pest Management Science* 65, 155–162.

Wilcox, W.F., Burr, J.A., Riegel, D.G. and Wong, F.P. (2003) Practical resistance to QoI fungicides in New York populations of *Uncinula necator* associated with quantitative shifts in pathogen sensitivities. (Abstr.) *Phytopathology* 93, S90.

Wille, P., Sierotzki, H., Stanger, C., Cleere, S., Burbidge, J., Hall, A., Windass, J. and Gisi, U. (2002) Qualitative and quantitative identification of SNPs in plant pathogens. *Modern Fungicides and Antifungal Compounds III.* AgroConcept, Bonn, Germany, pp. 131–139.

Ziogas, B.N., Baldwin, B.C. and Young, J.E. (1997) Alternative respiration: a biochemical mechanism of resistance to azoxystrobin (ICIA 5504) in *Septoria tritici. Pesticide Science* 50, 28–34.

Part III

Resistance Cases in Different Countries

13 Fungicide Resistance in *Plasmopara viticola* in France and Anti-Resistance Measures

Marie-France Corio-Costet
INRA, UMR SAVE (Health and Agroecology of Vineyard),
ISVV, Villenave d'Ornon, France

13.1 Introduction

Grapevine downy mildew [*Plasmopara viticola* (Berk. and Curt.) Berl. & de Toni] is a major disease causing widespread destruction in vineyards throughout the world. It was introduced into France in 1878, in the wake of the Phylloxera epidemic, and was discovered by Planchon in vineyards in Coutras (Bordeaux region). The following year, *P. viticola* was present in numerous French vineyards and in Italy (Galet, 1977). This highly destructive disease, combined with Phylloxera, led to poor harvests in all French vineyards during the period 1880–1890.

Ever since its early outbreak in France, management of this widespread disease of grapevine has been attempted with the use of chemicals. For a long time only products with copper salts were used (Bordeaux mixture), as developed by Millardet (Galet, 1977). It was to a certain extent the first anti-mildew product and, to our knowledge, *P. viticola* still shows no signs of resistance to it. However, the doses of copper discharged in the vineyard sometimes lead to problems of soil toxicity and the new European directives, therefore, recommend a significant decrease in its use. During the period 1950–1960, copper was widely used, then organic fungicides such as

dithiocarbamates and phthalimides appeared on the market, their main properties being to act on various cell targets (Table 13.1). They are generally considered to be preventive products because they do not penetrate the plant and are non-systemic (e.g. thiocarbamates). No case of resistance has been described for these products in French vineyards. The next step was products having a more targeted mode of action and with systemic properties (e.g. phenylamides, QoI fungicides, cell wall inhibitors) (Fig. 13.1; Table 13.1). Because of their penetration and bioavailability in the plant, most of them act preventively rather than curatively. These fungicides were very widely used for several decades and led more or less rapidly to the appearance of resistance, depending on the product involved (Gisi, 2002).

In France, there is a wide range of products that can be used in the fight against *P. viticola*. At present, this includes 58 approved products, covering eight different modes of action, representing 28 active ingredients (Table 13.1; Fig. 13.1). Under heavy disease pressure, from four to more than ten applications of these products may be necessary to control the disease. The use of these compounds raises various questions as to what strategy has been put in place to limit the

Table 13.1. Classification of compounds controlling *P. viticola* in France in 2010.

Mode of action	Chemical class	Common name of compounds used and date of market authorization	Resistance reported in *P. viticola*
Multi-site	Inorganics	Copper oxychloride, copper sulfate, cuprous oxide, copper hydroxide	No
	Dithiocarbamates	Mancozeb (1961), maneb (1958), metiram-Zn (1958), propineb (1963)	No
	Phthalimides	Captan (1952), folpel (1952)	No
	Phthalonitriles	Chlorothalonil (1964)	No
	Quinones	Dithianon (1963)	No
Plant elicitors	Phosphonates	Fosetyl-Al (1977), potassium phosphonate (2010)	No
Unknown	Cyanoacetamide oximes	Cymoxanil (1976)	Yes
RNA polymerase inhibitors (phenylamides)	Acylalanines	Benalaxyl (1981), mefenoxam or metalaxyl-M (1998)	Yes
Mitochondrial respiration inhibitors (QoIs)	Methoxy-acrylates (strobilurins)	Azoxystrobin (1992), pyrachlostrobin (2003)	Yes
	Oxazolidine-diones	Famoxadone (1996)	Yes
QiI + phosphonates	Cyano-imidazoles	Cyazofamid (2010)	No
Cell wall synthesis	Valinamide carbamates	Benthiavalicarb-isopropyl (2009)	Yes
		Iprovalicarb (1999)	Yes
		Valifenalate (2010)	?
	Mandelic acid amides	Mandipropamid (2009)	Yes
	Cinnamic acid amides	Dimethomorph (1988)	Yes
Microtubule biosynthesis	Benzamides	Zoxamide (2003)	No

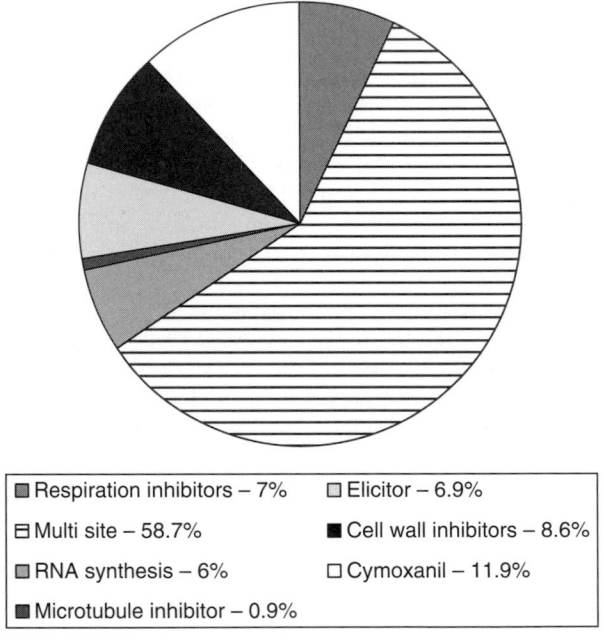

- ◨ Respiration inhibitors – 7%
- ⊟ Multi site – 58.7%
- ▨ RNA synthesis – 6%
- ▪ Microtubule inhibitor – 0.9%
- ▢ Elicitor – 6.9%
- ■ Cell wall inhibitors – 8.6%
- ☐ Cymoxanil – 11.9%

Fig. 13.1. Distribution of active ingredients in the list of authorized fungicides, presenting various modes of action, used on grapevines in France in 2010. (QoI fungicides represent 7% of the total of 58 authorized chemicals, which contain a total of 118 active ingredients in different formulations.)

appearance, dispersal and continuation of these resistance phenomena and how resistant strains can be managed. At present, management of resistance to fungicides with different modes of action, exhibiting inherent resistance risk, is done mainly by associating a multi-site fungicide with a risky fungicide (Couteux and Lejeune, 2010). A reasonable approach to disease management in the vineyard consists of adapting the programme of fungicide treatments to the risks in order to attain the goals fixed in terms of produce quality and quantity with a minimum of treatments.

13.2   Resistance to Phenylamides

These fungicides, which have been in use since the 1980s, inhibit ribosomal RNA synthesis (Fisher and Hayes, 1979), more specifically RNA polymerase, and have selective activity exclusively against Peronosporales. In addition, some phenylamides, such as benalaxyl or cyprofuram, may also affect uridine uptake into fungal cells (Davidse, 1995). The existence of cross-resistance among the strains resistant to various phenylamides is evident, and endogenous nuclear RNA polymerase activity of resistant strains is affected less than that of wild-type strains (Davidse, 1995). Although the major mechanisms of reported phenylamide resistance appeared to be a reduced sensitivity of RNA polymerase, some studies suggested or identified additional resistance mechanisms including reduced uptake of fungicides (Clerjeau *et al.*, 1985). More recently, the use of an active R-enantiomer of metalaxyl increased the efficiency of this fungicide with regard to the racemic mixture (R-S) (Nuninger *et al.*, 1996). As curative and preventive molecules, the wide use of phenylamides led to resistance build-up in *P. viticola*, described in French vineyards from 1981. Today, compounds such as benalaxyl and mefenoxam represent 7% of the approved fungicides in France (Fig. 13.1).

It was in autumn 1981, in the Bordeaux and Cognac regions of France, that, for the first time in Europe, strains of *P. viticola*

resistant to metalaxyl were detected in plots where efficacy of metalaxyl was being monitored (Staub and Sozzi, 1981; Clerjeau and Simone, 1982). The sensitive strains were controlled with concentrations of 0.01–0.1 mg/l, whereas the resistant strains were not, or were controlled only partially, even at higher doses of 10–100 mg/l of metalaxyl. At the discriminatory dose of 5 mg/l, it was found that resistance was distributed quite widely in the great majority of the Atlantic regions, which were also those most exposed to the disease and where phenylamide treatments had been used most frequently. In 1983, 40–70% of western vineyards (southwest, Cognac, Loire Valley) exhibited resistance to phenylamides. By contrast, eastern vineyards (Champagne, Bourgogne or southeast) showed only 2–25% of resistant strains (Clerjeau *et al.*, 1984). Resistance appeared rapidly in vineyards where there was more pathogen pressure. However, it was interesting to note that in 1978 in a Swiss vineyard never treated with phenylamides, resistant strains originating from naturally occurring isolates were found before the exposure of *P. viticola* populations to the fungicide (Bosshard and Schüepp, 1983). The fitness of resistant strains was good and their competitiveness appeared variable under laboratory conditions (Leroux and Clerjeau, 1985). However, an experiment performed at 30°C for 2 days, followed by 6 days at 22°C, showed disappearance of resistant isolates after three asexual cycles (Piganeau and Clerjeau, 1985), suggesting a lesser fitness of the resistant strains at high temperatures. In the vineyard, pathogen populations may contain individuals with sensitive, intermediate and resistant levels of response to phenylamides. The sensitivity range in *P. viticola* populations can be over 1000-fold, as evidenced by the samples collected in 1995 in vineyards in south-western France (Gisi, 2002). After a drastic decline in the late 1980s, the resistant population decreased between 1987 and 1998, from an initial 60–80% to about 25% in 1998 (Gisi, 2002), but at the same time intermediate populations increased from less than 5% to 38%. Data from 29 *P. viticola* isolates collected between 2003 and 2005 at the beginning of the growing season showed that 31% of populations in French vineyards were resistant.

The 29 isolates exhibited variable sensitivities to mefenoxam, with sensitive isolates having an ED_{50} lower than 0.05 mg/l, intermediate strains having an ED_{50} between 0.5 and 2 mg/l and resistant strains having an ED_{50} between 20 and 100 ml/l (Corio-Costet 2005, unpublished results).

The anti-resistance strategy developed consists of combining phenylamides with a multi-site fungicide and using in alternation with fungicides having different modes of action (e.g. association with mancozeb, folpel or fosetyl-aluminium) (Couteux and Lejeune, 2010). As phenylamides are never used alone, the loss of potential efficacy is not a systematic echo of the disease rate. Mixtures of single-site and contact fungicides have been formulated and they were expected to delay the build-up of resistant strains in fungal populations. Mixtures have a much lower selection pressure than single fungicides when the level of resistance is low. However, it seems that resistance to phenylamides is still present in populations of *P. viticola* in France. To limit the problem of the spread of resistance, the Fungicide Resistance Action Committee (FRAC, 2010) has recommended that fungicide applications with phenylamides should be restricted to two to four per season at not less than 14-day intervals (Brent and Holllomon, 2007a,b). In France, where phenylamide resistance remains strongly implanted in all regions, the interest of these products is limited directly to the type and the dose of the associated product (French National Memorandum on Mildew, 2010). The Memorandum on Mildew recommends the application of compounds containing phenylamides at a maximum of once or twice a year, and their use in nurseries is totally banned.

13.3 Resistance to Mitochondrial Inhibitors (QoI, QiI)

These fungicides inhibit the electron transfer of complex III containing two quinone binding sites, the quinol oxidization site (outer, Qo), involving a cytochrome *b*, and the quinone reduction site (inner, Qi),

controlled by ubiquinol-cytochrome *c* oxidoreductase (Hollomon *et al.*, 2005).

Quinone outside inhibitor (QoI)

QoI fungicides, which inhibit mitochondrial respiration at the cytochrome bc_1 enzyme complex, are one of the most important classes of fungicide (Jordan *et al.*, 1999). However, strong selection pressure following repeated QoI applications has led to the development of QoI-resistant populations, which has limited the success of these fungicides in grapes and other crops (Gisi *et al.*, 2002; Grasso *et al.*, 2006). Thus, only a few years after the introduction of QoI fungicides in 1996, *P. viticola*-resistant populations were detected not only in French vineyards (Magnien *et al.*, 2003) but also in the vineyards of other European countries and in the USA (Heaney *et al.*, 2000; Sierotzki *et al.*, 2005, 2008; Baudoin *et al.*, 2008; Corio-Costet *et al.*, 2008). In *P. viticola*, two mutations in the cytochrome *b* gene have been described (G143A, F129L) as being involved in resistance mechanisms, with G143A mutation in the majority of cases. The mutation resulting in the replacement of glycine by alanine at amino acid codon position 143 (G143A) has been widely recognized (Gisi *et al.*, 2002; Grasso *et al.*, 2006) and causes cross-resistance to all QoI fungicides (i.e. famoxadone, fenamidone, strobilurins) that are included in the same cross-resistance group (C3–11) described by FRAC.

These fungicides act efficiently in controlling infections at the first developmental stages of *P. viticola*, but have a low curative effect. Phylogenetic analysis of a large mitochondrial DNA fragment, including the cytochrome *b* gene (2.281 bp), across a wide range of *P. viticola* isolates, detected four major haplotypes belonging to two genetically distinct groups (I and II), in which the resistance allele carrying the G143A mutation could be present (Chen *et al.*, 2007). This indicated at least two origins of fungicide resistance in the grapevine downy mildew pathogen population. In France, haplotype group I (IR and IS) and group II (IIR and IIS) reached a frequency of 75 and 25%, respectively, among populations tested, but their distribution might vary (Chen *et al.*, 2007; Corio-Costet *et al.*, 2008). For instance, in Champagne (France) an average of 10% of haplotype group II has been detected, and the significance of the two mitochondrial haplotypes remains unclear (Corio-Costet *et al.*, 2006). As regards the resistant allele proportion IR and IIR within a population, the resistant haplotype IR ranged from 67.7 to 98.3% of resistant alleles, depending on the location (Corio-Costet *et al.*, 2006; Chen *et al.*, 2007). To estimate the persistence of QoI resistance and the survival capacity of isolates, one study showed that fitness of the resistant isolates was rather good during the asexual cycle (Corio-Costet *et al.*, 2011). On the other hand, several studies showed that after sexual reproduction, a decrease in the presence of resistant populations in the absence of pressure of selection was often described (Toffolatti *et al.*, 2007) and might be connected to the mitochondrial inheritance.

In French vineyards, all isolates of *P. viticola* collected from 2003 to 2007 showed that QoI resistance was persistent and that isolates grew with or without QoI selection pressure. In addition, G143A was present systematically in isolates where resistance was detected, and resulted in a resistance factor of more than 2000 (Chen *et al.*, 2007). Regional follow-up of resistance status carried out on *P. viticola* isolates at the beginning of the growing season in 2003 and 2004, in two regions where cultural practices and disease pressures were different, showed a similar development of resistance. Indeed, in the Bordeaux area, where each vine grower followed his or her own treatment programme, there was, on average, 29% resistant populations, which varied greatly depending on the plot of vineyard monitored, ranging from 0 to 70% (Corio-Costet *et al.*, 2008). In the Champagne region, where wine growers followed the recommendations of the CIVC (the Champagne Vine Trade Council), the situation was similar, with an average of 33% resistant isolates and variations in resistance frequency ranging from 0.01 to 77%. All the resistant isolates showed the mutation in G143A and the only difference between the two regions was the stronger presence of the mitochondrial haplotype II in the Bordeaux region (28%) than in

Champagne (10%) (Corio-Costet *et al.*, 2006). Diversity analysis of resistant populations with microsatellite markers (Delmotte *et al.*, 2006) showed that the genotypic diversity of *P. viticola* was less important when the resistance frequency was more than 30% and that the genotypic diversity tended to decrease at the end of the growing season when resistance frequency was highest in the Champagne and Bordeaux areas (Corio-Costet *et al.*, 2008).

The rapidity of the appearance of QoI resistance and the behaviour of resistant isolates in populations could suggest a good fitness value of resistant isolates during asexual cycles. Furthermore, recent monitoring has showed that in spite of a strong decrease in the use of strobilurin fungicides in France, the resistance of *P. viticola* to QoI fungicide is widespread and, in the majority of situations, the frequency of resistant populations is higher than 80% in more than three out of four cases. Comparison with 2003 observations reveals that the situation has become even worse when the use of QoIs for mildew control has been strongly reduced (Magnien *et al.*, 2009).

The 2010 Memorandum on Mildew states that the use of active substances belonging to the QoI family is no longer of interest due to the generalized and persistent presence of resistant strains. In this context, the efficacy of products containing QoIs is often related exclusively to the type and dose of the associated multi-site products (e.g. copper, mancozeb, cymoxanil, folpel, etc.).

Quinone inside inhibitor (QiI): cyazofamid

This fungicide exhibits a very high level of efficacy and preventive activity against oomycete pathogens. It inhibits the respiratory system, and more specifically in the mitochondrial complex III (Mitani *et al.*, 2001), and acts on the internal membrane of the mitochondria. In France, this contact fungicide was combined with a phosphonate and was approved in 2010 for the control of grapevine downy mildew. To date, no resistance to this fungicide has been described in *P. viticola* and no cross-resistance has been observed with QoI fungicides. However, reduced sensitivity to cyazofamid has been reported in *Phytophthora capsici* in the USA (Kousik and Keinanth, 2008). For the control of *P. viticola*, it is strongly recommended to limit its use to three applications, not consecutively, per growing season.

13.4   Resistance to Cell Wall Inhibitors

The carboxylic acid amide (CAA) fungicides, including dimethomorph, iprovalicarb, benthiavalicarb and mandipropamid, are uni-site fungicides inhibiting cell wall biosynthesis/assembly of Oomycetes (Young *et al.*, 2005). A cross-resistance exists among CAA fungicides (Gisi *et al.*, 2007; Blum *et al.*, 2010b). Recently, the target of mandipropamid has been described as being a 1,3-β-glucanase exerting a glucanosyltransferase activity (*PiCesA3*, *PvCesA3* genes) in cellulose biosynthesis in *P. infestans* and *P. viticola* (Blum *et al.*, 2010a,b). A mutation in the gene *PvCesA3* could be involved in resistance acquisition in field isolates (G1105S) of grapevine downy mildew fungus (Blum *et al.*, 2010b). The fact that this gene is involved may not, by itself, be sufficient to cause resistance, for there are certainly other recessive genes involved. Consequently, the resistance in *P. viticola* to CAAs might be moderated if appropriate strategies for product use were implemented in the field. FRAC recommends the use of CAAs in mixtures with multi-site fungicides or other non-cross-resistant fungicides.

When CAAs were first used (particularly dimethomorph on *P. viticola*), ED_{95} values were low in a range of 0.25–1.15 µg/ml of the active ingredient (Bissbort and Schlosser, 1991). This fungicide, with preventive and local systemic properties, showed very good activity for more than 10 years. In the same way, the ED_{50} values of iprovalicarb in *P. viticola* isolates during the period 1996–1998 showed a wide range from <1 to 30 µg/ml, with a mean of 2.0 µg/ml (Suty and Stenzel, 1999). In France, dimethomorph sensitivity assays performed in 2003 and 2004 on 63 isolates exhibited ED_{50} values in the range of 0.08–40 µg/ml, with a mean of 5.92 µg/ml of

active ingredient. In the same isolates collection, iprovalicarb sensitivity assays showed ED_{50} in the range of 0.3–500 µg/ml, with a mean of 78.7 µg/ml (Corio-Costet, 2005, unpublished results), suggesting a shift in the sensitivity of *P. viticola* isolates to iprovalicarb and dimethomorph. Monitoring carried out between 2005 and 2008 on more than 400 French *P. viticola* populations showed a decrease in the sensitivity of these populations, from 74 to 45%. This study also showed a more or less stable intermediate group during the monitoring period and an increase in resistant populations, with a shift from 17 to 39% (Magnien *et al.*, 2009). However, these results hide an important disparity in resistance behaviour according to region. As for the other fungicides, the Midi-Pyrénées region is the most affected by the fungicide resistance of *P. viticola*, as is the Mediterranean region. To the contrary, the vineyards of the Atlantic regions, along with those in the eastern regions, are little affected, with more than 50% still having sensitive populations. As there is, now, a majority of resistance-affected areas in most wine-producing regions of France, the efficacy of treatments based on CAA fungicides is connected directly to the type and dose of the associated product. The recommendation in the 2010 Memorandum on Mildew is to reduce the number of applications of CAA fungicides to a maximum of one or two per year, non-consecutively, and not to use them on declared outbreaks of the disease.

13.5   Resistance to Cymoxanil

Cymoxanil is the only representative fungicide of the cyano-acetamide group. It has been used to control *P. viticola* since 1976 (Table 13.1; Serres and Carraro, 1976). It has a strong curative activity against grape downy mildew, but is metabolized quickly in plant tissues (Douchet *et al.*, 1977; Belasco *et al.*, 1981). For this reason, cymoxanil has been used almost exclusively in mixtures with contact, i.e. mancozeb, and/or systemic fungicides, and greater protection may be achieved by using combinations than by alternation

(Samoucha and Gisi, 1987). The mode of action of this fungicide is not understood fully, but it is reported that several biochemical processes are disrupted, including the biosynthesis of nucleic acid and amino acids (cysteine, glycine, serine) (Despreaux *et al.*, 1981). With the appearance of resistance to phenylamides, cymoxanil rapidly has become a key component in limiting the development of *P. viticola*. Although the field efficacy of cymoxanil has remained good in France, as in other parts of Europe, reports have shown a reduction in the sensitivity of *P. viticola* populations (Genet and Vincent, 1999). Currently, mixtures containing cymoxanil associated with multi-site or systemic fungicides represent 11.9% of active ingredients approved in France (Fig. 13.1). In French populations of *P. viticola*, the ED_{50} varied in range from less than 10 μg/ml to more than 300 μg/ml, with the number of applications varying from 0 to 10 per growing season (Genet and Vincent, 1999). Populations collected in France between 1995 and 1997 showed that the majority (50%) were resistant, with an ED_{50} value between 100 and 300 μg/ml. The most sensitive populations (ED_{50} < 10 μg/ml) represented not more than 5% of the total populations and approximately 40% of the populations exhibited an ED_{50} lower than 100 μg/ml. The authors also observed that on plots of grapevine treated with fungicides other than cymoxanil, resistance was less prevalent, because 60% of the isolates had an ED_{50} lower than 100 μg/ml. In contrast, isolates collected in September on plots treated five times with cymoxanil presented only 20% of the isolates with an ED_{50} lower than 100 μg/ml (Genet and Vincent, 1999). In French vineyards, the sensitivity profile was similar in different regions, suggesting that the resistance level could result from the regular use of formulations containing cymoxanil associated with other fungicides. Since the mode of action of cymoxanil is not clearly understood, the mechanism of resistance is also unknown. However, the existence of different groups of sensitivity suggests that different genes might be involved in the acquisition of resistance and its maintenance. In France, the recommendation is not to implement an anti-downy mildew programme

based only on products containing cymoxanil, referring to the drift of sensitivity observed for several years (French National Memorandum on Mildew, 2011).

13.6 Multi-Site and Miscellaneous Fungicides

Multi-site fungicides

Multi-site inhibitors are non-systemic, generally used preventively and mostly interact unspecifically with many biochemical pathways. The best-known anti-mildew product was 'Bordeaux mixture', a mixture of copper sulfate ($CuSO_4$) and lime ($Ca(OH)_2$). Today, various derivatives including copper oxychloride, copper hydroxide, cuprous oxide, or copper sulfate are used in French vineyards. Copper ion exhibits its fungitoxicity by forming complexes, in particular with sulfhydryl and also with amino, carboxylic and hydroxylic groups of cell constituents (Buchenauer, 1985). These compounds generally are combined with uni-site fungicides, and 33% of the fungicide compounds marketed in France contain some copper salt (Couteux and Lejeune, 2010). To date, no resistance to copper fungicides has been reported in French populations of *P. viticola*.

Dithiocarbamates are also very old compounds and have been used to control *P. viticola* since 1941 (Galet, 1977), but some compounds such as zineb, mancopper and thiram are no longer used in French vineyards. However, dithiocarbamates combining Zn and/or Mn complexes, such as mancozeb, maneb, metiram-zinc and propineb, are still in use, combined with uni-site fungicides. To date, they are present in 46.5% of the fungicide products approved for use in France. Other compounds, such as phthalimides (captan, folpel), phthalonitrile (chlorothalonil) or a quinone (dithianon) are also used as multi-site fungicides, usually combined with a uni-site fungicide, with the exception of captan, which can still be used alone. Combining uni-site and multi-site products enables the association of contact compounds with products often possessing properties of translocation or redistribution via vapour, but also the association of products with preventive activities and those with potential curative activities (i.e. cymoxanil + folpel + famoxadone or fosetyl-Al + mandipropamid). The mode of action of these products is not clearly known and seems to affect different enzymes of fungal metabolism (Gisi, 2002).

Phosphonates

Fosetyl-Al, a phosphonate fungicide, has been used against grapevine downy mildew for more than 30 years. More recently, another phosphonate has just been authorized to partner cyazofamid. However, numerous fertilizers also contain phosphonates that are salts and esters of phosphorous acid (H_3PO_3), often mono- and dipotassium salts, sometimes referred to as potassium phosphite. Phosphonate fungicides are absorbed by plants and incorporated into cells as phosphite ions ($H_2PO_3^-$). To our knowledge, it is the only compound which exhibits ascending and descending systemic properties (symplastic ambimobility) and moves in both xylem and phloem. Thanks to these translocation properties, phosphonates such as fosetyl-Al associated with other fungicides possess a very interesting curative activity, but the fungicide action may be most effective following the pre-infectional period (Dercks and Creasy, 1989). Besides its fungicide activity, fosetyl-Al is also described as an elicitor of plant defences, since it increases phenolic biosynthesis in leaves, but only in the presence of P. viticola, and it is more efficient in the post-infection period than in the pre-infection period (Lafon et al., 1977; Williams et al., 1977; Raynal et al., 1980; Dercks and Creasy, 1989; Bouscaut, 2005). Results of its fungicidal efficacy on isolates of P. viticola seem variable, depending not on the cultivar but on the doses of inoculum, with a range of ED_{50} from 120 to 370 µg/ml (Clerjeau et al., 1981; Dufour et al., 2009). The doses necessary to control P. viticola isolates in laboratory conditions range from 1.2 g/l to more than 2 g/l (Gaulliard, 1985; Bouscaut, 2005). To date, we are not aware of any confirmed reports of pathogen resistance to phosphonate fungicides in grapevine downy mildew, although cases of field insensitivity have been reported in P. cinnamomi in Lawson's cypress and in Bremia lactucae on lettuce (Vegh et al., 1985; Brown et al., 2004). As the mode of action in target fungi probably involves several sites, and the host defences are involved in disease suppression, the risk of resistance with phosphonates seems low. However, detoxification and/or efflux systems (multi-drug resistance) are known to be effective in pathogens for protection against fungicides and plant defence compounds (Gisi, 2002) and the possibility of such a development should not be neglected with regard to P. viticola populations. In France, fosetyl-Al is always associated with either a multi-site fungicide (e.g. folpel, mancozeb, metiram-zinc), or a uni-site fungicide such as fenamidone (QoI), or a phenylamide. The efficacy of mixtures containing fosetyl-Al is optimal if these are used during the active growth of the grapevine. There is no particular limitation, apart from respecting the general principles of anti-mildew products.

Zoxamide

This benzamide is a recent fungicide that exhibits an anti-microtubule activity towards oomycete fungi and prevents nuclear division of zoospores, as well as inhibiting microtubule assembly (Young and Slawecki, 2001). The sensitivity of 47 isolates of P. viticola collected at the beginning of the growing season in 2003 from different vineyards where zoxamide had never been used showed that the ED_{50} of P. viticola isolates varied from 0.02 to 30 µg/ml (Corio-Costet, 2005, unpublished results). As with other fungicide baselines, we found highly sensitive isolates ($ED_{50} < 1$ µg/ml), intermediate isolates ($ED_{50} < 10$ µg/ml) and less sensitive isolates ($ED_{50} \geq 10$ µg/ml). From 2004 to 2006, sensitivity monitoring of 155 populations collected showed consistent results, with ED_{50} for the majority of sensitive isolates falling between 0.01 and 0.2 µg/ml. A population was described as sensitive when its ED_{50} was lower than or equal to 0.2 mg/l. The ED_{50} range of pathogen populations was included between 0.01 mg/l and 0.8 mg/l (Magnien *et al.*, 2009).

Used only in combination with a multi-site fungicide (mancozeb) in France, the risk that zoxamide may lead to resistance build-up in *P. viticola* populations seems to be low at the present time. Even after 6 years of moderate use, no sensitivity drift in *P. viticola* has been noticed with regard to this compound. At present, in the absence of any detection of sensitivity drift, the recommendation is to limit applications of zoxamide to a maximum of three non-consecutive treatments per season.

13.7   Anti-Resistance Measures

Grapevine downy mildew management in France

Grapevine downy mildew is a devastating disease in France and it is treated mainly by chemical means (organic or inorganic), decided according to the technical and economic constraints of each region. Treatments, which are generally multi-purpose and easy to schedule, are still carried out, in spite of certain inconveniences (e.g. cost, soil erosion, resistance). These kinds of systematic applications at fixed intervals of time are still used, but they do not take into account the biology of the pathogen and the epidemiology of the disease, nor the periods of susceptibility of the vine that might favour the appearance of fungicide resistance (Staub, 1991). But, there are also preventive methods which can limit the need for chemical treatment, such as planting the vineyard in healthy soil which has been drained beforehand, implementing a programme of rational fertilization, etc. These are practices which avoid predisposing the vine to pathogen attacks. Prophylactic methods such as desuckering and leaf reduction can restrict formation of the primary mildew inoculum in vineyards, and delay appreciably development of the disease in the vineyard. In French vineyards, some growers follow integrated pest management advice and apply good plant protection practices such as those described in the EPPO bulletin (2002).

Currently, other methods implementing a variable number of treatments, and decided at dates which do not follow a pre-established calendar, can help vine growers to adjust fungicide treatment. To be effective, these methods have to determine with a certain precision the period suitable for application that will prevent pathogen development. To implement this, agricultural warnings have been established, based on biological and phenological data, along with climatological observations, providing rational advice (Maurin, 1983). For several decades, models such as 'Milvit' or 'EPI-Mildiou' have offered a quantitative evaluation of the phytosanitary risks of *P. viticola*, which can enable a significant reduction in fungicide treatments and better positioning of treatments, avoiding curative treatments (Strizyk, 1983; Tran Manh Sung *et al.*, 1990; Magnien *et al.*, 1991). In France, the use of grapevine downy mildew models favours this less systematic approach to fungicide treatments. However, fear of getting less effective control of the disease has sometimes led some growers to use new products or mixtures presenting a curative effect, without carrying out proper alternations of the active ingredients, and this has facilitated the appearance of resistant strains (e.g. resistance to QoIs).

The usual stage at which the first fungicide application is recommended is when the shoots of the plant have 6–8 leaves. Application can be repeated once or twice before flowering, depending on the development of the plant. Around the flowering period, protection of bunches becomes most important. It is necessary to take into account the growth stage, the risk of infection, the local conditions, observations on infection of the grapes, the susceptibility of the cultivar and the weather conditions. Currently, the basic strategy for downy mildew control is to prevent establishment of any infection within the plant and to stop the spread of the disease. In order to control *P. viticola*, fungicides with three types of activity can be used, i.e. contact, locally systemic and systemic. A good practice is to apply contact fungicide sprays preventively and to use the others with curative effect when the weather conditions are especially favourable and the possibility of infection is higher. The quality and timing of application are critical and the efficacy of chemical control depends greatly on the quality of the application techniques.

Resistance and future risks

Resistance in *P. viticola* to fungicides can be found throughout French vineyards for phenylamides, CAA and QoI fungicides, and even for cymoxanil. However, there are sometimes contrasting situations which relate to regional characteristics (viz. climate favourable to epidemics, sensitivity of cultivars, cultural practices). As *P. viticola* is capable of acquiring and accumulating various resistant phenotypes after multiple selection pressures with uni-site products having different modes of action, a question arises: Is this multi-resistance the result of mutations accumulated in different target genes, or are we facing a phenomenon of non-specialized resistance similar to multi-drug resistance (MDR) systems? Among multi-resistant isolates found in French vineyards in 2003–2004, all QoI-resistant isolates had the specific resistant allele G143A. For isolates combining resistance to CAA and/or mefenoxam, not disposing molecular markers of resistance at present, the doubt persists. How will populations evolve if subjected systematically to the pressure of multiple fungicides?

Today, it is even possible to find, at the beginning of the growing season and before new seasonal treatments, downy mildew isolates presenting double or triple fungicide resistance (e.g. R-QoI + R-phenylamide, R-QoI + R-CAA or R-QoI + R-CAA + R-phenylamide) in vineyards in south-western France (Corio-Costet, unpublished results, 2004). In 2009, Magnien *et al.* showed that resistance to CAA fungicides was high in the Armagnac (south-west) and Provence areas (south-east). The resistance to QoI is generalized and seems to persist, and resistance also remains present for phenylamides, in spite of a slight improvement between 1999 and 2008. In view of the products available in France, the principle of prudence

should be not to use products containing a QoI fungicide and to moderate strongly the use of those containing CAAs and/or phenylamides.

In view of the difficulties which could occur, there is a possibility of using fungicides more moderately in key periods of vine development. In France, weather conditions suitable for downy mildew are closely monitored by field extension services and growers receive warnings of when or when not to treat. Optimum timing of the first application is critical. In general, the number of applications can vary from 4 to 10, depending on the region and the season. Recently, a formal model for integrated disease management decision making with regard to grapevine powdery and downy mildews has been described (Leger *et al.*, 2008, 2010) and used experimentally in the Bordeaux region, as a complement to advice from field extension services. It should enable treatments to be limited and restrict fungicide selection pressure, but the choice of fungicide used needs to be made carefully. Assessment of the model began in 2005 and it is being used at 29 sites today. It enabled an average economy of 28% on mildew treatment in 2008–2009, years with strong pathogen pressure, while maintaining production levels (e.g. only four treatments in Bordeaux, with satisfactory results). Generally, when available, vine growers use forecasting methods as a decision tool. When choosing a fungicide, its effect against other grapevine diseases should also be considered (e.g. the QoI fungicides used against powdery mildew may exert selection pressure on downy mildew pathogen populations).

In general, strategies currently in use to avoid fungicide resistance in *P. viticola* have been successful and resistance is normally not a problem in practice, except in one specific area (i.e. the Armagnac region in south-western France), and depends on the annual epidemic pressure. But, the appearance of isolates exhibiting multiple resistance to QoIs, phenylamides and CAAs could be a problem in the future and it might be necessary to develop strategies using alternative or complementary methods.

Resistance management outlook

A reasoned decision has to be made on how to use contact and systemic products against *P. viticola*. Locally systemic and systemic products have the great advantage of a high degree of efficacy and of a curative action. However, experience has shown that there is a considerable risk of the appearance of resistance to such products, viz. phenylamides, cymoxanil, QoIs and, more recently, CAAs. Accordingly, their use must be managed carefully. Thus, mixtures combining contact fungicides (multi-site) and systemic fungicides (often uni-site) are today indisputably necessary for managing fungicide resistance, and are shown to be better than alternation of products in the prediction model (Fry and Milgroom, 1990). However, the example of phenylamides shows that different combinations, in particular with multi-site fungicides, do not delay significantly the selection of resistant isolates in French vineyards (Leroux, 2000).

How can resistance be avoided? The resistance of pathogen populations depends on the inherent risk exerted by the active ingredient (i.e. selection pressure, sensitivity baseline, cell target) and on risk management in relation to the pathogen and to its cycle of development (genomic plasticity, sexual versus asexual cycle, dissemination). To this should be added the more general factors of weather, cultural practices and improper applications (under-dosing, treatment of one side of the row only). Since prediction of the risk of resistance for new fungicide groups remains difficult, monitoring of the sensitivity of pathogen populations is essential to assess resistance in the field and to follow the emergence and survival of resistant strains in the vineyard. At present, assessment of the risk of resistance is based on data provided by fungicide manufacturers, and this is integrated into the process of market authorization for a chemical compound (EU Directive 91/414). In addition to monitoring carried out by manufacturers and field extension services, the Memorandum on Mildew is available to indicate the basic principles to alert growers to the incurred risks and to propose methods of managing resistance involving, for example, a decrease in the number of treatments or a halt in the use of specific fungicides. To implement anti-resistance strategies successfully, vine growers, distributors and field extension services must all be in agreement.

As genetically modified organisms (GMOs) are not allowed in the grapevine in France, and as public opinion is not in favour of this biotechnology, the use of grapevine varieties resistant or partially resistant to *P. viticola*, cumulating different quantitative trait loci (QTL) of disease resistance, could be an interesting alternative. The first varieties resistant to *P. viticola* are due to be registered in the national French variety certification catalogue in the near future. A combination of reduced fungicide treatments with resistant varieties (total or partial) might be a solution for managing resistance to fungicides and overcoming varietal resistance. Another alternative would be to develop the use of practices facilitating the implementation of vine defence. Some biological extract elicitors, elicitor analogues, or organic compounds such as phosphonates may be authorized to accompany fungicide compounds, even to limit the quantity of fungicide and thus limit the risk of resistance to fungicides. This is the case for phosphonates, which are partners of, for example, cyazofamid, used in combination with various fungicides.

These alternative methods, even though complementary, could accompany the more conventional methods, whether chemical management or the use of inorganic products in organic vine growing. However, although these various combined methods may limit and/or enable potential management of the appearance of fungicide resistance, they will not prevent completely the selection pressures exercised on the pathogen populations and the search for new fungicide targets to limit outbreaks of disease remains a necessity. This will all depend strictly on the coordination that will need to be put in place between research institutes, field extension services, manufacturers, product distributors, vine growers and policy makers.

References

Baudoin, A.I., Olaya, G., Delmotte, F., Colcol, J.F. and Sierotzki, H. (2008) QoI resistance of *Plasmopara viticola* and *Erysiphe necator* in the Mid-Atlantic United States. *Plant Management Network. Plant Health Progress.* Doi:10.1094/PHP-2008, pp. 211–212.

Belasco, I.J., Han, J.C.Y., Chrzanowski, R.L. and Baude, F.J. (1981) Metabolism of ^{14}C cymoxanil in grapes, potatoes and tomatoes. *Pesticide Science* 12, 355–364.

Bissbort, S. and Schlosser, E. (1991) Sensitivity of *Plasmopara viticola* to dimethomoprh. *Mededelingen van de Faculteit Landbouwwetenschappen Rijksuniversiteit Gent* 56, 559–567.

Blum, M., Boehler, M., Randall, E., Young, V., Csukai, M., Kraus, S., *et al.* (2010a) Mandipropamid targets the cellulose synthase-like PiCesA3 to inhibit cell wall biosynthesis in the oomycete plant pathogen, *Phytophthora infestans. Molecular Plant Pathology* 11, 227–243.

Blum, M., Waldner, M. and Gisi, U. (2010b) A single point mutation in the novel *PcCesA3* gene confers resistance to the carboxylic acid amide fungicide mandipropamid in *Plasmopara viticola. Fungal Genetic Biology* 47, 499–510.

Bosshard, E. and Schüepp, H. (1983) Variability of selected strains of *Plasmopara viticola* with respect to their metalaxyl sensitivity under field conditions. *Zeitschrift Fur Pflanzenkrankheiten Und Pflanzenschutz-Journal of Plant Diseases and Protection* 90, 449–459.

Bouscaut, J. (2005) Diversité des systèmes de défenses induites de la vigne et efficacité sur le mildiou et l'oïdium. PhD in Oenology-Ampelology, Bordeaux II University, No. 1263, 210 pp.

Brent, K. and Hollomon, D.W. (2007a) *Fungicide Resistance in Crop Pathogens: How Can it be Managed?* FRAC Monograph No. 1 (2nd edition). CropLife International, Brussels, 55 pp.

Brent, K. and Hollomon, D.W. (2007b) *Fungicide Resistance: The Assessment of Risk.* FRAC Monograph No. 2 (2nd edition). CropLife International, Brussels, 52 pp.

Brown, S., Koike, S.T., Ochoa, O.E., Laemmlen, F. and Michelmore, R.W. (2004) Insensitivity to the fungicide fosetyl-aluminium in California isolates of the lettuce downy mildew pathogen, *Bremia lactucae. Plant Disease* 88, 502–508.

Buchenauer, H. (1985) General review of the mode of action of fungicides. In: Smith, I.M. (ed.) *Proceedings of the Centenary Meeting Bordeaux Mixture – Fungicides for Crop Protection 100 Years of Progress.* BCPC Publications, Croydon, UK, Monograph No. 31, 1, pp. 55–65.

Chen, W.-J., Delmotte, F., Richard Cervera, S., Douence, L., Greif, C. and Corio-Costet, M.-F. (2007) At least two origins of fungicide resistance in grapevine downy mildew populations. *Applied Environmental Microbiology* 73, 5162–5172.

Clerjeau, M. and Simone, J. (1982) Apparition en France de souches de mildiou (*Plasmopara viticola*) résistantes aux fongicides de la famille des anilides (métalaxyl, milfuram). *Le Progrès Agricole et Viticole* 99(3), 59–61.

Clerjeau, M., Lafon, R. and Bugaret, Y. (1981) Etude sur les propriétés et le mode d'action des nouveaux fongicides (cymoxanil, metalaxyl, milfuram, phosethyl-Al) antimildiou chez la vigne. *Phytiatrie-Phytopharmacie* 30, 215–234.

Clerjeau, M., Moreau, C., Piganeau, B. and Malato, G. (1984) Resistance of *Plasmopara viticola* to anilide-fungicides: evaluation of the problem in France. *Mededelingen van de Faculteit Landbouwwetenschappen Rijksuniversiteit Gent* 49, 179–184.

Clerjeau, M., Irhir, H., Moreau, C., Piganeau, B., Staub, T. and Diriwachter, G. (1985) Etude de la résistance croisée au métalaxyl et au cyprofurame chez *Plasmopara viticola*: évidence de plusieurs mécanismes de résistance indépendants. In: Smith, I.M. (ed.) *Proceedings of the Centenary Meeting of Bordeaux Mixture – Fungicides for Crop Protection 100 Years of Progress.* BCPC Publications, Croydon, UK, Monograph No. 31, 1, pp. 303–306.

Corio-Costet, M.-F., Delmotte, F., Martinez, F., Giresse, X., Raynal, M., Richart-Cervera, S., *et al.* (2006) Resistance of *Plasmopara viticola* to QoI fungicides: origin and diversity. *Proceedings of 8th International Conference on Pests and Diseases 2006.* AFPP (ed.), France, pp. 612–620.

Corio-Costet, M.-F., Martinez, F., Delmotte, F., Douence, L., Richart-Cervera, S. and Chen, W.-J. (2008) Resistance of *Plasmopara viticola* to QoI fungicides: origin and diversity. In: Dehne, H.W, Gisi, U., Kuck, K.H., Russell, P.E. and Lyr, H. (eds) *Modern Fungicides and Antifungal Compounds V.* DPG Selbstverlag, Braunschweig, Germany, pp. 107–112.

Corio-Costet, M.-F., Dufour, M.-C., Cigna, J., Abadie, P. and Chen, W.-J. (2011) Diversity and fitness of *Plasmopara viticola* isolates resistant to QoI fungicides. *European Journal of Plant Pathology* 12, 315–329.

Couteux, A. and Lejeune, V. (2010) *Index Phytosanitaire ACTA 2010.* ACTA Publications, Paris, 747 p.

Davidse, L.C. (1995) Phenylamide fungicides – biochemical action and resistance. In: Lyr, H. (ed.) *Modern Selective Fungicides: Properties, Applications, Mechanisms of Action.* Gustav Fischer Verlag, Jena, Germany, pp. 347–354.

Delmotte, F., Chen, W.J., Richard-Cervera, S., Greif, C., Papura, D., Giresse, X., *et al.* (2006) Microsatellite DNA markers for *Plasmopara viticola,* the causal agent of downy mildew of grapes. *Molecular Ecology Notes* 6, 379–381.

Dercks, W. and Creasy, L.L. (1989) Influence of fosetyl-Al on phytoalexin accumulation in the *Plasmopara viticola* grapevine interaction. *Physiological and Molecular Plant Pathology* 34, 203–213.

Despreaux, D., Fritz, R. and Leroux, P. (1981) Mode d'action biochimique du cymoxanil. *Phytiatrie-Phytopharmacie* 30, 245–255.

Douchet, J.P., Absi, M., Hay, S.J.B., Mutan, L. and Villani, A. (1977) European results with DPX3217 a new fungicide for the control of grape downy mildew and potato late blight. In: *Proceedings Brighton Crop Protection Conference – Pests and Diseases.* BCPC Publications, UK, pp. 535–540.

Dufour, M.-C., Druelle, L., Sauris, P., Taris, G. and Corio-Costet, M.-F. (2009) Efficacités de stimulateurs de défenses des plantes (BTH et phosphonates) sur l'oïdium et le mildiou de la vigne: impact de la diversité des pathogènes. *AFPP, 9ème Conférence Internationale sur les maladies des plantes* (Tours, 4–5 December). Association Française de protection des plantes (AFPP), Alfortville, France, pp. 526–535.

EPPO Panel (2002) Good plant protection practice. *OEPP/EPPO Bulletin* 32, 371–392.

Fisher, D.J. and Hayes, A.L. (1979) Mode of action of the systemic fungicides furalaxyl, metalaxyl and ofurace. *Pesticide Science* 13, 330–339.

FRAC (2010) FRAC Code List: Fungicides sorted by mode of action (including FRAC Code numbering) 1–10 (http://www.frac.info/).

French National Memorandum of Grapevine Downy Mildew (2011) (http://draaf.midi-pyrenees.agriculture.gouv.fr/IMG/pdf/note_nationale_Vigne_mildiou_oidium_2011_Vdef_cle01fdcf.pdf).

Fry, W.E. and Milgroom, M.G. (1990) Population biology and management of fungicide resistance. In: Green, M.B, LeBaron, H.M. and Moberg, W.K. (eds) *Managing Resistance to Agrochemicals.* American Chemical Society, Washington, DC, pp. 275–285.

Galet, P. (1977) Mildiou. In: Galet, P. (ed.) *Les maladies et les parasites de la vigne.* Paysan du midi, Montpellier, France, pp. 89–222.

Gaulliard, J.M. (1985) Efficacité du phoséthyl-Al contre les souches de *Plasmopara viticola* résistantes aux anilides (acylalanines). *OEPP/EPPO Bulletin* 15, 437–441.

Genet, J.L. and Vincent, O. (1999) Sensitivity of European *Plasmopara viticola* populations to cymoxanil. *Pesticide Science* 55, 129–136.

Gisi, U. (2002) Chemical control of downy mildew. In: Spencer-Phillips, P.T.N., Gisi, U. and Lebeda, A. (eds) *Advances in Downy Mildew Research.* Kluwer Academic Publishers, Dordrecht, The Netherlands, pp. 119–159.

Gisi, U., Sierotzki, H., Cook, H. and McCaffery, A. (2002) Mechanisms influencing the evolution of resistance to Qo inhibitor fungicides. *Pest Management Science* 58, 859–867.

Gisi, U., Waldner, M., Kraus, N., Dubuis, P.H. and Sierotzki, H. (2007) Inheritance of resistance to carboxylic acid amide (CAA) fungicides in *Plasmopara viticola. Plant Pathology* 56, 199–208.

Grasso, V., Palermo, S., Sierotzki, H., Garibaldi, A. and Gisi, U. (2006) Cytochrome b structure and consequences for resistance to Qo inhibitor fungicides in plant pathogens. *Pest Management Science* 62, 465–472.

Heaney, S.P., Hall, A.A., Davies, S.A. and Olaya, G. (2000) Resistance to fungicides in the QoI-STAR cross-resistance group: current perspectives. *Proceedings Brighton Crop Protection Conference – Pest and Diseases.* BCPC, Farham, Surrey, UK, pp. 755–764.

Hollomon, D.W., Wood, P.M., Reeve, S. and Miguez, M. (2005) Alternative oxidase and its impacts on the activity of Qo and Qi site inhibitors. In: Dehne, H.W, Gisi, U., Kuck, K.H., Russell, P.E. and Lyr, H. (eds) *Modern Fungicides and Antifungal Compounds IV.* BCPC, Hampshire, UK, pp. 55–62.

Jordan, D.B., Livingston, R.S., Bisaha, J.J., Duncan, K.E., Pember, S.O., Picollelli, M.A., *et al.* (1999) Mode of action of famoxadone. *Pesticide Science* 55, 105–108.

Kousik, C.S. and Keinanth, A.P. (2008) First report of insensitivity to cyazofamid among isolates of *Phytophthora capsici* from the southeastern United States. *Plant Disease* 92, 979.

Lafon, R., Bugaret, Y. and Bulit, J. (1977) New prospects for controlling grapevine downy mildew (*Plasmopara viticola*) with a systemic fungicide, aluminium ethylphosphite. *Phytiatrie-Phytopharmacie* 26, 19–40.

Leger, B.O., Cartolaro, P., Delière, L., Delbac, L., Clerjeau, M. and Naud, O. (2008) An expert-based crop
    protection decision strategy against grapevine's powdery and downy mildews epidemics: part 1,
    formalization. *IOBC/WPRS Bulletin* 36, 145–153.
Leger, B., Naud, O., Bellon-Maurel, V., Clerjeau, M., Delière, L., Cartolaro, P. and Delbac, L. (2010)
    GrapeMildeWS: a formally designed integrated pest management decision process against grapevine
    powdery and downy mildews. In: Manos, B., Paparrizos, K., Matsatsinis, N. and Papathanasiou, J.
    (eds) *Decision Support Systems in Agriculture, Food and the Environnement: Trends, Applications and
    Advances.* Information Science Reference, Hershey, New York, pp. 246–269.
Leroux, P. (2000) La lutte chimique contre les maladies fongiques de la vigne. *Phytoma* 533, 32–38.
Leroux, P. and Clerjeau, M. (1985) Resistance of *Botrytis cinerea* pers. and *Plasmopara viticola* (Berk. et Curt.)
    Berl. and de Toni to fungicides in French vineyards. *Crop Protection* 4, 137–160.
Magnien, C., Jacquin, D., Muckensturm, N. and Guillemard, P. (1991) Milvit, a descriptive quantitative
    model for the asexual phase of grapevine downy mildew. *OEPP/EPPO Bulletin* 21, 451–460.
Magnien, C., Micoud, A., Glain, M. and Remuson, F. (2003) QoI resistance of downy mildew-monitoring and
    tests 2002. In: *Proceedings of 7th International Conference on Pest and Diseases 2003.* AFPP, France, pp. 1–8.
Magnien, C., Micoud, A., Remuson, F. and Grosman, J. (2009) La résistance du mildiou de la vigne aux
    fongicides: résultats des plans de surveillance de la sous direction de la qualité et de la protection des
    végétaux de 2005 à 2008. In: *Proceeding of 9th International Conferences on Pests and Diseases.* Association
    Française de Protection des Plantes (AFPP), France, pp. 750–760.
Maurin, G. (1983) Application d'un modèle d'état potentiel d'infection à *Plasmopara viticola. OEPP/EPPO
    Bulletin* 13, 263–269.
Mitani, S., Araki, S., Takii, Y., Ohshima, T., Matsuo, N. and Miyoshi, H. (2001) The biochemical mode of
    action of the novel selective fungicide cyazofamid: specific inhibition of mitochondrial complex III in
    Pythium spinosum. Pesticide Biochemistry and Physiology 71, 107–115.
Nuninger, C., Watson, G., Leadbeater, N. and Ellgehausen, H. (1996) CGA 329351: introduction of the
    enantiomeric form of the fungicide metalaxyl. *Proceedings Brighton Crop Protection Conference: Pests
    and Diseases* Volume 1. BCPC, Alton, UK, pp. 41–46.
Piganeau, B. and Clerjeau, M. (1985) Influence différentielle de la température sur la germination de spo-
    rocyste et la sporulation des souches de *Plasmopara viticola* sensible et résistantes aux phénylamides.
    Fungicides for Crop Protection. BCPC Monograph 31, 327–330.
Raynal, L., Ravisé, A. and Bompeix, G. (1980) Action du tri-o-éthylphosphonate d'aluminium (phoséthyl-
    Al) sur la pathogénie de *Plasmopara viticola* et sur la stimulation des réactions de défenses de la vigne.
    Annales de Phytopathologie 12, 163–175.
Samoucha, Y. and Gisi, U. (1987) Systemicity and persistence of cymoxanil in mixture with oxadixyl and
    mancozeb against *Phytophthora infestans* and *Plasmopara viticola. Crop Protection* 6, 393–398.
Serres, J.M. and Carraro, G.A. (1976) DPX-3217, a new fungicide for the control of grape downy mildew,
    potato blight and other Peronosporales. *Mededelingen van de Faculteit Landbouwwetenschappen
    Rijksuniversiteit Gent* 42, 645–650.
Sierotzki, H., Kraus, N., Assemat, P., Stanger, C., Cleere, C., Windass, J., *et al.* (2005) Evolution of resistance
    to QoI fungicides in *Plasmopara viticola* populations in Europe. In: Dehne, H.W, Gisi, U., Kuck, K.H.,
    Russell, P.E. and Lyr, H. (ed.) *Modern Fungicides and Antifungal Compounds.* BCPC, Alton, UK,
    pp. 73–80.
Sierotzki, H., Kraus, N., Pepin, S., Ferandes, N. and Gisi, H. (2008) Dynamics of QoI resistance in *Plasmopara
    viticola.* In: Dehne, H.W., Gisi, U., Kuck, K.H., Russell, P.E. and Lyr, H. (eds) *Modern Fungicides and
    Antifungal Compounds V.* DPG Selbstverlag, Braunschweig, Germany, pp. 151–157.
Staub, T. (1991) Fungicide resistance: practical experience with antiresistance strategies and the role of
    integrated use. *Annual Review of Phytopathology* 29, 421–442.
Staub, T. and Sozzi, D. (1981) Résistance au Métalaxyl en pratique et les conséquences pour son utilisation.
    Phytiatrie-Phytopharmacie 30, 283–291.
Strizyk, S. (1983) *Modèle d'état potentiel d'infection.* Association technique viticole, France, pp. 1–46.
Suty, A. and Stenzel, K. (1999) Iprovalicarb sensitivity of *Phytophthora infestans* and *Plasmopara viticola*:
    determination of baseline sensitivity and assessment of the risk of resistance. *Pflanzenschutz-
    Nachrichten Bayer* 52, 71–82.
Toffolatti, S.T., Serrati, L., Sierotzki, H., Gisi, U. and Vercesi, A. (2007) Assessment of QoI resistance in
    Plasmopara viticola oospores. *Pest Management Science* 63, 194–201.
Tran Manh Sung, C., Strizyk, S. and Clerjeau, M. (1990) Simulation of the date of maturity of *Plasmopara
    viticola* oospores to predict the severity of primary infections in grapevine. *Plant Disease* 74, 120–124.

Vegh, I., Leroux, P., Leberre, A. and Lanen, C. (1985) Détection sur *Chamaecyparis lawsoniana* 'Ellwoodii' d'une souche de *Phytophthora cinnamomi Rands* résistante au phoséthyl-Al. *PHM-Revue Horticole* 262, 19–21.

Williams, D.J., Beach, B.G.W., Horriere, D. and Marechal, G. (1977) LS 74-783, a new systemic fungicide with activity against phycomycete diseases. *Proceedings Brighton Crop Protection Conference – Pests and Diseases*, Volume 2. BCPC, Alton, UK, pp. 565–573.

Young, D.H. and Slawecki, R.A. (2001) Mode of action of zoxamide (RH-7281), a new oomycete fungicide. *Pesticide Biochemistry and Physiology* 69, 100–111.

Young, D.H., Kemmit, G.M. and Owen, J. (2005) A comparative study of XR-539 and other oomycete fungicides: similarity to dimethomorph and amino acid amides in its mechanism of action. In: Dehne, H.W, Gisi, U., Kuck, K.H., Russell, P.E. and Lyr, H. (eds) *Modern Fungicides and Antifungal Compounds IV*. BCPC, Alton, UK, pp. 145–152.

14 QoI Resistance in *Plasmopara viticola* in Italy: Evolution and Management Strategies

Silvia Laura Toffolatti and Annamaria Vercesi
Department of Plant Production, Plant Pathology Section,
University of Milan, Milano, Italy

14.1 Introduction

Viticulture is an important sector of agricultural production in Europe and in particular in Italy, where the numerous cultivars and the high diversity of climatic and topographical conditions result in a wide assortment of wines and table grapes (Fregoni, 2005). The main cultivation areas are located in southern Italy, in Sicilia and Puglia, followed by the northern and central regions, namely Veneto, Piemonte, Emilia-Romagna and Toscana. In the Italian viticultural regions, downy mildew, caused by *Plasmopara viticola* (Berk. *et* Curt.) Berlese and De Toni, is considered one of the most devastating diseases, particularly where moderate temperatures and frequent rainfall occur during the grapevine vegetative season (Lafon and Bulit, 1981). Chemical control is, at the moment, the most efficient means to prevent severe downy mildew epidemics (Gisi and Sierotzki, 2008a).

Several chemical classes of fungicides, with different characteristics in terms of systemicity, specificity, duration of activity and risk of resistance, are available against *P. viticola* on the Italian market, according to the European Directive 91/414 for the authorization of plant protection products (http://www.salute.gov.it/fitosanitariwsWeb_new/FitosanitariServlet).

Copper, the dithiocarbamates mancozeb, metiram and propineb, and the phthalimide folpet, employed either solo or in mixture with penetrative fungicides, are multi-site, non-systemic, preventive active substances that inhibit pathogen development before its penetration into the host tissue by unspecifically disrupting several biochemical processes (Russell, 2005; Gisi and Sierotzki, 2008a). Recently, protectant active substances characterized by a single-site mode of action were introduced to the market: zoxamide, acting on the microtubule assembly during mitosis (Young and Slawecki, 2001) and the QiI (quinone inside inhibitor) cyazofamid that affects mitochondrial respiration by binding to the inner binding site for ubiquinone (Qi) located in cytochrome *b* (Mitani *et al.*, 2001). Systemic fungicides, introduced into the market in the 1970s, are taken up and translocated inside the plant tissues, either through the apoplast or through the symplast, and are usually single-site inhibitors able to interfere with a single or a few targets of the fungal cell (Gisi, 2002). An acropetal movement characterizes the phenylamides benalaxyl, metalaxyl and their optical isomers benalaxyl-M and metalaxyl-M, inhibitors of nucleic acid synthesis acting on RNA polymerase I, while the ethyl phosphonate fosetyl-Al is the only fungicide able to translocate both acropetally and basipetally.

The cytotropic and/or translaminar fungicides include cymoxanil, whose mechanism of action is unknown (Genet and Vincent, 1999); the carboxylic acid amides dimethomorph, mandipropamid, benthiavalicarb, iprovalicarb and valifenalate, which act on cell wall synthesis (Blum *et al.*, 2010); fluopicolide, which interferes with the spectrin-like proteins, modifying their spatial and cellular distribution (Toquin *et al.*, 2010); and finally the QoIs (quinone outside inhibitors) that inhibit mitochondrial respiration by binding to the outer (Qo) binding site for ubiquinol located in cytochrome *b* (complex III). QoIs include several compounds belonging to different chemical classes (Bartlett *et al.*, 2002): the strobilurins azoxystrobin, pyraclostrobin, kresoxim-methyl and trifloxystrobin, the imidazolinone fenamidone and the protectant oxazolidinedione famoxadone. While famoxadone and fenamidone are active exclusively against *P. viticola* on grapevine, the strobilurins are employed also against the causal agent of powdery mildew *Erysiphe necator* Schwein. In particular on grapevine, kresoxym-methyl and trifloxystrobin are used exclusively against the powdery mildew agent, even if they are also moderately active against *P. viticola* (Bartlett *et al.*, 2002).

Kresoxym-methyl and azoxystrobin have been employed in Italy since 1997, followed in 1999 and 2001 by famoxadone and trifloxystrobin. Two years later, fenamidone was introduced into the Italian market, and finally pyraclostrobin was registered in 2005. The solo use of QoI fungicides against *P. viticola* lasted only a few years, due to the occurrence of resistant strains of the pathogen in northern Italy, and was replaced by mixtures with both systemic and protectant fungicides. However, solo formulations of QoIs used mainly against *E. necator* are still applied in vineyards.

14.2   QoI Resistance in Italy

P. viticola populations containing individuals resistant to QoIs were found at first in France and in an Italian field trial site during 1999 (Heaney *et al.*, 2000). Beginning in 2000, surveys have been carried out in vineyards located in different Italian regions by the Universities of Bologna, Torino and Milano, supported mainly by Regional Phytosanitary Services, which are responsible for resistance management under the coordination of the Italian Ministry of Agriculture. The monitoring activity has been limited to a few regions where severe downy mildew epidemics usually occur and a high number of fungicide treatments are applied during the same grapevine vegetative season. The surveys, concentrated between 2000 and 2002, were carried out in vineyards located mainly in northern Italian regions, Emilia Romagna (Brunelli *et al.*, 2002; Toffolatti *et al.*, 2007), Lombardia (Collina *et al.*, 2004, 2007; Toffolatti *et al.*, 2007, 2008a, in press), Friuli Venezia Giulia (Collina *et al.*, 2004, 2007), Trentino Alto Adige and Piemonte (Gilardi *et al.*, 2003, 2006; Gullino *et al.*, 2004) and in southern Italy, in Puglia only (Toffolatti *et al.*, 2008a,b, in press). In addition, resistance data, gathered by several agrochemical companies in Italy and in other European countries, are discussed by the QoI Fungicides Working Group of the Fungicide Resistance Action Committee (FRAC) and published on the website as the resistance risk level (www.frac.info).

A difficult control of *P. viticola* epidemics was observed in some field trials and commercial vineyards treated with fungicides belonging to the QoI group in northern Italy (Brunelli *et al.*, 2002, 2008). Reduced field performances usually were associated with a high number of QoI applications as a solo formulation. In fact, the QoI fungicides available on the Italian market at the end of the 1990s/beginning of the 2000s, the strobilurins azoxystrobin, kresoxym-methyl and trifloxystrobin, often were applied more than three times per season (Gullino *et al.*, 2004), consecutively and not in mixtures with partners belonging to different resistance groups. The repeated application of the strobilurins was due probably to their broad spectrum of activity that allowed a simultaneous protection against both grapevine downy and powdery mildew agents (Faretra *et al.*, 1998; Morando *et al.*, 1998; Gullino *et al.*, 2000; Liguori *et al.*, 2000; Manaresi and Coatti, 2002), even if with different control levels (Bartlett *et al.*, 2002). The strong selection pressure led, in some vineyards, to an increased frequency of resistance over sensitive strains. In fact, populations characterized by reduced sensitivity were found in QoI-treated vineyards located in some northern Italian regions and in Puglia. However, the reduced sensitivity of *P. viticola* populations did not always result in poor disease control.

In the following paragraphs, the results of the surveys carried out to evaluate the sensitivity of *P. viticola* to QoIs in different Italian regions are presented.

Emilia Romagna

Resistant populations in both experimental and commercial fields located in Emilia Romagna were found from 2000 to 2004 (Brunelli *et al.*, 2002; Collina *et al.*, 2004, 2007; Toffolatti *et al.*, 2007). In 2000, for the first time after the introduction of the strobilurins,

a reduced disease control, associated with the presence of resistant populations, was signalled in Italian commercial vineyards and confirmed by field trials carried out for 2 consecutive years in a vineyard located in the province of Bologna that had been treated with strobilurins from 1995 until 1999, with good results. Based on leaf-disc assays, the sporulated surface on untreated leaves of four of the seven *P. viticola* populations sampled during 2000 in highly infected commercial vineyards (six located in the province of Ravenna, one in the province of Bologna) were analogous to those observed at the field dose of azoxystrobin (250 mg/l). The minimum inhibitory concentration (MIC) was higher than 500 mg/l in all the cases (Table 14.1). Interestingly, a reduced sensitivity characterized the *P. viticola* populations sampled from two vineyards, the first located in the Forlì-Cesena province, where QoIs applied for 3 years controlled the disease efficiently, and the second in the Ravenna province, never treated with QoIs (Table 14.1).

Table 14.1. Results of sensitivity assays carried out on leaf discs sprayed with azoxystrobin in different Italian regions.

Region	Province	Year	QoI	Vineyards	ED_{50} (mg/l)	MIC (mg/l)	Reference
Piemonte		2000	Never applied	13	n.d.	0.3–3	Gullino *et al.* (2004)
	Alessandria	2002	No	11	0.1–1	0.3–10	Gilardi *et al.* (2003)
			Yes	3	1–3	1–10	
	Cuneo[a]	2002	No	1	1	30	
				2	100; >300	>300	
			Yes	4	30–100	1; >300	
				6	>300	>300	
		2002	Never applied	12	n.d.	≤3	Gullino *et al.* (2004)
Trentino Alto Adige		1999	Unknown	18	n.d.	1.3–5.3	Gullino *et al.* (2004)
				3	n.d.	11–21	
		2001	Never applied	1	3	100	Gullino *et al.* (2004); Gilardi *et al.* (2006)
				1	30	300	
				2	10–300	>1000	
			Yes	6	≤30	>300	
				1	30–100	>1000	
				3	100–300	>1000	
				9	>300	>1000	

Continued

Table 14.1. Continued.

Region	Province	Year	QoI	Vineyards	ED_{50} (mg/l)	MIC (mg/l)	Reference
			Unknown	2	10–100	≥1000	Gilardi *et al.* (2006)
				2	>1000	>1000	
		2002	Never applied	1	3	100	Gullino *et al.*
				2	0.1; >300	>300	(2004); Gilardi *et al.* (2006)
			Yes	2	0.1–100	300	
				2	100–300	>300	
				1	>300	>300	
			Unknown	1	<1	100	Gilardi *et al.* (2006)
				4	10–30	≥300	
				2	100–300	>300	
				6	>300	>300	
		2004	Unknown	3	0.1–1	>300	Gilardi *et al.* (2006)
				8	>300	>300	
Friuli		2001	Yes	4	10–300	≥1000	Gullino *et al.*
Venezia		2002	Unknown	1	0.3	>300	(2004); Gilardi
Giulia				13	>300	>300	*et al.* (2006)
Emilia	Bologna[b]	2000, 2001	Yes	2	n.d.	>2000	Brunelli *et al.* (2002)
Romagna	Ravenna	2000	Never applied	1	n.d.	>500	
			Yes	6[a]	n.d.	>500	
	Reggio Emilia	2001	Yes	1	n.d.	>100	
	Forlì-Cesena	2001	Never applied	1	n.d.	1	
		2000	Yes	1	n.d.	>1000	

Notes: [a]Vineyards at high disease incidence; [b]one vineyard is an experimental site; n.d., not determined.

In vitro assays carried out from 2000 until 2009 on oospores sampled from three different vineyards where QoIs had been applied from 1998 until 2004, showed an increased presence of resistant oospores in 2001 and 2002 (68 and 50%), compared to that observed in 2000 (7%), 2003 and 2004 (25 and 30%). Oospores collected from adjacent plots, untreated and treated with QoI fungicides within the same vineyard, were characterized by analogous resistance percentages (Table 14.2). From 2005, following the complete interruption of QoI use, the percentages of both resistant oospores and mutated allele decreased progressively towards the values observed in 2000.

Molecular assays carried out by Collina and co-workers (2004, 2007) and Toffolatti and co-workers (2007, 2010) between 2002 and 2004 showed percentages of mutated allele frequently higher than 50% in populations sampled in vineyards where *P. viticola* epidemics were controlled poorly by QoIs (Table 14.2).

Piemonte

Based on dose–response results (ED_{50} and MIC lower than 3 and 10 mg/l of azoxystrobin, respectively), fully sensitive *P. viticola* populations were found in vineyards never treated with QoIs in both 2000 and 2002. Analogous data were found in treated and untreated sites located in the province of Alessandria (Table 14.1). To the contrary, in the province of Cuneo, sampled in 2002, ED_{50} was higher than 10 mg/l in most cases.

Table 14.2. Average percentage of resistant oospores (RO) and mutated allele (MA) calculated on samples collected in Emilia Romagna, Puglia, Veneto and Lombardia.

Region	Province	Year	QoI	Vineyards	RO (%)	MA (%)	References
Emilia	Ravenna	2000	Untreated	1	1.5	–	Toffolatti *et al.*
Romagna			QoI	1	7	–	(2007)
		2001	QoI	3	68	–	
		2002	Untreated	2	45	–	
			QoI	2	58	–	
		2003	QoI	1	25	42	
		2004	Untreated	3	36	46	
			QoI	3	30	45	
		2005	Untreated	6	21	24	Toffolatti *et al.*
		2006	Untreated	5	14	24	(2008a)
		2007	Untreated	2	7	–	Toffolatti *et al.*
		2008	Untreated	1	5	–	(in press)
		2009	Untreated	2	1.8	–	
Puglia	Bari	2004	Untreated	1	3	7	Toffolatti *et al.*
			Solo	14	57	63	(2010)
			Mixture	3	5	7	
		2005	Untreated	3	11	11	
			Solo	4	25	32	
			Mixture	2	47	51	
		2006	Untreated	2	11	11	
			Solo	4	55	53	
			Mixture	4	33	49	
		2007	Untreated	1	7	–	Toffolatti *et al.*
			Solo	4	31	–	(in press)
			Mixture	1	4	–	
		2008	Untreated	0	–	–	
			Solo	1	79	–	
			Mixture	1	2	–	
		2009	Untreated	1	47	–	
			Solo	1	78	–	
			Mixture	1	61	–	
Veneto	Verona	2000	Never applied	1	0	–	Toffolatti *et al.*
		2001		1	0	–	(2007)
		2002		1	0	–	
		2003		1	0	0	
		2004	Never applied	1	5	9	Toffolatti *et al.*
	Padova	2005		2	3	9	(2010)
	Verona	2006		2	9	16	
Lombardia	Pavia	2005	Never applied	1	12	10.5	Toffolatti *et al.*
		2006		1	2	3	(2010)
		2007		1	0.5	–	Toffolatti *et al.*
	Sondrio	2007	Mixture	1	1	–	(in press)

Trentino-Alto Adige

The vineyards monitored in 1999 in Trentino-Alto Adige were characterized by fully sensitive *P. viticola* populations: leaf-disc assays carried out on bulk populations sampled in 19 commercial vineyards showed that the pathogen sporulation was, in most cases, suppressed completely at 1.3–5.3 mg/l of azoxystrobin (Table 14.1). The situation changed in 2001 and 2002, when most of the bulk populations sampled showed EC_{50} and

MIC values exceeding 10 and 300 mg/l, respectively, following the application of more than three QoI treatments in the same season (Gullino *et al.*, 2004). No data on the percentage of resistant strains inside each population are available, but some problems in controlling *P. viticola* epidemics have occurred (Gullino *et al.*, 2001).

Friuli Venezia Giulia

EC_{50} and MIC values higher than 10 and 300 mg/l, respectively, were detected in 2001 and 2002 in the 18 vineyards monitored in Friuli Venezia Giulia (Table 14.1).

Most of the *P. viticola* populations collected from 2000 to 2005 by Collina and co-workers (2007) in 84 vineyards were able to grow and sporulate on leaf discs sprayed with increasing doses of azoxystrobin (30, 60 and 250 mg/l) and were considered resistant to QoIs. However, the percentage of the mutated allele associated with resistance in the same populations was lower than 25% in 18 out of 28 vineyards. The higher proportion of sensitive over resistant individuals inside the population could explain the good field performance of QoIs observed in most cases.

Lombardia and Veneto

Sensitivity assays on oospores have been carried out in three vineyards never treated with QoIs, two located in Veneto, in the provinces of Verona (2000–2006) and Padova (2005–2006), and one in Lombardia, in the province of Pavia (2005–2007). The oospore populations were characterized by low percentages of resistant spores and mutated allele, ranging from 0 to 12 and 16%, respectively (Table 14.2). Very few oospores, 1% of the total population collected in the province of Sondrio from a vineyard treated just once with a QoI in mixture with a partner belonging to a different resistance group, were able to germinate at 10 mg/l azoxystrobin.

Low percentages of mutated allele (1–10%) were detected in populations of oospores and sporangia collected in the province of Pavia

(Toffolatti *et al.*, 2007, 2010) and Brescia (Collina *et al.*, 2007) between 2000 and 2005. To the contrary, in the province of Sondrio, a high frequency of point mutation G143A, ranging from 50 to 100%, was detected in sporangia populations collected in four out of seven vineyards.

Puglia

Sensitivity monitoring has been carried out on *P. viticola* oospore populations sampled from 2004 until 2009 in the province of Bari. The average percentage of individual oospores resistant to QoIs in different populations collected in untreated vineyards was generally lower than 11%, except for a single field in 2009, when 47% was recorded. It has to be pointed out that in this particular vineyard, high rates of resistant oospores (79%) associated with solo QoI treatments had been recorded in the previous year. The populations sampled from vineyards treated with QoIs, against both downy and powdery mildew, with a total number of applications often higher than recommended, showed variable rates of resistant oospores that were generally higher where the fungicides were applied as solo formulation (mean value 54%) than in mixture (mean value 25%). The percentage of mutated allele strictly followed the resistant oospore values (Table 14.2).

14.3 Characteristics of Resistant Strains and Evolution of QoI Resistance

Mitochondrial DNA inheritance is essentially non-Mendelian and transmission to offspring is predominantly uniparental and, in the particular case of Oomycetes, maternal (Gisi *et al.*, 2002). QoI resistance shows a segregation pattern of 1:3, slightly different from the expected 0:1 ratio, and is homoplasmic: single isolates are either fully sensitive and possess G143 allele, or fully resistant with A143 allele (Blum and Gisi, 2008; Gisi and Sierotzki, 2008b). The possible evolution of resistance in vineyards is influenced strongly by the fitness of resistant strains.

Studies carried out by measuring parameters such as infection ability and sporangia size and production suggest that QoI-resistant strains of *P. viticola* possess fitness properties as high as sensitive strains on untreated, detached leaves (Lafarge *et al.*, 2008). However, a decline in the frequency of resistant strains is observed in mixed populations after several generations on untreated leaves. The relative fitness of resistant subpopulations in competition with sensitive ones during asexual cycles was determined by dose–response assays on three populations, the first composed of sensitive individuals, the second of resistant and the third of a 1:1 mixture of sensitive and resistant individuals, kept on untreated vine seedlings for ten generations. Results showed that while purely sensitive and resistant strains maintained the same EC_{50}, low and high, respectively, for ten generations, the mixed population showed decreasing EC_{50} starting from the eighth generation and values analogous to full sensitive strains at the ninth (Heaney *et al.*, 2000). Similar declines in resistance levels, but starting from the fourth generation, were observed in a population containing 5% resistant individuals (Genet *et al.*, 2006). Based on the germinability on an azoxystrobin-amended medium and the percentage of mutated allele, a reduction in the frequency of resistant individuals following the suspension of QoI use has also been demonstrated in oospore populations sampled from Italian vineyards located in Emilia Romagna and Puglia (Toffolatti *et al.*, 2010). Therefore, in the absence of selection pressure, sensitive phenotypes seem to overcome resistant ones.

14.4 Methods for the Detection of QoI Resistance in *P. viticola*

With the purpose of finding fast and reliable tests able to detect QoI-resistant strains in the pathogen population, several different methods, based on the response of *P. viticola* to increasing concentrations of the chemicals or on the detection of the allele leading to G143A

and/or F129L substitutions, were developed. QoI sensitivity of *P. viticola* population can be tested by biological assays performed *in vivo*, by evaluating the symptoms of the disease, expressed as a percentage of the infected leaf area, on treated grape seedlings or vine cuttings inoculated with sporangial suspensions obtained from symptomatic leaves sampled in vineyard (Brunelli *et al.*, 2002). Due to the standardization of the method, *in vitro* tests, carried out on leaf discs kept in Petri dishes or microtitre plates (www.frac.info), are usually employed (Gullino *et al.*, 2004). The results are expressed as EC_{50} (median effective concentration) and ED_{50} (median effective dose), but also as minimum inhibitory concentration (MIC), and compared with the values of sensitive reference strains (Russell, 2004). Collina and co-workers (2004, 2007) identified resistant populations by their ability to sporulate on leaf discs treated with QoI discriminatory doses. The biological assays carried out on bulk samples provide qualitative information on QoI resistance. The composition of the sampled populations, i.e. the relative frequency of resistant strains over sensitive ones, is estimated exclusively on single strains isolated from single lesions or sporangia. Another biological test developed for the assessment of QoI resistance is based on the germinability of the oospores at the discriminatory concentration of the fungicide and allows the calculation of the percentage of spores in the population resistant to the fungicide (Toffolatti *et al.*, 2007, 2010). Heaney and co-workers (2000) stated that the reference concentration for resistant field isolates was 10 mg/l.

Since the discovery that QoI resistance is associated mainly with the point mutation leading to G143A substitution in cytochrome *b* protein, molecular diagnostics, based on PCR techniques, have been developed and are currently available. Allele-specific amplification using specific primers for wild-type and mutated alleles by real-time PCR combines detection to the quantification of the point mutation leading to resistance (Sierotzki and Gisi, 2003; Collina *et al.*, 2007; Toffolatti *et al.*, 2007). Collina and co-workers (2007) observed that percentages of mutated allele higher than 50% usually were associated with poor disease control.

14.5   Mechanism of QoI Resistance

A major function of mitochondria is the conversion of energy released by metabolic processes in the mitochondrial matrix into the energy currency of the cell, ATP (Vonck and Schäferb, 2009). Inhibition of respiration therefore has severe consequences for cell maintenance and proliferation.

The components of the electron transport chain in eukaryotes are the transmembrane protein complexes I, NADH-ubiquinone reductase; II, succinate-ubiquinone reductase; III, ubiquinol-cytochrome c reductase or cytochrome bc_1 complex; and IV, cytochrome c oxidase. Respiratory complexes I and II pass electrons from the redox substrates NADH and succinate on to complex III through ubiquinone or coenzyme Q, and from complex III to IV through cytochrome c, generating an electro-osmotic proton gradient across the membrane that is used by ATP synthase (complex V) to synthesize ATP (Griffin, 1994; Schägger, 2001). Cytochrome bc_1, a membrane-bound homodimeric complex composed of two monomeric units of 10–11 different polypeptides (Fisher and Meunier, 2005), possesses two spatially separated binding sites, called Qo and Qi from their relative positions in the inner mitochondrial membrane, for ubiquinol and for its oxidized form, ubiquinone. At Qo, ubiquinol (UQH_2) is oxidized to ubiquinone (UQ), whereas at Qi, ubiquinone is reduced to semi-quinone anion (UQ^-) (Wood and Hollomon, 2003). A complete turnover of the enzyme requires the oxidation of two ubiquinol molecules, resulting in reduction of the semi-quinone (Wenz *et al.*, 2007).

Most of the QoI fungicides used in agriculture are synthetic analogues of strobilurin A, a β-methoxyacrylate (MOA) derivative synthesized by Basidiomycetes fungi such as *Strobilurus tenacellus* (Pers.) Singer (Clough and Godfrey, 1998; Balba, 2007). The strobilurins used on grapevine are characterized by different toxophore groups: methoxyacrylate (azoxystrobin), methoxycarbamate (pyraclostrobin) and methoxyiminoacetate (kresoxymmethyl and trifloxystrobin). Famoxadone and fenamidone show the same mode of action as strobilurins but have different chemical structures (Bartlett *et al.*, 2002).

Crystal structures of bovine cytochrome bc_1 in complex with inhibitors such as MOA azoxystrobin showed that these inhibitors were bound in the hydrophobic Qo pocket with the toxophore inserted between the residues Phe128 and Tyr131 of the C helix, forming a hydrogen bond between the methyl ester oxygen of the fungicide and the amide group of Glu271 (Esser *et al.*, 2004). Analogous results, involving the same amino acids but at a different position (Phe129 and Glu272), were obtained in yeast (Gisi *et al.*, 2002; Wenz *et al.*, 2007). Phe129 presumably supports the methoxy vinyl group of the fungicide by favourable van der Waals interactions and stabilizes the pyrimidyl ring by aromatic–aromatic interactions (Gisi *et al.*, 2002). A single point mutation in the cytochrome b gene leading to a substitution of Phe129 for a different amino acid can lead to severe respiratory deficiencies, as happens with Lys and Arg in yeast, or to a decreased sensitivity towards the fungicides, as in the case of Leu (F129L). Mutation F129L, in fact, removes the stabilizing interactions between the protein and the inhibitor (Fisher and Meunier, 2008). F129L mutation, found in resistant strains of several plant pathogens, such as *Pyricularia grisea* (Sacc.) and *Pythium aphanidermatum* (Edson) Fitzp., is not frequently detected in field isolates of *P. viticola* (Sierotzki *et al.*, 2005).

The phenyl group of MOAs that carries and orients the toxophore properly in bovine cytochrome b is wedged between the Pro270 in the PEWY sequence and Gly142 of the cd1 helix. The substitution of Gly with an Ala in this position (G143A mutation in fungi) reduces the affinity of the azoxystrobin binding: Ala is large enough to interfere sterically with the inhibitor binding, but is sufficiently small as not to cause a performance penalty during ubiquinol oxidation (Esser *et al.*, 2004). G143A mutation occurs in, among others (Gisi and Sierotzki, 2008b), *P. viticola*, *Venturia inaequalis* (Cooke) Wint., *Alternaria alternata* (Fr.) Keissl., *Pyrenophora tritici-repentis* (Died.) Drechsler and *Blumeria graminis* (DC.) Speer.

Several other mutations in two regions of the cytochrome *b* gene, corresponding to amino acid positions 127–147 and 275–296, have been found in different organisms (Gisi *et al.*, 2002), but since they reduce the pathogen fitness as a consequence of respiratory deficiency, they are detected at very low frequency in the field. Between F129L and G143A mutations, the latter is associated to particularly high resistance factors and found at higher frequency in field populations of *P. viticola* (Sierotzki *et al.*, 2005).

Resistance mechanisms, such as the presence of alternative oxidation and efflux transporters, described, for example, for *Botrytis cinerea* Pers. and *Mycosphaerella graminicola* (Fuckel) J. Schröt., mainly in response to fungicides belonging to different chemical classes (Leroux *et al.*, 2002; Wood and Hollomon, 2003; de Waard *et al.*, 2006), have not been reported, to the best of our knowledge, for *P. viticola*.

14.6   QoI Resistance Management

The first recommendations for a rational use of QoI fungicides that could limit the occurrence of resistance in practice were presented in Italy in 2000. Due to the existence of cross-resistance among the different chemicals of this class, it was generally suggested to apply QoIs not consecutively but in alternation with fungicides with a different mechanism of action and for a maximum number of three sprays per year (Faretra and Gullino, 2000). In the particular case of *P. viticola*, the indications of FRAC on the use of QoIs not as solo applications but in mixture with appropriate partners belonging to a different resistance group were recommended (Grasso *et al.*, 2005; Gilardi *et al.*, 2006). As demonstrated by experimental assays, in comparison with mixtures, solo treatments exert a higher selection pressure on *P. viticola* populations and should, therefore, be avoided. A reduction of the sensitivity level was observed in a population containing 5% resistant individuals starting from the fourth propagation cycle on grapevine seedlings treated with famoxadone and azoxystrobin (Genet *et al.*, 2006). On the contrary, the application of QoIs in mixture did

not change the sensitivity level, as a consequence of a delayed selection pressure. The effect of the QoI formulation (solo or in mixture) on the distribution of resistant individuals and mutated allele (G143A) has also been studied on the oospores: while the populations sampled from solo-treated vineyards showed very variable rates of resistant oospores and mutated allele (0–100%), with a 50% portion of the samples ranging from 15–20 to 80–90%, those collected where QoIs had been applied in mixture showed reduced percentages of the same parameters (0–75%), with the middle half of the data enclosed between 15 and 45% (Toffolatti *et al.*, 2010).

Field trials carried out in Italian vineyards by using different QoI fungicides against *P. viticola* (azoxystrobin, famoxadone, fenamidone and pyraclostrobin) in mixtures also showed that a good disease control could be achieved by using anti-resistance strategies (Mutton *et al.*, 2004; Morando *et al.*, 2006; Brunelli *et al.*, 2008). In Emilia Romagna, several trials were carried out from 2001 until 2007 in an experimental farm located in the province of Bologna where *P. viticola* populations were characterized by a reduced sensitivity in 2000 (Brunelli *et al.*, 2002). On the plots where QoIs were applied solo, the percentage of infected leaves and bunches was, in general, significantly higher than on the plots sprayed with mixtures, where a good disease control was achieved (Brunelli *et al.*, 2008). A poor disease control of solo formulations compared to the mixtures was also observed during field trials carried out in 2002 and 2003 in a vineyard located in Friuli Venezia Giulia where QoI-resistant populations of *P. viticola* were found in 2001 (Gullino *et al.*, 2004; Mutton *et al.*, 2004). No reduction in the field performance of the fungicide class was ever observed in fields located in Piemonte, where QoIs were always applied three times per season in mixtures from 2002 until 2005 (Morando *et al.*, 2006).

14.7   Conclusions

A highly variable distribution of *P. viticola* sensitivity to QoIs is illustrated by the published data concerning the few regions monitored in Italy. In general, due to the high number of treatments, often as solo formulations, QoI-resistant populations were detected in most of the vineyards sampled in the early 2000s. In Emilia Romagna, Trentino Alto Adige and Friuli Venezia Giulia, the presence of resistant strains was seldom associated with a reduced disease control. In other vineyards, the field performance of the fungicides was good, despite the presence of less sensitive populations, probably because the frequency of the resistant individuals was lower than the sensitive ones. Where anti-resistance strategies such as the use of mixtures and limitation of QoI treatments to three per season were applied, downy mildew was controlled efficiently and the frequency of resistant oospores often decreased.

An extensive monitoring activity on fungicide resistance should be carried out, therefore, in Italian vineyards in order to gather information on the sensitivity status of *P. viticola* populations and to promote the correct use of the chemical classes available, not only in terms of disease management but also in terms of anti-resistance strategies. Vine growers using QoIs against both downy and powdery mildew agents should take particular care in programming the treatments, to avoid an excessive number of QoI applications, especially if solo formulations are employed. Preservation of the activity of all the available classes of fungicides is particularly crucial in Italy, where numerous chemical sprays are performed to prevent severe *P. viticola* epidemics. From this point of view, the reduction in the frequency of resistant individuals following the interruption of QoI use suggests that a shift towards full sensitivity can be achieved and, therefore, that this important and effective class of chemicals can still be utilized following proper resistance management.

References

Balba, H. (2007) Review of strobilurin fungicide chemicals. *Journal of Environmental Science and Health Part B Pesticides, Food Contaminants, and Agricultural Wastes* 42, 441–451.

Bartlett, D.W., Clough, J.M., Godwin, J.R., Hall, A.A., Hamer, M. and Parr-Dobrzanski, B. (2002) The strobilurin fungicides. *Pest Management Science* 58, 649–662.

Blum, M. and Gisi, U. (2008) Inheritance of fungicide resistance in *Plasmopara viticola*. In: Dehne, H.W., Deising, H.B., Gisi, U., Kuck, K.H., Russell, P.E. and Lyr, H. (eds) *Modern Fungicides and Antifungal Compounds V*. DPG Selbstverlag, Braunschweig, Germany, pp. 101–104.

Blum, M., Boehler, M., Randall, E., Young, V., Csukai, M., Kraus, S., *et al.* (2010) Mandipropamid targets the cellulose synthase-like PiCesA3 to inhibit cell wall biosynthesis in the oomycete plant pathogen, *Phytophthora infestans. Molecular Plant Pathology* 11, 227–243.

Brunelli, A., Collina, M., Guerrini, P. and Gianati, P. (2002) Ridotta sensibilità di *Plasmopara viticola* ai fungicidi QoI in Emilia Romagna. *Atti Giornate Fitopatologiche*, Volume 2. CLUEB, Bologna, Italy, pp. 279–288.

Brunelli, A., Gianati, P., Portillo, I., Sedda, G. and Collina, M. (2008) Effectiveness on grapevine downy mildew of QoIs applied in mixture or alternation with different fungicides in vineyards with resistance of *Plasmopara viticola. Atti Giornate Fitopatologiche*, Volume 2. CLUEB, Bologna, Italy, pp. 253-260.

Clough, J.M. and Godfrey, C.R.A. (1998) The strobilurin fungicides. In: Hutson, D. and Miyamoto J. (eds) *Fungicidal Activity – Chemical and Biological Approaches to Plant Protection*. John Wiley and Sons, Chichester, UK, pp. 109–148.

Collina, M., Landi, L., Guerrini, P., Branzati, M.B. and Brunelli, A. (2004) QoI resistance of *P. viticola* in Italy: biological and quantitative real-time PCR approaches. *Atti Giornate Fitopatologiche*, Volume 2. CLUEB, Bologna, Italy, pp. 203–204.

Collina, M., Landi, L., Branzati, M.B. and Brunelli, A. (2007) Studies on *Plasmopara viticola* QoI resistance by biological assays and quantitative PCR real-time. *Italus Hortus* 14, 242–246.

De Waard, M.A., Andrade, A.C., Hayashi, K., Schoonbeek, H., Stergiopoulos, I. and Zwiers, L.-H. (2006) Impact of fungal drug transporters on fungicide sensitivity, multidrug resistance and virulence. *Pest Management Science* 62, 195–207.

Esser, L., Quinn, B., Li, Y.-F., Zhang, M., Elberry, M., Yu, L., *et al.* (2004) Crystallographic studies of quinol oxidation site inhibitors: a modified classification of inhibitors for the cytochrome bc_1 complex. *Journal of Molecular Biology* 341, 281–302.

Faretra, F. and Gullino, M.L. (2000) La resistenza ai fungicidi nella protezione delle colture. *Informatore Fitopatologico* 10, 52–58.

Faretra, F., Santomauro, A., Polizzi, G., Catara, A., D'Ascenzo, D., Grande, C., *et al.* (1998) The control of powdery mildew with azoxystrobin and kresoxim-methyl in vineyards located in central and southern Italy. *Atti Giornate Fitopatologiche*, Volume 2. CLUEB, Bologna, Italy, pp. 569–574.

Fisher, N. and Meunier, B. (2005) Re-examination of inhibitor resistance conferred by Qo-site mutations in cytochrome *b* using yeast as a model system. *Pest Management Science* 61, 973–978.

Fisher, N. and Meunier, B. (2008) Molecular basis of resistance to cytochrome bc_1 inhibitors. *FEMS Yeast Research* 8, 183–192.

Fregoni, M. (2005) *Viticoltura di qualità*, 2nd edn. Tecniche Nuove, Milano, Italy.

Genet, J.-L. and Vincent, O. (1999) Sensitivity of European *Plasmopara viticola* populations to cymoxanil. *Pesticide Science* 55, 129–136.

Genet, J.L., Jaworska, G. and Deparis, F. (2006) Effect of dose rate and mixtures of fungicides on selection for QoI resistance in populations of *Plasmopara viticola*. *Pest Management Science* 62, 188–194.

Gilardi, G., Benzi, D., Cravero, S., Garibaldi, A. and Gullino, M.L. (2003) Sensibilità a diverse famiglie di fungicidi in popolazioni di Plasmopara viticola in Piemonte. *Notiziario sulla Protezione delle Piante* 17, 49–54.

Gilardi, G., Bertetti, D., Grasso, V. and Gullino, M.L. (2006) Un aggiornamento sulla resistenza ai principali fungicidi impiegati in viticoltura. *Informatore Fitopatologico* 4, 30–38.

Gisi, U. (2002) Chemical control of downy mildews. In: Spencer-Phillips, P.T.N., Gisi, U. and Lebeda, A. (eds) *Advances in Downy Mildew Research*. Kluwer Academic Publishers, Dordrecht, The Netherlands, pp. 119–159.

Gisi, U. and Sierotzki, H. (2008a) Fungicide modes of action and resistance in downy mildews. *European Journal of Plant Pathology* 122, 157–167.

Gisi, U. and Sierotzki, H. (2008b) Molecular and genetic aspects of fungicide resistance in plant pathogens. In: Dehne, H.W., Deising, H.B., Gisi, U., Kuck, K.H., Russell, P.E. and Lyr, H. (eds) *Modern Fungicides and Antifungal Compounds V*. DPG Selbstverlag, Braunschweig, Germany, pp. 53–61.

Gisi, U., Sierotzki, H., Cook, A. and McCaffery, A. (2002) Mechanisms influencing the evolution of resistance to Qo inhibitor fungicides. *Pest Management Science* 58, 859–867.

Grasso, V., Garibaldi, A. and Gullino, M.L. (2005) Resistance to QoI fungicides: state of the art and practical problems. *Informatore Fitopatologico* 7/8, 33–39.

Griffin, D.H. (1994) *Fungal Physiology*. Wiley-Liss, New York.

Gullino, M.L., Leroux, P. and Smith, C.M. (2000) Uses and challenges of novel compounds for plant disease control. *Crop Protection* 19, 1–11.

Gullino, M.L., Gilardi, G., Stefanelli, G., Mescalchin, E. and Garibaldi, A. (2001) Detection of populations of *Plasmopara viticola* resistant to fungicides inhibiting mitochondrial respiration (QoI STAR) in vineyards located in North-Eastern Italy. *Informatore Fitopatologico* 12, 86–87.

Gullino, M.L., Gilardi, G., Tinivella, F. and Garibaldi, A. (2004) Observations on the behaviour of different populations of *Plasmopara viticola* resistant to QoI fungicides in Italian vineyards. *Phytopathologia Mediterranea* 43, 341–350.

Heaney, S.P., Hall, A.A., Davies, S.A. and Olaya, G. (2000) Resistance to fungicides in the QoI-STAR cross resistance group: current perpectives. In: *Proceedings of Brighton Crop Protection Conference, Pests and Diseases*. BCPC, Farnham, UK, pp. 755–762.

Lafarge, D., Abadie, P., Cigna, J., Douence, L., Dufour, M.C. and Corio-Costet, M.-F. (2008) Competitive fitness and adaptation of QoI-resistant *Plasmopara viticola* strains. In: Dehne, H.W., Deising, H.B., Gisi, U., Kuck, K.H., Russell, P.E. and Lyr, H. (eds) *Modern Fungicides and Antifungal Compounds V*. DPG Selbstverlag, Braunschweig, Germany, pp. 181–185.

Lafon, R. and Bulit, J. (1981) Downy mildew of the vine. In: Spencer D.M. (ed.) *The Downy Mildews*. Academic Press, New York, pp. 601–614.

Leroux, P., Fritz, R., Debieu, D., Albertini, C., Lanen, C., Bach, J., *et al.* (2002) Mechanisms of resistance to fungicides in field strains of *Botrytis cinerea*. *Pest Management Science* 58, 876–888.

Liguori, R., Bertona, A., Bassi, R., Filì, V., Filippi, G., Saporiti, M., *et al.* (2000) Trifloxystrobin (CGA 279202): new broad spectrum fungicide. *Atti Giornate Fitopatologiche*, Volume 2. CLUEB, Bologna, Italy, pp. 3–8.

Manaresi, M. and Coatti, M. (2002) F500 (pyraclostrobin): a newest broad-spectrum strobilurin fungicide. *Atti Giornate Fitopatologiche*, Volume 2. CLUEB, Bologna, Italy, pp. 119–124.

Mitani, S., Araki, S., Takii, Y., Ohshima, T., Matsuo, N. and Miyoshi H. (2001) The biochemical mode of action of the novel selective fungicide cyazofamid: specific inhibition of mitochondrial complex III in *Phythium spinosum*. *Pesticide Biochemistry and Physiology* 71, 107–115.

Morando, A., Lembo, S., Morando, D. and Morando, M. (1998) Trials of downy mildew control on grapevine with new fungicides. *Atti Giornate Fitopatologiche*, Volume 2. CLUEB, Bologna, Italy, pp. 525–530.

Morando, A., Lavezzaro, S. and Gallesio, G. (2006) Four years of downy mildew control trials by using QoI fungicides. *Atti Giornate Fitopatologiche*, Volume 2. CLUEB, Bologna, Italy, pp. 219–226.

Mutton, P., Boccalon, W., Bressan, S., Marchi, G. and Mucignat, D. (2004) Activity of azoxystrobin and famoxadone in vineyards with *Plasmopara viticola* resistant to QoI fungicides. *Atti Giornate Fitopatologiche*, Volume 2. CLUEB, Bologna, Italy, pp. 197–202.

Russell, P.E. (2004) *Sensitivity Baselines in Fungicide Resistance Research and Management*. CropLife International, Brussels.

Russell, P.E. (2005) A century of fungicide evolution. *Journal of Agricultural Science* 143, 11–25.

Schägger, H. (2001) Respiratory chain supercomplexes. *IUBMB Life* 52, 119–128.

Sierotzki, H. and Gisi, U. (2003) Molecular diagnostics for fungicide resistance in plant pathogens. In: Voss, G. and Ramos, G. (eds) *Chemistry of Crop Protection*. Wiley-VCH, Weinheim, Germany, pp. 71–89.

Sierotzki, H., Kraus, N., Assemat, P., Stanger, C., Cleere, S., Windass, J., *et al.* (2005) Evolution of resistance to QoI fungicides in *Plasmopara viticola* in Europe. In: Dehne, H.W., Gisi, U., Kuck, K.H., Russel, P.E. and Lyr H. (eds) *Modern Fungicides and Antifungal Compouds IV*. BCPC, Alton, UK, pp. 73–80.

Toffolatti, S.L., Serrati, L., Sierotzki, H., Gisi, U. and Vercesi, A. (2007) Assessment of QoI resistance in *Plasmopara viticola* oospores. *Pest Management Science* 63, 194–201.

Toffolatti, S.L., Prandato, M., Serrati, L. and Vercesi, A. (2008a) Investigations on the *Plasmopara viticola* strains resistant to QoIs in northern Italy. *Atti Giornate Fitopatologiche*, Volume 2. CLUEB, Bologna, Italy, pp. 247–252.

Toffolatti, S.L., Prandato, M., Serrati, L., Sierotzki, H., Gisi, U. and Vercesi, A. (2008b) Monitoring QoI resistance in *Plasmopara viticola* oospore populations. In: Dehne, H.W., Deising, H.B., Gisi, U., Kuck, K.H., Russell, P.E. and Lyr, H. (eds) *Modern Fungicides and Antifungal Compounds V*. DPG Selbstverlag, Braunschweig, Germany, pp. 159–165.

Toffolatti, S.L., Prandato, M., Serrati, L., Sierotzki, H., Gisi, U. and Vercesi, A. (2010) Evolution of QoI resistance in *Plasmopara viticola* oospores. *European Journal of Plant Pathology* 129, 331–338.

Toffolatti, S.L., Serrati, L. and Vercesi, A. CAA, phenylamide and QoI resistance assessment in *Plasmopara viticola* oospores. In: *Modern Fungicides and Antifungal Compounds VI* (in press).

Toquin, V., Barja, F., Sirven, C., Gamet, S., Mauprivez, L., Peret, P., *et al.* (2010) Novel tools to identify the mode of action of fungicides as exemplified with fluopicolide. In: Gisi, U., Chet, I. and Gullino, M.L. (eds) *Recent Developments in Management of Plant Diseases*. Springer, Dordrecht, The Netherlands, pp. 19–36.

Vonck, J. and Schäferb, E. (2009) Supramolecular organization of protein complexes in the mitochondrial inner membrane. *Biochimica et Biophysica Acta* 1793, 117–124.

Wenz, T., Covian, R., Hellwig, P., MacMillan, F., Meunier, B., Trumpower, B.L., *et al.* (2007) Mutational analysis of cytochrome *b* at the ubiquinol oxidation site of yeast complex III. *The Journal of Biological Chemistry* 282, 3977–3988.

Wood, P.M. and Hollomon, D.W. (2003) A critical evaluation of the role of alternative oxidase in the performance of strobilurin and related fungicides acting at the Qo site of Complex III. *Pest Management Science* 59, 499–511.

Young, D.H. and Slawecki, R.A. (2001) Mode of action of zoxamide (RH-7281), a new oomycete fungicide. *Pesticide Biochemistry and Physiology* 69, 100–111.

15 Fungicide Resistance in Italian Agriculture and Strategies for its Management

Maria Lodovica Gullino, Domenico Bertetti and Angelo Garibaldi
Agroinnova, University of Torino, Grugliasco, Italy

15.1 Introduction

Fungicides are and will remain essential for maintaining healthy crops and reliable, high quality yields in countries such as Italy, characterized by a high quality, intensive horticultural production. Fungicides are key components of integrated crop management and their effectiveness must be sustained for as long as possible (Leadbeater and Gisi, 2010). Pathogen resistance to fungicides is widespread worldwide. As a consequence, the performance of many recently developed fungicides has been affected to some degree. In Italy, however, fungicide resistance could be much worse. Most classes of fungicides are still effective in many situations, due probably to several reasons (presence of relatively small-scale farms which use different classes of fungicides, widespread use of IPM, adoption of cultural practices, etc.) which eventually lead to a more limited use of fungicides in comparison with other countries.

In some cases (i.e. phenylamides in the early 1980s), the early adoption of countermeasures proved to be beneficial in avoiding the development of resistance problems, in comparison to what was observed in other areas. This chapter will review the situation of fungicide resistance in Italy in the past 10 years, by taking into account the most used classes of fungicides and the most widespread diseases on economically important crops. The status of fungicide resistance in Italy is also summarized under Table 15.1. In order to facilitate readers, classes of fungicides will be used for describing the phenomenon of resistance development in different pathogens.

15.2 Situation of Fungicide Resistance in Italy

Benzimidazoles

Although very few components of this class of fungicides are still registered for use, resistance to them is widespread in many pathogens, such as *Botrytis cinerea*, causal agent of grey mould on several crops (grapevine, tomato, strawberry, kiwi, several ornamental crops) (Garibaldi *et al.*, 1987; Garibaldi and Gullino, 1990; Gullino, 1992; Gullino and Garibaldi, 2007). Benzimidazole-resistant strains are quite stable and remain in pathogen populations years after their use has been dismissed. However, a recent survey carried out in Italian vineyards showed a reduction in the frequency of benzimidazole-resistant mutants of *B. cinerea* (Bertetti *et al.*, 2008). The genetic basis of resistance to benzimidazoles in *B. cinerea* has been widely exploited (Faretra and Pollastro, 1991). Benzimidazole resistance

Table 15.1. Practical cases of fungicide resistance in Italy.

Fungicide	Pathogen (disease)	Crop(s)	Year of detection
Benzimidazoles	*Venturia inaequalis* (apple scab)	Apple	1972
	Botrytis cinerea, Botrytis spp. (grey mould)	Grapevine, vegetables, ornamentals	1970s
Dodine	*Venturia inaequalis* (apple scab)	Apple	1983
Dicarboximides	*Botrytis cinerea* (grey mould)	Grapevine, vegetables, ornamentals	1980
	Stemphylium vesicarium (pear brown spot)	Pear	2010
	Sclerotinia homeocarpa (dollar spot)	Turf	2001
Phenylamides	*Plasmopara viticola* (grape downy mildew)	Grapevine	1990
	Bremia lactucae (lettuce downy mildew)	Lettuce	1996
	Phytophthora nicotianae var. *parasitica* (root rot)	Ornamentals	1986
	Pythium spp. (blight)	Turf	2009
Cymoxanil	*Plasmopara viticola* (grape downy mildew)	Grapevine	1997
QoI	*Plasmopara viticola* (grape downy mildew)	Grapevine	2000
	Venturia inaequalis (apple scab)	Apple	2003
	Stemphylium vesicarium (pear brown spot)	Pear	2006
Anilynopyrimidines	*Venturia inaequalis* (apple scab)	Apple	2004
DMIs	*Venturia inaequalis* (apple scab)	Apple	1987
	Erysiphe necatrix (grape powdery mildew)	Grapevine	1990
Oxycarboxin	*Uromyces caryophyllinus*	Carnation	1986

is also present in populations of *B. cinerea* and *Penicillium* spp., agents of postharvest fruit rots on apple, pear (Romano *et al.*, 1983; Bertetti *et al.*, 2003) and citrus (Gullino *et al.*, 1986). Benzimidazole resistance is still widespread in several strains of different *formae speciales* of *Fusarium oxysporum*, as well as in *Penicillium* spp. causal agents of bulb rots on several bulb crops (Garibaldi and Gullino, 1990). Also, strains of *Sclerotinia homeocarpa* resistant to benzimidazoles have been detected on Italian golf courses (Mocioni *et al.*, 2001).

Dicarboximides

In Italy in the 1980s this group of fungicides replaced benzimidazoles in the control of grey mould on many crops. Resistance developed quickly, first on tomato and grapevine, and later on other crops (Gullino, 1992). In most cases, resistance to dicarboximides developed in *B. cinerea* strains that were already benzimidazole resistant (Gullino and Garibaldi, 1986; Gullino, 1992). Dicarboximide-resistant strains of *B. cinerea* were less stable than benzimidazole-resistant strains. The genetic basis of resistance to dicarboximides in *B. cinerea* has been widely exploited (Faretra and Pollastro, 1991).

Only a few representatives of this chemical class are still used in Italy. Resistance to iprodione has been described recently in *S. homeocarpa*, causal agent of dollar spot of turfgrass in Italy (Mocioni *et al.*, 2011).

Resistance to dicarboximides is widespread in Italian isolates of *Stemphylium vesicarium*, casual agent of brown spot of pear, a severe disease present in the Po valley since the early 1990s (Brunelli *et al.*, 1997; Alberoni *et al.*, 2005). Four phenotypes of *S. vesicarium* were recognized, according to their *in vitro* responses to dicarboximides: sensitive, low, moderately and highly resistant. Cross-resistance between dicarboximides and aromatic hydrocarbons is present in all resistant phenotypes, while cross-resistance to phenylpyrroles is present only in dicarboximide highly resistant strains (Alberoni *et al.*, 2010b). The *S. vesicarium* phenotypes with different levels of sensitivity to dicarboximides are characterized by specific mutations in the histidine kinase gene, causing unfavourable changes in osmotic and oxidative stress response and fitness, compared to the sensitive phenotype. The resistant phenotypes may, therefore, be stable in the field and their decline within the populations may need a long period of suspension of dicarboximide application (Alberoni *et al.*, 2010a).

Anylinopyrimidines

Strains of *B. cinerea* resistant to this class of fungicides have been detected in vineyards (Gullino *et al.*, 2000; Santomauro *et al.*, 2000), showing a low level of resistance. Anylinopyrimidines are still providing a satisfactory level of control of grey mould of grapevine and other crops under practical conditions. Resistance to anilinopyrimidines is also present in Italian isolates of *Venturia inaequalis*, inducing a reduction of efficacy of this class of fungicides under field conditions (Fiaccadori *et al.*, 2007).

Fludioxonil

In the case of this fungicide, resistance in *B. cinerea* can develop *in vitro*, without causing practical problems under field conditions (Bertetti *et al.*, 2008).

Fenhexamid

This fungicide, belonging to the class of sterol biosynthesis inhibitor fungicides, shows a novel mode of action inhibiting the enzymatic complex 3-keto reductase involved in C-4 demethylation, and is interesting for the control of grey mould, due to the absence of positive cross-resistance with other groups of fungicides. Field isolates with reduced sensitivity to fenhexamid have been reported in France, Switzerland and Germany, but so far, loss of fungicide effectiveness under field conditions has not been observed. In laboratory mutants of *B. cinerea*, two levels of resistance have been observed. To avoid the risk of selection in the field of resistant strains different from those observed up to now, appropriate anti-resistance strategies are suggested also for fenhexamid (De Guido *et al.*, 2010).

Boscalid

In the case of this relatively new botryticide, resistance has been detected only *in vitro* (Bertetti *et al.*, 2008; De Miccolis *et al.*, 2010).

Phenylamides

With this class of fungicides, introduced in Italy later in comparison with other European countries such as France and the Netherlands, it was possible to demonstrate in practice the importance of the early adoption of anti-resistance strategies in the case of high-risk chemicals. In the case of grapevine, resistance in *Plasmopara viticola* to phenylmides developed in France as well as in other countries 1 year after their commercialization. Due to the resistance problems observed in other viticultural areas, phenylamides were commercialized in Italy only in mixture with multi-site fungicides. Resistance developed only in the mid-1990s (Mezzalama *et al.*, 1995) without causing severe practical problems. Isolates of *Bremia lactucae*, causing lettuce downy mildew, resistant to this class of fungicides have been reported in Italy (Cobelli *et al.*, 1998). Resistance to mefenoxam has been observed in strains of *Pythium* spp., agents of Pythium blight on turfgrass (Mocioni *et al.*, 2010a).

Cymoxanil

In the case of this fungicide, Italy was the first country where resistance developed in *P. viticola*, leading to a reduced effectiveness of this chemical in vineyards (Gullino *et al.*, 1997). Several factors did contribute to the development and spread of cymoxanil-resistant populations of *P. viticola*. Probably the most important was the extensive use of such fungicide in curative sprays: some growers sprayed cymoxanil ten or more times in 1994 (Gullino *et al.*, 1997). Resistance to cymoxanil was later detected in Germany (Harms *et al.*, 2000), as well as in France (Leroux, 2002).

Demethylation inhibitors

This class of fungicides has been, and still is, used largely on horticultural crops against a range of pathogens. Resistance developed in several pathogens after some years of use, causing a slight decrease in their efficacy year by year. Resistance developed in *Erysiphe necatrix* on grapevine (Garibaldi *et al.*, 1990) and on *Venturia inaequalis* in apple (Fiaccadori *et al.*, 1987).

QoI fungicides

QoI-resistant populations of *P. viticola* have been present in Italian vineyards since 2000 (Gullino *et al.*, 2001, 2004; Toffolatti *et al.*, 2007). Recently, a population of *Pseudoperonospora cubensis*, agent of downy mildew of cucurbits, resistant to QoI fungicides was observed in northern Italy (Gilardi Giovanna, personal communication).

Resistance to QoI fungicides has been detected also in *S. vesicarium* in pear orchards in 2006 (Collina *et al.*, 2007). The molecular analysis of mitochondrial cytochrome *b* gene of some monospore isolates with different levels of sensitivity confirmed the presence of the mutation causing G143A substitution in all the resistant isolates (Alberoni *et al.*, 2010a).

Resistance to azoxystrobin has been observed in strains of *Pythium* spp., agents of Pythium blight on turfgrass, as a consequence of the illegal use of this fungicide, not registered on turf (Mocioni *et al.*, 2010b).

15.3   Strategies for Resistance Management

Although not always successful, fungicide resistance management undoubtedly has prevented or delayed potentially more serious losses of disease control than those which actually have occurred in Italian agriculture. When practical resistance develops, the importance of avoiding ineffective treatments is now widely recognized and the public sector and industry work together in order to develop and implement anti-resistance strategies. Since very few new fungicide groups will emerge from the industrial laboratories in the future, due to stringent European legislation (Leadbeater and Gisi, 2010), it will be vital to conserve for a long time the activity of those already on the market. An important part of the assessment of new fungicides is resistance risk and resistance management.

Monitoring

Monitoring is crucial to determine whether resistance is the cause in cases of lack of disease control under field conditions, as well as to check if resistance management strategies are effective. Monitoring must start early, to gain valuable baseline data before commercial use begins. Results must be interpreted carefully, to avoid misleading conclusions. Unfortunately, during the past two decades, limited public funding for fungicide-resistance research led to the lack of fully published monitoring results when it was carried out by industry. The information deriving from monitoring, including baseline data, has long-term value and is now published more frequently in journals, or summarized by the Fungicide Resistance Action Committee (FRAC) in publications and on their website (www.frac.info).

Both bioassays and real-time PCR are useful and powerful techniques for detecting and quantifying the presence of resistant phenotypes in the field. When resistance resulted from specific DNA changes in resistant isolates, various PCR diagnostic methods were developed and used to monitor that resistance. Management of QoI anti-resistance strategies relies almost entirely on PCR diagnostics, and similar methods can be applied to monitor resistance to benzimidazoles, dicarboximides and demethylation inhibitors. It is important that researchers are aware of advances in real-time PCR and array technologies and also that sufficient resources are allocated to the laboratories involved in routine monitoring to keep their technicians, as well as instrumentation, up to date in order to obtain the best benefits from those developments.

However, bioassay protocols, which can also be improved (Fraaije *et al.*, 2005), must remain a component of monitoring programmes, since resistance may emerge through selection of different target site mutations, or completely different mechanisms (Brent and Hollomon, 2007).

Anti-resistance strategies

Mixtures may be useful in resistance management but, as shown by Shaw (2007), this requires quite specific conditions concerning the initial frequency of resistant strains at introduction of the fungicide, fitness costs and perhaps also synergy between the effects of the mixture components (Gisi *et al.*, 1985; Van den Bosch and Gilligan, 2008). Well-documented cases of synergistic interactions are mixtures of phenylamides combined with multi-site fungicides (Garibaldi *et al.*, 1985;

Gisi *et al.*, 1985; Vigo *et al.*, 1986; Gisi, 1996). An effective resistance management for fungicides is only possible if the discovery and development of novel modes of action of pesticides and a safe and effective diversity of products in the market are maintained.

Mixtures also deserve renewed attention for their insurance capacity in the sense that large yield losses may be prevented when one component in the fungicide mixture suddenly fails. This is especially relevant in situations where the pathogen causes large epidemics in some but not all years, to protect the crop against the risk of damage in adverse years. Resistance may arise and increase in frequency without being noticed during a series of non-epidemic years followed by a year when a severe epidemic develops and control fails. Systems need to be in place to calculate optimum composition for such fungicide mixtures. In the case of mixture usage, it is important to avoid under-dosage of the mixing partner. This would also require criteria for optimum dosage and optimum mixtures, which might change during the phase of invasion and spread of the resistant strain (Van den Bosch and Gilligan, 2008).

Mathematical models provide an alternative means to test hypotheses about strategies that are designed to delay or reduce the risk of build-up of fungicide-resistant strains in pathogen populations.

Acknowledgements

This work was carried out with grants from Regione Piemonte through Agroinnova's Centro di Saggio.

References

Alberoni, G., Collina, M., Pancaldi, D. and Brunelli, A. (2005) Resistance to dicarboximide fungicides in *Stemphylium vesicarium* from Italian pear orchards. *European Journal of Plant Pathology* 113, 211–219.

Alberoni, G., Cavallini, D., Collina, M. and Brunelli A. (2010a) Characterisation of the first *Stemphylium vesicarium* isolates resistant to strobilurins in Italian pear orchards. *European Journal of Plant Pathology* 126, 453–457.

Alberoni, G., Collina, M., Lanen, C., Leroux, P. and Brunelli, A. (2010b) Field strains of *Stemphylium vesicarium* with resistance to dicarboximide fungicides correlated with changes in a two-component histidine kinase. *European Journal of Plant Pathology* 128, 171–184.

Bertetti, D., Kejji, S., Garibaldi, A. and Gullino, M.L. (2003) Valutazione della sensibilità di alcuni agenti di marciume su pomacee in post-raccolta nei confronti di diversi fungicidi. *Informatore Fitopatologico* 53(6), 57–59.

Bertetti, D., Garibaldi, A. and Gullino, M.L. (2008) Resistance of *Botrytis cinerea* to fungicides in Italian vineyards. *Communications in Agricultural and Applied Biological Sciences* 73(2), 273–282.

Brent, K.J. and Hollomon, D.W. (2007) *Fungicide Resistance in Crop Pathogens: How Can it be Managed*. FRAC Monograph No 1, (2nd edn). CropLife International, Brussels, 55 pp.

Brunelli, A., Gherardi, I. and Adani, N. (1997) Ridotta sensibilità di *Stemphylium vesicarium*, agente della maculatura bruna del pero, ai fungicidi dicarbossimidici. *Informatore Fitopatologico* 47(9), 44–48.

Cobelli, L., Collina, M. and Brunelli, A. (1998) Occurrence in Italy and characteristics of lettuce downy mildew (*Bremia lactucae*) resistant to phenylamide fungicides. *European Journal of Plant Pathology* 104, 449–455.

Collina, M., Alberoni, G. and Brunelli, A. (2007) First occurrence of strobilurin-resistant isolates of *Stemphylium vesicarium* in an Italian pear orchard. *Communications in Agricultural and Applied Biological Sciences, Ghent University* 72(4), 735–737.

De Guido, M.A., De Miccolis Angelini, R.M., Pollastro, S., Santomauro, A. and Faretra, F. (2010) Selection and genetic analysis of laboratory mutants of *Botryotinia fuckeliana* resistant to fenhexamid. *Journal of Plant Pathology* 89, 203–210.

De Miccolis Angelini, R.M., Habib, W., Rotolo, C., Pollastro, S. and Faretra F. (2010) Selection, characterization and genetic analysis of laboratory mutants of *Botryotinia fuckeliana* (*Botrytis cinerea*) resistant to the fungicide boscalid. *European Journal of Plant Pathology* 128, 185–199.

Faretra, F. and Pollastro, S. (1991) Genetic basis of resistance to benzimidazole and dicarboximide fungicides in *Botryotinia fuckeliana* (*Botrytis cinerea*). *Mycological Research* 95, 943–951.

Fiaccadori, R., Gielink, A.J. and Dekker, J. (1987) Sensitivity to inhibitors of sterol biosynthesis in isolates of *Venturia inaequalis* from Italian and Dutch orchards. *Netherlands Journal of Plant Pathology* 93, 285–287.

Fiaccadori, R., Cicognani, E., Collina, M. and Brunelli, A. (2007) Study of the sensitivity of *Venturia inaequalis* to anilinopyrimidine fungicides in Italy. *Communications in Agricultural and Applied Biological Sciences, Ghent University* 72(4), 997–1001.

Fraaije, B.A., Burnett, F.J., Clark, W.S., Motteran, J. and Lucas, J.A. (2005) Resistance development to QoI inhibitors in populations of *Mycosphaerella graminicola* in the UK. In: Dehne, H.W., Gisi, U., Kuck, K.H., Russel, P.E. and Lyr, H. (eds) *Modern Fungicides and Anti-fungal Compounds*. BCPC, Alton, UK, pp. 63–71.

Garibaldi, A. and Gullino, M.L. (1990) Disease management of ornamental plants: a never-ending challenge. *Meded. Fac. Landbouww. Rijksuniv. Gent* 55, 189–201.

Garibaldi, A., Romano, M.L. and Gullino, M.L. (1985) Synergism between fungicides with different mechanisms of action against acylalanine-resistant strains of *Phytophthora* spp.. *EPPO Bulletin* 15, 545–551.

Garibaldi, A., Migheli, Q. and Gullino, M.L. (1987) Evaluation of the efficacy of several fungicides against *Cryptocline cyclaminis* on cyclamen. *Mededelingen Faculteit Landbouwwetenschappen Rijksuniversiteit Gent* 52, 859–865.

Garibaldi, A., McKenzie, L. and Gullino, M.L. (1990) Comparsa in Italia di una popolazione di *Uncinula. necatrix* che presenta ridotta sensibilità verso alcuni inibitori della biosintesi degli steroli. *Atti Giornate Fitopatologiche* 2, 141–150.

Gisi, U. (1996) Synergistic interaction of fungicides in mixtures. *Phytopathology* 86, 1273–1279.

Gisi, U., Binder, H. and Rimbach, E. (1985) Synergistic interactions of fungicides with different modes of action. *Transactions of British Mycological Society* 85, 299–306.

Gullino, M.L. (1992) Chemical control of *Botrytis* spp. In: *Recent Advances in Botrytis Research*. Pudoc, Wageningen, The Netherlands, pp. 217–222.

Gullino, M.L. and Garibaldi, A. (1986) Fungicide resistance monitoring as an aid to tomato grey mould management. *Proceedings of the British Crop Protection Conference* 1, 277–505.

Gullino, M.L. and Garibaldi, A. (2007) Critical aspects in management of fungal diseases of ornamental plants and directions in research. *Phytopathologia Mediterranea* 46, 135–149.

Gullino, M.L., Mirandola, R. and Garibaldi, A. (1986) Valutazione della sensibilità di popolazioni di *Penicillia*, agenti di marciumi degli agrumi, nei confronti di fungicidi utilizzabili in post-raccolta. *Atti Convegno Agrumicoltura, Cagliari* 729–734.

Gullino, M.L., Mescalchin, E. and Mezzalama, M. (1997) Sensitivity to cymoxanil in populations of *Plasmopara viticola* in northern Italy. *Plant Pathology* 46, 729–736.

Gullino, M.L., Bertetti, D., Monchiero, M. and Garibaldi, A. (2000) Sensitivity to anilinopyrimidines and phenylpyrroles in *Botrytis cinerea* in north-Italian vineyards. *Phytopathologia Mediterranea* 39, 433–446.

Gullino, M.L., Gilardi, G., Stefanelli, G., Mescalchin, E. and Garibaldi, A. (2001) Presenza di popolazioni di *Plasmopara viticola* resistenti ai fungicidi inibitori della respirazione mitocondriale (QoI Star) in vigneti dell'Italia nord-orientale. *Informatore Fitopatologico* 51(12), 86–87.

Gullino, M.L., Gilardi, G., Tinivella, F. and Garibaldi, A. (2004) Observations on the behaviour of different populations of *Plasmopara viticola* resistant to QoI fungicides in Italian vineyards. *Phytopathologia Mediterranea* 43, 341–350.

Harms, M., Fuhr, I., Lorenz, D. and Buchenauer, H. (2000) Investigation on the sensitivity of *Plasmopara viticola* to cymoxanil in the vine growing area of Palatinate, Germany. *IOBC/WPRS Bulletin* 23(4), 41–44.

Leadbeater, A. and Gisi, U. (2010) The challenges of chemical control of plant diseases. In: Gisi, U., Chet, I. and Gullino, M.L. (eds) *Recent Developments in Management of Plant Diseases*. Springer, Dordrecht, The Netherlands, pp. 3–17.

Leroux, P. (2002) Classification des fongicides agricoles et résistance. *Phytoma – La défense des végetaux* 566, 36–40.

Mezzalama, M., Aloi, C., Frausin, C., Ortez, A. and Gullino, M.L. (1995) Indagine sulla sensibilità alle fenilammidi in popolazioni di *Plasmopara viticola* in vivaio. *Vignevini* 22 (Suppl. 4), 16–18.

Mocioni, M., Gennari, M. and Gullino, M.L. (2001) Reduced sensitivity of *Sclerotinia homeocarpa* to fungicides on some Italian golf courses. *International Turfgrass Society Research Journal* 9, 701–704.

Mocioni, M., Gullino, M.L. and Garibaldi, A. (2010) Fungicide failure to control Pythium blight on turfgrass. Proc. AFPP 2me Conference sur l'entretien des espaces verts, jardins, gazons, forets, zone aquatiques et autres zones non agricoles, Angers, 28–29 November 2009. Association Francaise de Protection des Plantes, Paris.

Mocioni, M., Gullino, M.L. and Garibaldi, A. (2011) Sensitivity of *Sclerotinia homeocarpa* populations to demethylation-inhibiting (DMI) fungicides and iprodione in Italy. *Phytopathologia Mediterranea*, in press.

Romano, M.L., Gullino, M.L. and Garibaldi, A. (1983) Evaluation of the sensitivity to several fungicides of post-harvest apple pathogens in North-western Italy. *Mededelingen Faculteit Landbouwwetenschappen Rijksuniversiteit Gent* 48, 591–602.

Santomauro, A., Pollastro, S., De Guido, M.A., Pollastro, S. and Faretra, F. (2000) A long-term trial on the effectiveness of new fungicides against grey mould on grapevine and on their influence on the pathogen's population. *XII International Botrytis Symposium*, Reims, 2000 (abstract).

Shaw, M.W. (2007) Is there such a thing as a fungicide resistance strategy? A modeler's perspective. *Annals of Applied Biology* 78, 37–44.

Toffolatti, S.L., Serrati, L., Sierotzki, H. and Gisi, U. (2007) Assessment of QoI resistance in *Plasmopara viticola* oospores. *Pesticide Management Science* 63, 194–201.

Van den Bosch, F. and Gilligan, C.A. (2008) Models of fungicide resistance dynamics. *Annual Review of Phytopathology* 46, 123–147.

Vigo, M., Romano, M.L. and Gullino, M.L. (1986) Caratteristiche biologiche di isolati di *Phytophthora nicotianae* var. *parasitica* resistenti alle fenilammidi ottenuti in laboratorio. *Atti Giornate Fitopatologiche* 347–358.

16 Challenge of Fungicide Resistance and Anti-Resistance Strategies in Managing Vegetable Diseases in the USA

Margaret T. McGrath

Department of Plant Pathology and Plant-Microbe Biology, Cornell University, Long Island Horticultural Research and Extension Center, New York, USA

16.1 Introduction

Fungicide resistance has become a major issue when considering disease management programmes for vegetable crops. This is a consequence of the quantity of fungicides now commercially available that are listed as having a moderate to high intrinsic risk of resistance development, and the importance of these fungicides for disease control (Table 16.1). Many of these fungicides are important because they are mobile in the plant, imparting inherently greater activity compared to contact fungicides, and they have lower risk of adverse impact on non-target organisms and the environment. These fungicides have a specific single-site mode of action; thus are active on a critical pathway of the pathogen and therefore are generally more at risk for resistance development than other fungicides. Resistance is a major component of a disease management programme because the primary goal is to delay resistance development, rather than to manage resistant strains after they have been detected. Management practices need to be implemented when an at-risk fungicide is registered for commercial use rather than after resistance has been detected in the pathogen.

The general strategy recommended to farmers for managing resistance is to reduce the need for fungicides by using an integrated management programme and to use at-risk fungicides on a limited basis and in combination with other fungicides. A common anti-resistance fungicide programme for vegetable crop diseases consists of alternating applications among all available effective at-risk fungicide modes of action (MOA) for the disease and tank mixing these fungicides with contact protectant fungicides that are multi-site inhibitors. Cross-resistance is assumed to be a common phenomenon for fungicides in the same MOA class with high intrinsic risk of resistance; thus, the emphasis is on alternating among fungicide MOA classes. Simple alternations are recommended much more commonly than block alternations, which consist of more than one application of an at-risk fungicide chemistry before switching to another. Occasionally, mixtures of fungicides with high-risk MOA are recommended as an anti-resistance strategy. High label rates often are recommended when resistance is anticipated to be quantitative, to minimize the opportunity for selection of strains with intermediate levels of sensitivity. Some diseases can be managed effectively without including an at-risk fungicide in every application, such as

Table 16.1. Fungicides currently registered in the USA for diseases of vegetable crops that are prone to developing resistance, listed by chemistry group.

FRAC code[a]	Common name	Trade name	Resistance risk[b]	Crop groups labelled
1	Thiophanate-methyl	Topsin M	H	Beans, cucurbits, watermelon, potato
1	Thiobendazole	Mertect	H	Carrot, sweet potato
2	Iprodione	Rovral	M-H	Beans, cole crops, garlic, lettuce, onion, potato
3	Myclobutanil	Rally	M	Cucurbits
3	Triflumizole	Procure	M	Cucurbits
3	Tebuconazole	Folicur	H	Onion
3 + 9	Difenconazole + cyprodinil	Inspire Super	H	Cucumber, garlic, leek, muskmelon, onion, pumpkin, winter squash, watermelon
4	Metalaxyl	MetaStar	H	Cucumber, asparagus, beet, garlic, horseradish, lettuce, muskmelon, peas, pepper, pumpkin, winter squash, spinach, summer squash, tomato, watermelon
4	Mefenoxam	Ridomil Gold, Ultra Flourish	H	Cucumber, asparagus, beet, cole crops, carrot, aubergine, garlic, greens, horseradish, lettuce, muskmelon, onion, parsnips, peas, pepper, pumpkin, winter squash, radish, spinach, summer squash, tomato, watermelon, potato
4 + M1	Mefenoxam + copper	Ridomil Gold Copper	M-H	Pepper
4 + M5	Mefenoxam + chlorthalonil	Ridomil Gold Bravo	H	Cole crops
7	Flutolanil	Moncut	L-M	Potato
7	Boscalid	Endura	M-H	Potato, tomato
7 + 11	Boscalid + pyraclostrobin	Pristine	H	Carrot, cucumber, garlic, leek, muskmelon, onion, pumpkin, winter squash, summer squash, watermelon
11	Azoxystrobin	Quadris	H	Cucurbits, asparagus, beans, beet, cole crops, celery, garlic, greens, horseradish, lettuce, muskmelon, okra, onion, parsnip, peas, pepper, potato, pumpkin, radish, spinach, sweetcorn, tomato, watermelon, winter squash
11	Trifloxystrobin	Flint, Gem	H	Aubergine, cucurbits, pepper, potato, tomato
11	Pyraclostrobin	Cabrio, Headline	H	Aubergine, beans, beet, cole crops, carrot, cucumber, garlic, greens, horseradish, leek, muskmelon, onions, parsnip, pepper, potato,

Continued

Table 16.1. Continued.

FRAC code[a]	Common name	Trade name	Resistance risk[b]	Crop groups labelled
				radish, spinach, sweetcorn, tomato, watermelon
11	Fenamidone	Reason	H	Muskmelon, onion, potato, spinach, tomato, watermelon
11 + 3	Azoxystrobin + difenconazole	Quadris Top	H	Aubergine, cucumber, muskmelon, pumpkin, summer squash, watermelon, winter squash
11 + 3	Azoxystrobin + propiconazole	Quilt	M-H	Sweetcorn
11 + 3	Trifloxystrobin + propiconazole	Stratego	M-H	Sweetcorn
12	Fludioxonil	Scholar	M-H	Sweet potato
13	Quinoxyfen	Quintec	H	Muskmelon, pepper, pumpkin, watermelon, winter squash
14	PCNB	Blocker, Terraclor	L-M	Pepper, potato, tomato
21	Cyanofamid	Ranman	M-H	Tomato, potato
22 + M3	Zoxamide + mancozeb	Gavel	L-M	Cucumber, muskmelon, potato, summer squash, tomato, watermelon
25	Streptomycin	Agri-Mycin, Agri-Strep	H	Pepper (greenhouse)
27	Cymoxanil	Curzate	L-M	Cucumber, muskmelon, potato, pumpkin, summer squash, tomato, watermelon, winter squash
11 + 27	Famoxadone + cymoxanil	Tanos	H	Lettuce, potato, tomato
28	Propamocarb HCL	Previcur Flex	L-M	Cucumber, lettuce, muskmelon, potato, pumpkin, summer squash, tomato, watermelon, winter squash
30	Triphenyltin hydroxide	Super tin	L-M	Potato
40	Dimethomorph	Forum	L-M	Aubergine, beans, cucurbits, garlic, greens, leeks, lettuce, muskmelon, pepper, potato, pumpkin, summer squash, tomato, watermelon, winter squash
40	Mandipropamid	Revus	H	Lettuce
40 + 3	Mandipropamid + difenoconazole	Revus Top	L-H	Tomato
43	Fluopicolide	Presidio	H	Cucumber, muskmelon, pepper, pumpkin, spinach, summer squash, tomato, watermelon, winter squash

Note: [a]FRAC, Fungicide Resistance Action Committee; [b]resistance risk as determined by FRAC: L, low risk, M, moderate risk and H, high risk for fungicide resistance to develop.

early in disease development when conditions are not favourable for development of the pathogen, late in disease development when a high level of control is no longer necessary, or other situations where protectant fungicides are adequate. Maintaining effective control is an important aspect of resistance management because the potential to select resistant strains increases with the size of the pathogen population. Further reducing use of an at-risk fungicide is recommended when strains of the pathogen with reduced sensitivity or resistance have been detected.

Labels for most fungicides with moderate to high risk of resistance development contain information and use restrictions for managing resistance. To assist farmers with product selection, FRAC (Fungicide Resistance Action Committee) has assigned code numbers for each fungicide MOA group and most fungicide registrants specify the FRAC code on products labelled in the USA, displayed prominently at the top of the cover page. This is not required. Most registrants also provide on product labels use directions for resistance management and specify use restrictions to ensure resistance management is implemented. Restrictions include specifying applications with protectant fungicides and limiting the number of consecutive applications and the maximum number of applications. Since the

label is a legal document, farmers are legally required to follow these restrictions. Some at-risk fungicides are marketed in combination products with a protectant fungicide or another at-risk fungicide to ensure an anti-resistance strategy is employed.

16.2   Challenges of Fungicide Resistance and its Management

Fungicide resistance has created many challenges to managing diseases of vegetable crops effectively, and there have been many challenges to managing fungicide resistance effectively.

Lack of tools has been a major challenge to managing fungicide resistance effectively. Until recently, typically only one at-risk fungicide chemistry has been available or effective for farmers to use to manage a particular pathogen, making it impossible to implement the common anti-resistance strategy of alternating among fungicide chemistry. Contact protectant fungicides are not suitable alternation partners for pathogens that infect and develop on plant tissue which cannot easily be reached directly with fungicide spray, such as the underside of leaves (Fig. 16.1). On the other hand, protectant

Fig. 16.1. Control of powdery mildew appears to be very good in this pumpkin crop based on there being no symptoms on upper leaf surfaces (A); however, powdery mildew is very severe on the lower surface (B), which is not readily visible without turning leaves over. It is critical, for management success, to control powdery mildew on the lower surface because this is where the pathogen develops best and, without adequate control, leaves will die prematurely. These images are illustrative of what can occur when only protectant fungicides are applied, due to the challenge of delivering spray material directly to the lower surface, or when the pathogen is resistant to the mobile fungicides applied.

fungicides are valuable as tank-mix partners for at-risk fungicides to control any resistant strains that they contact. The protectant fungicides currently registered in the USA (chlorothalonil, mancozeb and copper) are labelled very broadly; however, they cannot be used on all vegetable crops. There are a few crops, including lettuce, for which there no longer are any protectant fungicides registered since the manufacturer of maneb cancelled its registration voluntarily in September 2008. Host plant resistance is an important tool for an integrated management programme; but this tool also has challenges because resistant cultivars are not available for as many diseases as are fungicides. Pathogens are also able to adapt to genetic resistance, and it is a large, time-consuming process to incorporate resistance into multiple cultivars encompassing the diversity of types in demand for several fresh-market vegetable crops.

Farmers may be reluctant to implement a resistance management programme when the use of an at-risk fungicide in an integrated fungicide management programme can increase the cost and/or decrease effectiveness because other fungicides are more expensive and/or less effective. This is more likely to be a potential issue with a wind-dispersed pathogen that cannot survive without living plant tissue, such as the cucurbit downy mildew pathogen. Farmers may perceive that their actions will have relatively minimal impact compared to the impact of the fungicide programmes used by farmers where the disease has developed earlier in the year, and any resistant strain selected in their crops will not survive in their fields until the next season.

Minimizing use of an at-risk fungicide is challenging when it is applied for other diseases or uses, such as yield enhancement with FRAC Code 11 fungicides. Not only will there be need to continue using the fungicide after resistance has started to develop, but these other uses also may extend the treatment period and could include applications when the population of a resistance-prone pathogen is not at a size necessitating control with the at-risk fungicide, thereby increasing selection pressure for resistance. Additionally, selection for resistant strains could occur when other

agricultural products with similar active ingredients are used, such as plant growth supplements containing phosphonic acid, which are similar to FRAC Code 33 fungicides.

Selection pressure for fungicide resistance could be greater when an IPM action threshold is used to initiate a spray programme compared to a preventive schedule. Initial infection would not be suppressed under such an IPM programme. A solution is to apply a protectant fungicide on a preventive schedule, with the at-risk fungicide applied beginning early in disease development. An IPM programme could be an asset for resistance management if routine scouting ensures disease is detected early in development. Under a calendar-based schedule, fungicide applications could start when a pathogen is well established.

Predicting resistance has proved challenging. While it is well established that it is possible to predict whether or not a fungicide has a risk of developing resistance based on if it has single-site mode of action, it is not as straightforward to predict details such as how quickly the resistance will develop, the type of resistance (qualitative or quantitative), the degree of cross-resistance and how long the fungicide will continue to provide control after resistance is detected. For example, when the first FRAC Code 11 fungicide was commercialized in the USA, it was predicted that the relative risk for this group was low (compared to the benzimidazoles), it would take several years for resistance to develop and it would be quantitative. Additionally, the cucurbit powdery mildew pathogen was expected to be the first vegetable pathogen in which resistance would develop. However, the risk proved to be high, qualitative resistance developed, and quickly, with control failure occurring during the 4th year of commercial use for cucurbit powdery mildew. Resistance was detected prior to this in *Didymella bryoniae*. Limited useful information has been obtained about the potential for resistance to a particular fungicide to develop in a pathogen by attempting to induce resistance under controlled conditions, sometimes with the aid of a mutagenic agent. These studies provide some indication of the potential, but the specifics, such as the type of genetic change, have been different from what occurred subsequently under field conditions.

Farmers lacking knowledge about fungicide resistance can affect its management as well as disease control. Those who have not learned that the primary goal of resistance management is to delay resistance development may assume the goal is to manage resistance after it has developed, and thus will not understand that they need always to use a resistance management programme. Fungicides often remain labelled for uses after resistance has developed. Growers who are not knowledgeable about the current situation regarding fungicide resistance for the target pathogen may select a product whose efficacy has been affected by resistance. Examples include products with mefenoxam for downy mildew in lettuce, late blight in potato and tomato, or downy mildew in cucurbits; products with QoI chemistry for powdery and downy mildews in cucurbits; and products with methyl benzimidazole carbamate (MBC) chemistry for powdery mildew in cucurbits.

It can be difficult to detect when a pathogen has developed fungicide resistance in a commercial production field because the other tools used in an integrated management programme (e.g. other fungicides and resistant cultivars) can provide sufficient control to compensate for the ineffective fungicide, and there are no other treatments for comparison in contrast to a replicated research field, where treatments may include the affected fungicide evaluated alone and fungicide treatments not affected by resistance, as well as a non-treated control. Additionally, there are other causes for control being below expectation, including application errors, accelerated fungicide degradation, improper timing relative to crop growth stage or disease development and conditions being ideal for disease development and/or unfavourable for chemical control (e.g. frequent rain). On the other hand, there have been situations where using a fungicide on a resistant pathogen population has resulted in the disease becoming more severe than if no fungicide was applied. Possible explanations include fungicide spray contributing to leaf wetness, pressure of the spray moving pathogen spores and the fungicide suppressing other pathogens or beneficial fungi.

The information that would provide a good foundation for dealing with fungicide resistance is generally not available for vegetable crop pathogens. The resources needed to conduct the required research are enormous. First, the pathogen's baseline sensitivity to each at-risk fungicide is needed to form a benchmark to detect changes in sensitivity after the fungicide has been used in commercial production. Extensive monitoring of pathogen populations is needed to detect initial changes, thereby enabling changes to be made to the management programme in an effort to thwart further development of resistance and avoid control failure. Monitoring is also needed after resistance develops to determine the extent of occurrence and the frequency of resistant isolates in order to make sound recommendations on fungicide use. Additionally, anti-resistance strategies cannot be evaluated until after resistance has developed. Thus, it is not possible to determine whether a simple or block alternation would be more effective, or whether a mixture would be even better.

Most occurrences of fungicide resistance in the USA have been detected after resistant strains have become sufficiently common to cause an observable control failure. Additionally, the focus of research generally is on determining whether control failure is due to pathogen resistance; thus, whether resistance is occurring elsewhere is often not investigated. A factor that can hinder, or at least delay, examining pathogen isolates collected in another state is the requirement in the USA for a permit for their movement, which is issued only to investigators with the necessary facilities to ensure the pathogen cannot escape from the work area. When control failure has already occurred, it may not be possible to change fungicide recommendations in order to extend the useful life of the fungicide, in contrast with when resistance is detected early in development through a monitoring programme. This raises the question of how often resistance development might go undetected in the future because, for many diseases, farmers now have several different fungicides to use in a resistance management programme; therefore, there could be little and perhaps no detectable impact on overall control when one fungicide in the programme is failing to control the pathogen. Thus, there is now a greater need for resistance monitoring by assaying pathogen populations and fungicide evaluations with at-risk fungicides used alone.

Knowledge of the appropriate methodology to use is critical to the success of monitoring efforts. This is especially important with *in vitro* assay procedures. Whether to examine germination, mycelial expansion, sporulation, or another stage in pathogen development needs to be based on an understanding of the biological activity of the fungicide. Another consideration is whether to use the formulated product or technical grade material. Inert ingredients in the product might affect the assay results, but pure fungicide often has low solubility in water, necessitating first dissolving it in a solvent like acetone.

16.3 Occurrence of Fungicide Resistance in Pathogens of Vegetables in the USA

Cucurbits

Downy mildew

Downy mildew re-emerged as an important disease in the USA in 2004. For many years it was controlled effectively with resistant cucumber cultivars and fungicides. Other cucurbit crops were affected sporadically, and often not until late in the growing season. Control failures that occurred beginning in 2004 were associated with the presence of pathogen strains virulent on resistant cucumber cultivars and resistant to FRAC Code 11 fungicides. Since 2004, downy mildew has been occurring more commonly in the USA. A forecasting system (http://cdm.ipmpipe.org/) has proven to be a valuable tool for timing applications of at-risk fungicides when conditions are predicted to be favourable for disease onset. This is a true forecasting system that predicts where downy mildew will occur

based on the location of current outbreaks in the region, forecasted wind trajectories from these locations and favourability of environmental conditions for spore production, spore survival during transport and for infection. A customized alert system has been implemented to enable farmers to receive a text message on their cell phone or by electronic mail notifying them when downy mildew has been detected nearby and there is a significant risk, as defined by the farmer. By using this system, farmers can avoid the need for curative applications and avoid unnecessary applications being made when conditions are not favourable, most importantly when environmental conditions are favourable for infection (e.g. rain) but wind trajectories are not forecast to go from known sources to locations where downy mildew is not yet present.

Presently, there are targeted fungicides with inherent risk of resistance development in eight FRAC groups labelled for managing cucurbit downy mildew in the USA. These are FRAC Codes 11 (azoxystrobin, famoxadone and others), 21 (cyazofamid), 22 (zoxamide), 27 (cymoxanil), 28 (propamocarb hydrochloride), 33 (phosphorous acid fungicides), 40 (dimethomorph and mandipropamid) and 43 (fluopicolide) (Table 16.1).

Gummy stem blight/black rot

D. bryoniae has demonstrated an ability to develop resistance. In the USA, resistance has been found to benomyl, thiophanate-methyl, strobilurins (particularly azoxystrobin) and boscalid. Resistance to FRAC Code 1 fungicides (benomyl and thiophanate-methyl) was detected first in Florida in 1989, then in New York in 1990 and South Carolina in 1991 (Keinath and Zitter, 1998). Resistant strains were found to be more common in South Carolina, where these fungicides were more commonly used due to greater disease pressure, than in New York through a survey conducted from 1992 to 1996. Cross-resistance was documented: almost all isolates that were resistant to benomyl were also resistant to thiophanate-methyl. Resistance to azoxystrobin in the eastern USA was first detected 2 years after its first commercial use in 1998 (Keinath, 2009). Resistant populations of

D. bryoniae can cause more disease in the presence than in the absence of the fungicide.

Presently, there are targeted fungicides with inherent risk of resistance development in four FRAC groups labelled for managing gummy stem blight in the USA. These are FRAC Codes 3 (tebuconazole, difenconazole), 7 (boscalid), 9 (cyprodinil) and 12 (fludioxonil) (Table 16.1).

Powdery mildew

In the USA, the cucurbit powdery mildew pathogen *Podosphaera xanthii* has demonstrated an ability to develop resistance, often quickly, to fungicides prone to resistance due to their single-site mode of action (McGrath, 2001). This pathosystem serves as a good illustration of the challenges of fungicide resistance (McGrath and Shishkoff, 2001). Host plant resistance is the only other management practice for powdery mildew to use in an integrated management programme. While there is good potential utility of using resistant cultivars in an integrated programme for managing fungicide resistance, it has limitations due to resistance not being incorporated into all horticultural types and also pathogen ability to evolve to overcome genetic resistance as well as fungicides.

MBC fungicides, also known as benzimidazoles (FRAC Code 1), were the first chemical class of fungicides with a single-site mode of action used for *P. xanthii* (McGrath, 2001). Resistance developed very quickly to benomyl, the first fungicide in this group. In the USA, benomyl-resistant strains were detected in 1967, the first year of field evaluations at US university facilities. This was the first documented case of resistance in the USA. At that time, global experiences of fungicide resistance were limited and thus the potential impact on control and the need for management were not recognized. Benomyl formulated as Benlate was registered in 1972 for commercial use on cucurbit crops in the USA. The first case of control failure in the field occurred the next year. Another MBC fungicide, thiophanate-methyl, formulated as Topsin M, is still labelled for cucurbit powdery mildew and thus available for use in production fields. It is not recommended because resistant strains continue to be found widely and commonly, despite the limited use of thiophanate-methyl for other diseases. (McGrath *et al.*, 1996; McGrath, 2006) Resistance to this group of fungicides is qualitative, thus pathogen strains are sensitive or fully resistant. And cross-resistance occurs among the fungicides, thus resistant strains are insensitive to all fungicides in the group.

The next chemical class developed for cucurbit powdery mildew was the demethylation inhibitor (DMI) fungicides (FRAC Code 3) (McGrath, 2001). The first active ingredient in this group was triadimefon. It was registered for cucurbit powdery mildew in the USA in April 1984. The first reported control failure documented through university fungicide efficacy experiments occurred just 2 years later. Control failure became widespread during the early 1990s. Resistance to DMIs is quantitative, thus pathogen strains exhibit a range in sensitivity. While cross-resistance exists among fungicides in this group, there are inherent differences in activity. The next DMI fungicide developed, myclobutanil, was effective in university experiments when used at a high concentration against pathogen strains fully resistant to triadimefon and where this fungicide was ineffective. Myclobutanil was granted registration in the USA in 2000. For 2 years prior to 2000 an emergency exemption from registration (FIFRA Section 18) was granted for myclobutanil in some states because neither benomyl nor triadimefon, the only mobile fungicides registered for this use at the time, were adequately effective due to resistance. The degree of DMI insensitivity in the *P. xanthii* population continued to shift during the 1990s. As a result, myclobutanil applied at its lowest label rate no longer controlled powdery mildew as well as at the highest rate. USA federal (Section 3) registration was granted for another new DMI, triflumizole, in 2002. Subsequently, sensitivity to the DMI fungicides remained fairly stable through 2010. Myclobutanil and triflumizole have provided effective control of powdery mildew in most field efficacy evaluations conducted over those years until 2009 (McGrath and Hunsberger, 2011). None of the DMI fungicides developed recently, which are difenoconazole, tebuconazole and metconazole, have exhibited greater inherent activity than the DMIs currently registered, unlike the situation with myclobutanil, being substantially more active than triadimefon.

Quinone outside inhibitor (QoI) fungicides (FRAC Code 11) were the next chemical class developed for cucurbit powdery mildew (McGrath, 2001). Azoxystrobin was registered in the USA in spring 1999. It could be used in some states in 1998 where an emergency exemption was granted. Additional QoIs were registered in late 1999 (trifloxystrobin) and 2002 (pyraclostrobin). Resistance to QoIs was first detected in the USA in 2002 (McGrath and Shishkoff, 2003). Control failures were reported from several states throughout the USA; resistant strains were confirmed to be present in Georgia, North Carolina, Virginia and New York. Impact on control was dramatic, with failure occurring where QoIs were highly effective the previous year, reflecting the qualitative nature of resistance to this group of fungicides. Resistant strains of *P. xanthii* have been common in the USA based on bioassays conducted recently in several states. QoI fungicides are no longer recommended for cucurbit powdery mildew because resistant strains are common, they are fully resistant due to the qualitative nature of the resistance and there is cross-resistance among QoI fungicides. Resistance to QoI and also to MBC fungicides has been detected at the start of cucurbit powdery mildew development where tested. There continues to be selective pressure to maintain QoI resistance in the *P. xanthii* pathogen population in the USA because the only fungicide available with a new active ingredient, boscalid, also contains a QoI fungicide.

Carboximide (FRAC Group 7) was the fourth chemical class of mobile fungicides available for managing cucurbit powdery mildew in the USA that was at risk for resistance development. The first product, which was registered in 2003, contained boscalid plus pyraclostrobin. Pathogen strains have exhibited a range in sensitivity to boscalid. Strains fully resistant to this fungicide were first detected in 2008 (McGrath, unpublished). These strains were able to tolerate label rates (500 ppm) in a leaf-disc bioassay (McGrath and Fox, 2010). Control failure in a fungicide evaluation conducted in 2009 in

New Jersey was associated with their presence (C.A. Wyenandt, personal communication) and in 2010 in New York (McGrath and Hunsberger, 2011). It is feasible, but not yet known, that the new carboximides expected to be registered for this use in the near future are sufficiently different chemically from boscalid that their efficacy will not be compromised due to cross-resistance with boscalid.

Quinoline (FRAC Group 13) is the most recent chemical class to become available for use in the USA. Quinoxyfen was registered for use on melon in 2007 and on pumpkin and winter squash in 2009. It has been highly effective in university fungicide evaluations (e.g. McGrath and Fox, 2009). There are a few additional fungicides in development with high inherent activity based on the baseline sensitivity of *P. xanthii*.

Presently, there are targeted fungicides with inherent risk of resistance development in five FRAC groups labelled for managing cucurbit powdery mildew in the USA. These are FRAC Codes 1 (thiophanate-methyl), 3 (triflumizole and others), 7 (boscalid), 11 (azoxystrobin and others) and 13 (quinoxyfen) (Table 16.1).

Lettuce

Downy mildew

The downy mildew pathogen *Bremia lactucae* occurring in the USA is complex and prone to adapting to fungicides as well as resistant cultivars. Multiple races (pathotypes) have been described. Metalaxyl (FRAC Code 4) was the first targeted at-risk fungicide used commercially for this disease. The product Ridomil 2E was registered in the USA for this disease in 1983. Beforehand, farmers in California, the major production area for lettuce in the USA, had been struggling since the early 1970s to manage downy mildew with protectant fungicides that were only moderately effective (copper fungicides) or had very long preharvest intervals (14 days for EBDC fungicides), plus cultural practices to minimize leaf wetness when a new pathotype was evident that rendered ineffective the major resistance gene (*Dm5/8*) in most resistant cultivars (Schettini *et al.*, 1991). At that time, farmers lacked the tools to manage resistance since there was not another mobile fungicide to use in alternation or an effective resistant cultivar and, furthermore, there was limited knowledge generally about managing fungicide resistance. Control failures occurred in California in 1987, the fifth season metalaxyl was used (Schettini *et al.*, 1991), and in Florida in 1989 during the fourth use season (Raid *et al.*, 1990). A strategy of integrating fungicide applications and genetic resistance was considered a viable management approach based on experience in England where control failure due to metalaxyl resistance had occurred earlier (in 1983, the sixth year of use) (Schettini *et al.*, 1991). For at least 4 years after resistance became widespread, downy mildew was managed effectively in England with metalaxyl plus a protectant fungicide applied to lettuce cultivars with a resistance gene to which all metalaxyl-resistant strains were found to be avirulent. Another resistance gene was bred into lettuce when virulent metalaxyl-resistant strains were detected. This integrated approach was considered likely to be more durable in California, where there were only asexually propagating pathotypes, in contrast with England, where the pathogen was reproducing sexually and variation was extensive. The approach has proved viable and continues to be used in California; however, pathogen adaptation to resistant cultivars necessitates periodic changes to the specific recommendations to farmers (S.T. Koike, personal communication). Metalaxyl resistance remained common despite limited use. All 134 isolates collected in 1999 and 2000 in California were resistant (Brown *et al.*, 2004). Unexpectedly, farmers have observed over the past few years that they are able to obtain control of downy mildew with mefenoxam (the R-enantiomer of metalaxyl and active ingredient now in use); isolates have been tested and found to be sensitive (S.T. Koike, personal communication).

Fosetyl-aluminium (FRAC Code 33), marketed under the brand name Aliette, was the next fungicide with single-site mode of action used for downy mildew in lettuce, beginning after metalaxyl became ineffective (Brown *et al.*, 2004). It has a low resistance risk in contrast with metalaxyl, which has a high risk. Control failures occurred with Aliette in 1998 in California. These events were associated with insensitivity in the pathogen. Insensitivity was detected in multiple pathotypes and in isolates from all the regions sampled. Although these insensitive isolates were able to sporulate on seedlings treated with rates twice the normal field dosage, downy mildew was suppressed 50% by Aliette applied to older plants. Insensitive isolates exhibit cross-resistance to phosphonic acid, which has a similar active ingredient. Use of plant growth supplements containing phosphonic acid might have contributed to the development of insensitivity to fosetyl-aluminium, especially if doses were sublethal for the pathogen. Insensitivity did not become widespread in California, but rather has evidently declined based on recent observation (S.T. Koike, personal communication).

The EBDC fungicide maneb was an important tool for managing fungicide resistance development in *B. lactucae*. Farmers routinely used this protectant fungicide early in crop development, then switched to at-risk fungicides. This tool was lost when in September 2008 the manufacturer cancelled registration of maneb in the USA. Lettuce was one of the very few vegetable crops that was not on the mancozeb label; thus, product substitution was not an option until 2011 when the label was expanded.

Presently, there are fungicides with inherent risk of resistance development in six FRAC groups labelled for managing lettuce downy mildew in the USA. These are FRAC Codes 4 (mefenoxam), 11 (azoxystrobin, famoxadone and others), 27 (cymoxanil), 28 (propamocarb hydrochloride), 33 (fosetyl-aluminium) and 40 (dimethomorph and mandipropamid) (Table 16.1). Azoxystrobin is not recommended due to its low inherent activity for this pathogen. Fungicides to which *B. lactucae* has developed resistance (FRAC Codes 4 and 33) have a role in an integrated management programme today in locations where the pathogen population is dominated by sensitive pathotypes.

Drop

Sclerotinia minor readily developed resistance *in vitro* to iprodione, a dicarboximide fungicide (FRAC Code 2); however, no iprodione-resistant isolates of *S. minor* were found under field conditions (Hubbard *et al.*, 1997).

Resistant strains in this study were selected by plating isolates on media amended with successive fungicide doses. This resistance generated under laboratory conditions was found to impose a fitness cost, and virulence of the resistant strains declined over time relative to the wild-type strains, with some strains becoming avirulent. Most iprodione-resistant strains exhibited cross-resistance to vinclozolin, another dicarboximide fungicide. Neither resistance to iprodione nor to boscalid, a carboximide fungicide (FRAC Code 7), was found during a recent evaluation of more than 200 isolates from 200 commercial production fields in California (Subbarao and Wu, unpublished data, cited in Klose *et al.*, 2010). Resistance to dicarboximide fungicides has been detected in *S. minor* isolates and other *Sclerotinia* species obtained from other plants, including peanut and creeping bentgrass. Poor control of lettuce drop with dicarboximide fungicides could be due to the application not being made at the ideal time and crop stage. Accelerated soil degradation of fungicides following repeated use was found to be another plausible explanation. Relative persistence in soils was related to fungicide efficacy for Rovral (iprodione), Botran (dicloran), Switch (cyprodinil plus fludioxonil) and Endura (boscalid), with Endura being the most effective and having the greatest residual soil fungicide concentrations.

Presently, there are several at-risk fungicides in the USA labelled for lettuce drop (iprodione, boscalid, cyprodinil plus fludioxonil, and dicloran). They are used in combination with cultural, biological and physical control practices (Klose *et al.*, 2010). Baseline sensitivity to boscalid, fenhexamid, fluazinam, fludioxonil and vinclozolin has been determined for *S. minor* as well as *S. sclerotiorum*, which also causes drop outside of California (Matheron and Porchas, 2004).

Potato

Early blight

Alternaria solani developed resistance to QoI (FRAC Code 11) fungicides quickly in the USA and by means of a different mutation from most other QoI-resistant pathogens. The

first fungicide in this group, azoxystrobin, was registered for use on potatoes in 1999. It was used intensively in the Midwestern region of the USA, with 4–6 applications to a crop being common, because of the superior control achievable over the protectant fungicides that were being used and because conditions were favourable for disease development (Pasche and Gudmestad, 2008). In the second year of use (2000), reduced sensitivity to azoxystrobin was documented in isolates of *A. solani* collected from Nebraska in 2000 (Pasche *et al.*, 2004). Consequently, use of QoI fungicides declined substantially. These isolates were found to have the F129L mutation in which leucine replaced phenylalanine at position 129, rather than the G143A mutation, in which alanine replaced glycine at amino acid position 143 (Pasche *et al.*, 2004, 2005). An important difference between these is that the G143A mutation provides cross-resistance among QoI fungicides, while the F129L mutation provides a differential effect on fungal sensitivity to QoI fungicides. *A. solani* isolates with the F129L mutation have resistance factors that are 10- to 15-fold for azoxystrobin and pyraclostrobin compared to wild-type isolates, and efficacy of these fungicides is consequently affected, while resistance factors are only 2- to 3-fold for trifloxystrobin, famoxadone and fenamidone, and their efficacy is not affected based on results from growth chamber studies. However, these fungicides lacked the intrinsic activity of azoxystrobin and pyraclostrobin and thus they were never as efficacious on a sensitive *A. solani* population. The net result is that these fungicides now all have similar efficacy.

Resistance was found to be widespread in the USA and to affect the field performance of all QoI fungicides (Pasche and Gudmestad, 2008). Almost all (96.5%) of the 4238 isolates of *A. solani* collected from 2002 to 2006 in 11 potato-producing states were demonstrated to have reduced sensitivity to QoI fungicides and/or to contain the F129L mutation. These results reveal the F129L mutation is stable, and it is present under conditions that are less conducive for the pathogen and thus there is less selection pressure by QoI fungicides. Comparing results from fungicide evaluations conducted in 2000 and 2001, when the

A. solani population was dominated by wild-type isolates, to those conducted from 2002 to 2006, when F129L mutant isolates dominated, revealed that resistance has affected the field performance of all QoI fungicides. However, resistance has not rendered the QoI fungicides completely ineffective; thus, they continue to be a component of an effective early blight management programme when used with other fungicides, such as boscalid and pyrimethanil, and alternated with protectant fungicides, such as mancozeb and chlorothalonil.

While resistance is a phenomenon with single-site mode of action fungicides, intensive use of protectant fungicides can lead to reduced sensitivity. Following repeated use of chlorothalonil in commercial potato fields, isolates of A. solani collected at the end of the season from five of the seven fields were significantly less sensitive to chlorothalonil than isolates collected at the start (Holm et al., 2003).

Presently, there are fungicides with inherent risk of resistance development in three FRAC groups labelled for managing early blight in the USA. These are FRAC Codes 3 (difenoconazole), 7 (boscalid) and 11 (azoxystrobin, famoxadone and others) (Table 16.1).

Fusarium dry rot

Resistance to thiabendazole, a FRAC Code 1 fungicide, has developed in Fusarium species associated with dry rot of potato in the USA (Hanson et al., 1996). At the time, it was the only fungicide registered for control of this postharvest disease. It had been used extensively since the early 1970s, before control failures were reported about 20 years later. Most isolates collected in 1990 and 1991 were resistant (Desjardins et al., 1993). In a survey of tubers collected during 1992 and 1993 primarily from the north-eastern USA, resistance was detected in 41% of the Fusarium isolates obtained. Resistant isolates were from most locations and all tuber types (seed, tablestock and processing). The species were F. sambucinum, F. solani, F. oxysporum, F. acuminatum and F. culmorum (Hanson et al., 1996). There was no evidence of a fitness cost (reduced growth or pathogenicity) in F. sambucinum or

F. oxysporum; however, resistant F. solani isolates exhibited slower growth in culture compared to sensitive isolates.

Thiabendazole remains the only fungicide with inherent risk of resistance development registered today for managing Fusarium dry rot in the USA.

Late blight

The first cases of metalaxyl failing to control Phytophthora infestans in the USA occurred in 1989 in Washington state (Deahl et al., 1993). Subsequently, resistance to phenylamide fungicides (FRAC Code 4) was confirmed in this area. Severe epidemics were reported in other areas during the next few years (Goodwin et al., 1998). Reports had come from throughout the USA by the end of 1993. The economic impact was substantial. These epidemics were caused primarily by two new resistant genotypes of P. infestans, US-7 and US-8, that had migrated from Mexico to the USA. Populations of this pathogen in the USA have been characterized by their clonality and absence of sexual reproduction. Before this event, P. infestans populations in the USA were dominated by the US-1 clonal lineage. This genotype was highly sensitive to phenylamide fungicides; therefore, late blight was controlled effectively by metalaxyl during the 1970s and 1980s. Several additional genotypes have been detected since the 1990s, mostly in tomato (Goodwin et al., 1998; Hu et al., 2010). A few have been sensitive to phenylamide fungicides, which raises the question of whether this chemistry could be used again; however, the potential impact on control could be great if the pathogen population included resistant strains. US-8 has been the dominate genotype in potato.

P. infestans has a low potential to develop practical (field) resistance to dimethomorph, a carboxylic acid amide (FRAC Code 40) fungicide (Stein and Kirk, 2004). This assessment is based on detecting a biological cost associated with dimethomorph resistance induced under laboratory conditions using ethidium bromide/UV light mutagenesis and repeated culturing on dimethomorph-amended medium. Most resistant isolates exhibited reduced growth rates on non-amended medium and reduced frequency of infection of leaf and tuber tissue. The resistance mechanism may be quantitative and possibly multigenic based on the low amount of resistance that developed to dimethomorph compared with previously published phenylamide resistance. If this is the case, then resistance in the field is predicted to develop through directional selection and occur in small increments.

Another factor contributing to the low potential for fungicide resistance to develop is that use of at-risk fungicides is limited due to the fact that late blight has been occurring sporadically, especially outside major potato production areas. Also, there has been no evidence of sexual reproduction, although there have been a few occasions when both mating types have occurred together.

Presently, there are targeted fungicides with an inherent risk of resistance development in five FRAC groups labelled for managing late blight in potato in the USA. These are FRAC Codes 21 (cyazofamid), 22 (zoxamide), 27 (cymoxanil), 28 (propamocarb hydrochloride) and 40 (dimethomorph and mandipropamid) (Table 16.1). Another at-risk fungicide, fluopicolide (FRAC Code 43), is registered for use on tomato, while another targeted fungicide with low resistance risk, fluazinam (FRAC Code 29), is registered for use on potato.

Leak

Pythium ultimum, the causal agent for leak, has developed resistance to the phenylamide fungicides (FRAC Code 4) metalaxyl and mefenoxam in the USA (Taylor *et al.*, 2002). Among the 213 isolates of *P. ultimum* tested during a 4-year study (1997–2000), only 3.7% were resistant. Most of these resistant isolates were collected in Minnesota (5 of 7) during the final year of the study. The other two resistant isolates were from Idaho and Washington. Only sensitive isolates were found in the seven other states, plus one Canadian province. Isolates varied markedly in their sensitivities to mefenoxam, with EC_{50} values ranging from below 0.05 µg/ml to > 100 µg/ml. EC_{50} was at most 0.15 µg/ml

for 68% of the isolates, with most in the 0.06–0.1 range.

Fungicides with inherent risk of resistance development that are registered today for managing leak in the USA are mefenoxam (FRAC Code 4) and cyazofamid (FRAC Code 21).

Pink rot

Resistance to the phenylamide fungicides (FRAC Code 4) metalaxyl and mefenoxam in *Phytophthora erythroseptica*, the causal agent for the soilborne disease pink rot, is considered to be widespread in the USA (Taylor *et al.*, 2002). This chemistry was effective during the 1980s. No evidence of resistance was detected in this pathogen through surveying potato storages during the late 1980s and early 1990s (Stack *et al.*, 1993). Subsequently, resistance was detected in Maine in 1994 (Lambert and Salas, 1994), New York in 1995 (Goodwin and McGrath, 1995), then Idaho in 1998 and Minnesota in 2000 (Taylor *et al.*, 2002). During a 4-year study (1997–2000), resistance was not detected in Colorado, Delaware, Nebraska, North Dakota, South Dakota, Washington or Wisconsin (Taylor *et al.*, 2002). Isolates varied markedly in their sensitivities to mefenoxam, with EC_{50} values ranging from below 0.05 µg/ml to > 100 µg/ml, which was similar to *P. ultimum*. EC_{50} was at most 0.15 µg/ml for 68% of the isolates, with most in the 0.06–0.1 range. Through the 1990s there were no other fungicides registered for use to control this pathogen, and no resistant cultivars. Cultural management practices include crop rotation, avoiding wet soil conditions and harvesting properly. Resistant populations of *P. erythroseptica* are pathogenically fit and can cause more disease in the presence than in the absence of the fungicide (Taylor *et al.*, 2006).

An investigation of the inheritance of mefenoxam resistance in *P. erythroseptica*, a homothallic pathogen, revealed that it did not fit any of the previously described models proposed for the genus *Phytophthora* (Abu-El Samen *et al.*, 2005). Most likely, it is inherited quantitatively and is under the control of more than one major gene, plus some minor genes of additive effect. With this understanding of the inheritance, it was predicted that mefenoxam resistance in *P. erythroseptica* would continue to increase in magnitude over time and that sensitive isolates would continue to exist, never disappearing entirely even if mefenoxam was not used in all potato production fields. Quantitative shifts in the intermediately resistant progeny toward higher levels of insensitivity were observed, which supported the prediction.

Today there are two new fungicide groups available for managing pink rot: FRAC Code 21 (cyazofamid; Ranman) and 33 (phosphorous acid), plus mefenoxam remains registered for this use.

Silver scurf

Helminthosporium solani, which causes silver scurf on potato tubers, has developed resistance to FRAC Code 1 fungicides (benomyl, thiabendazole and thiophanate-methyl) in several potato growing regions (Merida and Loria, 1994). Applying this chemistry was the main control practice until outbreaks of silver scurf were associated with fungicide resistance in the mid-1990s. The first documented occurrence was in New York (Merida and Loria, 1990). Silver scurf cannot be controlled in storage once the majority of *H. solani* isolates have developed fungicide resistance (Frazier *et al.*, 1998). Resistance of *H. solani* to benomyl and thiophanate-methyl was found to be common (76% of 238 isolates collected from three locations in 1998) in Tulelake, California, where only thiophanate-methyl had been used, thereby documenting cross-resistance (Cunha and Rizzo, 2003). No difference in virulence was detected between sensitive and resistant isolates (Merida and Loria, 1994). With no difference in fitness, resistant strains could persist in the population even if benzimidazoles were no longer used. Resistance to this fungicide chemistry has been described as a serious problem and has raised concern about resistance developing to other seed treatment chemicals with single-site modes of action if used in a widespread, exclusive manner (Frazier *et al.*, 1998).

Fungicides with an inherent risk of resistance development that are registered today for managing silver scurf in the USA are thiophanate-methyl (FRAC Code 1),

azoxystrobin (FRAC Code 11) and fludioxonil (FRAC Code 12).

Tomato

Early blight

In 1999, the first fungicide in a chemical group with some risk of resistance development was registered in the USA for early blight in tomato, and also potato. It was azoxystrobin (FRAC Code 11). At the time, resistance risk was thought to be low for this chemical group and the pathogen was thought to have a low risk of developing resistance. The fungicide programme recommended then was azoxystrobin (formulated as Quadris) alternated with the protectant fungicide chlorothalonil. Resistance was confirmed in 2008 during the second production season that QoI fungicides provided unsatisfactory levels of early blight control in a major tomato production area in North Carolina (Ivors *et al.*, 2010). The F129L mutation in cytochrome *b* was found in 36% of the *A. solani* isolates examined. Resistance has also developed in the other early blight pathogen, *Alternaria tomatophila*, in New York, where it has been found to be more common on tomato than *A. solani* (T.A. Zitter, personal communication). Elsewhere in the USA, occurrence of resistance to FRAC Code 11 fungicides in these pathogens has not been examined in tomato crops. Resistant *A. solani* strains have been detected in potato (see the relevant section). Four more fungicides in this chemical group have been registered in the USA for early blight since 1999: pyraclostrobin, trifloxystrobin, famoxadone and fenamidone. These fungicides are still in commercial use because resistance is not thought to be widespread, and they are also labelled for other diseases, notably Septoria leaf spot and late blight. Presently, there are fungicides in four other chemical groups with inherent risk of resistance development that can be used in alternation with FRAC Code 11 fungicides: difenconazole (FRAC Code 3), boscalid (FRAC Code 7), cyprodinil (FRAC Code 9), pyrimethanil (FRAC Code 9) and propamocarb (FRAC Code 28). Difenconazole is marketed as a mixture with azoxystrobin (Quadris Top) and mandipropamid (Revus Top).

Cyprodinil is sold combined with fludioxonil (FRAC Code 12; Switch).

Resistance to FRAC Code 11 fungicides has not been detected in a related pathogen, *Alternaria dauci*, which causes Alternaria leaf blight in carrot. This was investigated in 2004 by testing isolates from an important carrot production area in Wisconsin where azoxystrobin had been used in rotation with chlorothalonil (Rogers and Stevenson, 2010). There have been no reports since then of reduced performance of FRAC Code 11 fungicides in commercial carrot fields in Wisconsin (A. Gevens, personal communication).

Multiple crops

Grey mould

Botrytis cinerea is a common saprophytic pathogen with a broad host range. Efficacy of dicarboximide fungicides is limited due to resistance. Resistance has also developed to FRAC Code 11 fungicides.

Phytophthora blight

P. capsici causes blight in a diversity of crops. The main ones affected are pepper and cucurbit crops. Aubergine, tomato, lima beans and snap beans are also susceptible. Resistance has developed to the phenylamide fungicides (FRAC Code 4) metalaxyl and mefenoxam (Lamour and Hausbeck, 2000; Parra and Ristaino, 2001). Pathogen strains with insensitivity to cyazofamid have also been detected in the USA (Kousik and Keinath, 2008).

Presently, there are targeted fungicides with inherent risk of resistance development in six FRAC groups labelled for managing Phytophthora blight in the USA. These are FRAC Codes 21 (cyazofamid), 22 (zoxamide), 27 (cymoxanil), 33 (phosphorous acid), 40 (dimethomorph and mandipropamid) and 43 (fluopicolide) (Table 16.1). They are not all registered for use on all susceptible crops.

16.4 Conclusion

Fungicide resistance has become a major issue when considering disease management programmes for vegetable crops in the USA, because fungicides with an inherent risk of resistance developing have become an important management tool due to their superior efficacy compared to protectant fungicides and the number of at-risk fungicides now available. Presently, there are fungicides in 18 chemical groups registered for diseases on vegetable crops in the USA. Resistance has developed in pathogens causing several of the most important diseases of vegetable crops in the USA, including powdery mildew, downy mildew, gummy stem blight and Phytophthora blight in cucurbit crops; downy mildew and drop in lettuce; early blight and late blight in tomato; early blight, late blight, leak, pink rot, Fusarium dry rot and silver scurf in potato. Experience with occurrence and management of fungicide resistance has revealed there are many associated challenges. Often when resistance developed, the affected fungicide was the only at-risk fungicide registered for the disease. Lack of fungicides to use in alternation programmes is no longer a major constraint to management for many diseases.

References

Abu-El Samen, F.M., Secor, G.A. and Gudmestad, N.C. (2003) Genetic variation among asexual progeny of *Phytophthora infestans* detected with RAPD and AFLP markers. *Plant Pathology* 52, 314–325.

Abu-El Samen, F.M., Oberoi, K., Taylor, R.J., Secor, G.A. and Gudmestad, N.C. (2005) Inheritance of mefenoxam resistance in selfed populations of the homothallic oomycete *Phytophthora erythroseptica* (Pethybr.), cause of pink rot of potato. *American Journal of Potato Research* 82, 105–115.

Brown, S., Koike, S.T., Ochoa, O.E., Laemmlen, F. and Michelmore, R.W. (2004) Insensitivity to the fungicide fosetyl-aluminum in California isolates of the lettuce downy mildew pathogen, *Bremia lactucae*. *Plant Disease* 88, 502–508.

Cunha, M.G. and Rizzo, D.M. (2003) Development of fungicide cross resistance in *Helminthosporium solani* populations from California. *Plant Disease* 87, 798–803.

Deahl, K.L., DeMuth, S.P., Pelter, G. and Ormrod, D.J. (1993) First report of resistance of *Phytophthora infestans* to metalaxyl in Eastern Washington and Southwestern British Columbia. *Plant Disease* 77, 429.

Desjardins, A.E., Christ-Harned, E.A., McCormick, S.P. and Secor, G.A. (1993) Population structure and genetic analysis of field resistance to thiabendazole in *Gibberella pulicaris* from potato tubers. *Phytopathology* 83, 164–170.

Frazier, M.J., Shetty, K.K., Kleinkopf, G.E. and Nolte, P. (1998) Management of silver scurf (*Helminthosporium solani*) with fungicide seed treatments and storage practices. *American Journal of Potato Research* 75, 129–135.

Goodwin, S.B. and McGrath, M.T. (1995) Insensitivity to metalaxyl among isolates of *Phytophthora erythroseptica* causing pink rot of potato in New York. *Plant Disease* 79, 967.

Goodwin, S.B., Smart, C.D., Sandrock, R.W., Deahl, K.L., Punja, Z.K. and Fry, W.E. (1998) Genetic change within populations of *Phytophthora infestans* in the United States and Canada during 1994 to 1996: role of migration and recombination. *Phytopathology* 88, 939–949.

Hanson, L.E., Schwager, S.J. and Loria, R. (1996) Sensitivity to thiabendazole in *Fusarium* species associated with dry rot of potato. *Phytopathology* 86, 378–384.

Holm, A.L., Rivera, V.V., Secor, G.A. and Gudmestad, N.C. (2003) Temporal sensitivity of *Alternaria solani* to foliar fungicides. *American Journal of Potato Research* 80, 33–40.

Hu, C., Perez, F.G., Donahoo, R., McLeod, A., Myers, K.L., Ivors, K.L., *et al.* (2010) Genetic structure of *Phytophthora infestans* population in eastern North America, 2002–2009. *Phytopathology* 100, S52.

Hubbard, J.C., Subbarao, K.V. and Koike, S.T. (1997) Development and significance of dicarboximide resistance in *Sclerotinia minor* isolates from commercial lettuce fields in California. *Plant Disease* 81, 148–153.

Ivors, K., Lacey, L., Milks, D. and Olaya, G. (2010) Detecting resistance to QoI fungicides in *Alternaria solani* isolates collected from tomatoes in North Carolina. *Phytopathology* 100, S55.

Keinath, A.P. (2009) Sensitivity to azoxystrobin in *Didymella bryoniae* isolates collected before and after field use of strobilurin fungicides. *Pest Management Science* 65, 1090–1096.

Keinath, A.P. and Zitter, T.A. (1998) Resistance to benomyl and thiophanate-methyl in *Didymella bryoniae* from South Carolina and New York. *Plant Disease* 82, 479–484.

Klose, S., Wu, B.M., Ajwa, H.A., Koike, S.T. and Subbarao, K.V. (2010) Reduced efficacy of Rovral and Botran to control *Sclerotinia minor* in lettuce production in the Salinas Valley may be related to accelerated fungicide degradation in soil. *Crop Protection* 29, 751–756.

Kousik, C.S. and Keinath, A.P. (2008) First report of insensitivity to cyazofamid among isolates of *Phytophthora capsici* from the southeastern United States. *Plant Disease* 92, 979.

Lambert, D.H. and Salas, B. (1994) Metalaxyl insensitivity of *Phytophthora erythroseptica* isolates causing pink rot of potato in Maine. *Plant Disease* 78, 1010.

Lamour, K.H. and Hausbeck, M.K. (2000) Mefenoxam insensitivity and the sexual stage of *Phytophthora capsici* in Michigan cucurbit fields. *Phytopathology* 90, 396–400.

McGrath, M.T. (2001) Fungicide resistance in cucurbit powdery mildew: experiences and challenges. *Plant Disease* 85, 236–245.

McGrath, M.T. (2005) Evaluation of fungicide programs for managing pathogen resistance and powdery mildew of pumpkin. *Fungicide and Nematicide Tests* 60, V049.

McGrath, M.T. (2006) Occurrence of fungicide resistance in *Podosphaera xanthii* and impact on controlling cucurbit powdery mildew in New York. In: Holmes, G.J. (ed.) *Proceedings of Cucurbitaceae 2006*, Asheville, NC, USA, 17–21 September 2006. Universal Press, Raleigh, North Carolina, pp. 473–482.

McGrath, M.T. and Fox, G.M. (2009) Efficacy of fungicides for managing cucurbit powdery mildew and treatment impact on pathogen sensitivity to fungicides, 2008. *Plant Disease Management Reports* 3, V125.

McGrath, M.T. and Fox, G.M. (2010) Efficacy of fungicides for managing cucurbit powdery mildew and treatment impact on pathogen sensitivity to fungicides, 2009. *Plant Disease Management Reports* 4, V147.

McGrath, M.T. and Hunsberger, L.K. (2011) Efficacy of fungicides for managing cucurbit powdery mildew and pathogen sensitivity to fungicides, 2010. *Plant Disease Management Reports* 5, V104.

McGrath, M.T. and Shishkoff, N. (2001) Resistance to triadimefon and benomyl: dynamics and impact on managing cucurbit powdery mildew. *Plant Disease* 85, 147–154.

McGrath, M.T. and Shishkoff, N. (2003) First report of the cucurbit powdery mildew fungus (*Podosphaera xanthii*) resistant to strobilurin fungicides in the United States. *Plant Disease* 87, 1007.

McGrath, M.T., Staniszewska, H., Shishkoff, N. and Casella, G. (1996) Fungicide sensitivity of *Sphaerotheca fuliginea* populations in the United States. *Plant Disease* 80, 697–703.

Matheron, M.E. and Porchas, M. (2004) Activity of boscalid, fenhexamid, fluazinam, fludioxonil, and vinclozolin on growth of *Sclerotinia minor* and *S. sclerotiorum* and development of lettuce drop. *Plant Disease* 88, 665–668.

Merida, C.L. and Loria, R. (1990) First report of resistance *of Helminthosporium solani* to thiabendazole in the United States. *Phytopathology* 80, 1027.

Merida, C.L. and Loria, R. (1994) Comparison of thiabendazole-sensitive and -resistant *Helminthosporium solani* isolates from New York. *Plant Disease* 78, 187–192.

Parra, G. and Ristaino, J.B. (2001) Resistance to mefenoxam and metalaxyl among field isolates of *Phytophthora capsici* causing Phytophthora blight of bell pepper. *Plant Disease* 85, 1069–1075.

Pasche, J.S. and Gudmestad, N.C. (2008) Prevalence, competitive fitness and impact of the F129L mutation in *Alternaria solani* from the United States. *Crop Protection* 27, 427–435.

Pasche, J.S., Wharam, C.M. and Gudmestad, N.C. (2004) Shift in sensitivity of *Alternaria solani* to QoI fungicides. *Plant Disease* 88, 181–187.

Pasche, J.S., Piche, L.M. and Gudmestad, N.C. (2005) Effect of the F129L mutation in *Alternaria solani* on fungicides affecting mitochondrial respiration. *Plant Disease* 89, 269–278.

Raid, R.N., Datnoff, L.E., Schettini, T. and Michelmore, R.W. (1990) Insensitivity of *Bremia lactucae* to metalaxyl on lettuce in Florida. *Plant Disease* 74, 81.

Rogers, P.M. and Stevenson, W.R. (2010) Aggressiveness and fungicide sensitivity of *Alternaria dauci* from cultivated carrot. *Plant Disease* 94, 405–412.

Schettini, T.M., Legg, E.J. and Michelmore, R.W. (1991) Insensitivity to metalaxyl in California populations of *Bremia lactucae* and resistance of California lettuce cultivars to downy mildew. *Phytopathology* 81, 64–70.

Stack, R.W., Salas, B., Gudmestad, N.C. and Secor, G.A. (1993) The lack of evidence for metalaxyl resistance in *Phytophthora erythroseptica,* the cause of pink rot of potato. *Phytopathology* 83, 886.

Stein, J.M. and Kirk, W.W. (2004) The generation and quantification of resistance to dimethomorph in *Phytophthora infestans. Plant Disease* 88, 930–934.

Taylor, R.J., Salas, B., Secor, G.A., Rivera, V. and Gudmestad, N.C. (2002) Sensitivity of North American isolates of *Phytophthora erythroseptica* and *Pythium ultimum* to mefenoxam (metalaxyl). *Plant Disease* 86, 797–802.

Taylor, R.J., Pasche, J.S. and Gudmestad, N.C. (2006) Biological significance of mefenoxam resistance in *Phytophthora erythroseptica* and its implications for the management of pink rot of potato. *Plant Disease* 90, 927–934.

17 Fungicide Resistance in India: Status and Management Strategies

Tarlochan S. Thind

Department of Plant Pathology, Punjab Agricultural University, Ludhiana, India

17.1 Introduction

Resistance to fungicides has become a serious issue in the management of crop diseases and has threatened the potential of some highly effective commercial products. Unlike insecticides, where resistance problems are known to occur much earlier, the practical problems of fungicide resistance emerged much later, in the 1970s and thereafter. The incidence of resistance to fungicides has remained restricted to mainly systemic fungicides that operate against single biochemical targets, also known as single-site inhibitors (Dekker, 1985; Brent, 1995). These site-specific systemic fungicides were introduced in the mid-1960s onwards and include several major groups of fungicides such as benzimidazoles, pyrimidines, phenylamides, sterol biosynthesis inhibitors, dicarboximides, phenylamides, etc. During the past decade, more novel compounds with different modes of action, notably phenylpyrroles, anilinopyrimidines, strobilurins, spiroxamines, phenylpyridylanines, quinolines, etc., have been developed having bioefficacy against diverse plant diseases. Several of these modern selective fungicides have become vulnerable to the risk of resistance development in target pathogens in different countries (Brent and Hollomon, 2007).

As compared to developed countries, not much work has been done on the problem of fungicide resistance in India. This could be attributed to the lack of awareness among Indian workers about the importance of the problem and the non-availability of trained scientific manpower in this field. Most of the earlier studies done on fungicide resistance in India pertained to acquired resistance using mutagens or training (pressurization) methods under laboratory conditions, without looking into their possible implications in practical disease control. However, during the past two decades, cases of resistance development in field situations have also been reported from different parts of the country (Thind, 2002, 2008).

17.2 Use of Fungicides in India

The use of fungicides in India is quite low compared to developed countries. Overall, fungicide use is less than insecticide and herbicide use. The consumption of fungicides in 2009 was 8307 MT compared to 26,756 MT of insecticides and 6040 MT of herbicides, with a market share of 19% compared to 61% of insecticides and 17% of herbicides. The current fungicide market in India is worth Rs 4.3 billion.

Apart from conventional compounds like sulfur, dithiocarbamates, mercurials, phthalimides and copper-based compounds, etc., several of the site-specific fungicides of the groups

Table 17.1. Site-specific fungicides registered in India for use either alone or in combinations with multi-site contact fungicides (as on 14 September 2010).

Fungicide group	Name of fungicide
Oxathiins	Carboxin, oxycarboxin
Benzimidazoles	Benomyl, carbendazim
Guanidines	Dodine
Thiophanates	Thiophanate-methyl
Phosphorothiolates	Edifenphos, iprobenfos
Dicarboximides	Iprodione
Acylalanines	Metalaxyl, metalaxyl-M (mefenoxam)
Cyano-acetamide oximes	Cymoxanil
Cinnamic acid derivatives	Dimethomorph
Trizoles	Propiconazole, penconazole, myclobutanil, triadimefon, bitertanol, hexaconazole, difenoconazole, tebuconazole, flusilazole
Morpholines	Tridemorph
Pyrimidines	Fenarimol
Melanin biosynthesis inhibitors	Tricyclazole, carpropamid
Dithiolanes	Isoprothiolane
Strobilurins	Azoxystrobin, kresoxim-methyl, trifloxystrobin
Oxazolidinones	Famoxadone
Imidazoles	Fenamidone
Valinamides	Iprovalicarb
Anilides	Thifluzamide
Antifungal antibiotics	Aureofungin, kasugamycin, validamycin

Source: Central Insecticides Board (www.cibrc.nic.in).

like benzimidazoles, oxathiins, thiophanates, organophosphorus, triazoles and related sterol inhibitors, phenylamides, strobilurins and other recently developed compounds are being used in India for controlling different diseases on a number of crops. As on 14 September 2010, 52 fungicides belonging to different groups were registered for use in India (Table 17.1). In addition, formulations of combination products containing systemic and contact fungicides are also registered.

Cropwise consumption of fungicides in India is maximum on pome fruit (12.7%), followed closely by potatoes (12.2%), rice (12.0%), tea (9.4%), coffee (8.0%), chillies (7.6%), grapevines (6.9%), other fruit (5.9%), other vegetables (4.6%) and other crops, which account for about 75% of the total fungicides used in India (Thind, 2002).

17.3 Acquired Resistance in the Laboratory

Various workers in India have reported under laboratory conditions several cases of adaptive resistance to many fungicides including multi-site action compounds, but the possible implications in disease control have not been indicated. Various methods such as adaptation to increasing fungicide concentrations, exposure to UV radiation and chemical mutagens have been employed to study resistance development in diverse fungi to dithiocarbamates, copper-based compounds, oxathiins, benzimidazoles, organophosphorus, phenylamides, alkyl phosphonates, morpholines and antifungal antibiotics in the laboratory (Thind, 1995).

Gangawane (1997) has given an account of such laboratory cases of resistance studied in India. Various workers in India have reported nearly 18 fungi that have developed laboratory resistance towards 24 fungicides, but without much practical significance as the studies lack meeting the essential parameters of resistance. Normally, resistance acquired through serial transfers on fungicide-amended medium is not stable. Some examples of laboratory resistance studied by Indian workers are: *Gloeosporium ampelophagum* to copper sulfate (Reddy and Apparao, 1967);

Helminthosporium oryzae to mancozeb and copper oxychloride (Gupta and Dass, 1971); *Rhizoctonia solani* to carboxin (Grover and Chopra, 1971); *Rhizoctonia bataticola* to maneb, thiram, kocide and carbendazim (Anil Kumar and Sastri, 1979); *Fusarium oxysporum* f. sp. *lycopersici, Cephalosporium sacchari, G. ampelophagum* to fentin acetate and *Pyricularia oryzae, G. ampelophagum, F. oxysporium* to copper oxychloride and zineb (Reddy *et al.*, 1979); *Colletotrichum capsici* to mancozeb and copper oxychloride (Thind and Jhooty, 1980); *Helminthosporium maydis* to mancozeb (Bains and Mohan, 1982); *Aspergillus flavus* to captan, carbendazim and thiophanate-methyl (Gangawane and Reddy, 1985); *Macrophomina phaseolina* to dichloran and quintozene (Pan and Sen, 1980); *Alternaria solani* to mancozeb (Ramaswamy and Anilkumar, 1991); and *Fusarium udum* to benomyl (Kamble and Gangawane, 1994). Acquired resistance to antifungal antibiotic cycloheximide was reported in *Sclerotium rolfsii* by Sullia and Maria (1985). The resistance in the latter case was probably due to the conversion of cycloheximide into isocycloheximide, which is less toxic. More cases of laboratory resistance as reported by various Indian workers have been discussed by Thind (2002).

Most of these laboratory resistance cases are of academic interest and do not reflect a similar situation occurring in practice. Some workers have reported that fungicide tolerance is accompanied by a change in pathogen virulence. Grover and Chopra (1971) observed a reduced pathogenicity of carboxin- and oxycarboxin-adapted isolates of *Rhizoctonia* species towards *Phaseolus mungo*. Later, Pan and Sen (1980) reported that adaptation of four isolates of *M. phaseolina* to dichloran, PCNB, carboxin and carbendazim reduced the virulence of these isolates to soybean significantly over their parental type. Similar findings on reduced virulence of the adapted isolates were reported by Thind and Jhooty (1980) in the case of mancozeb and copper oxychloride-adapted isolates of *C. capsici* when inoculated on chilli fruit.

Usually, the adaptive type of resistance obtained under laboratory conditions is temporary in nature. The tolerant isolates soon regain their sensitive response after the fungicide is withdrawn from the culture media.

In most cases, tolerance is lost gradually as the isolates are cultured serially on fungicide-free medium. Ramamoorthy (1991) observed that mutants of *F. oxysporum* f. sp. *lycopersici* adapted on carbendazim-amended medium were not stable in resistance and were non-pathogenic, but EMS and UV mutants were found to be stable in resistance even after ten generations and could cause infection on tomato plants. Laboratory mutants showed reduction in sporulation, spore germination and mycelial growth. Annamalai and Lalithakumari (1992) have reported that about 50% of laboratory mutants (obtained through chemical mutagens) of *Bipolaris oryzae*, the cause of brown leaf spot of rice, were found to retain their resistance to edifenphos after ten transfers on fungicide-free medium. In contrast, adapted strains gradually lost their tolerance when transferred on fungicide-free medium and completely reverted back to a parental level of sensitivity after four transfers.

Laboratory mutants of *Venturia inaequalis* to two sterol inhibitors, viz. bitertanol and fenapanil, were developed by adaptation and EMS mutation (Revathi *et al.*, 1992). Adapted mutants exhibited much higher resistance to test fungicides than the EMS mutants. Both types of mutants completely lost pigmentation and sporulation. The conidial size of bitertanol mutants showed significant reduction in size over sensitive strains. Using chemical mutagenesis (MNNG), laboratory strains of *V. inaequalis* with high resistance level to penconazole were also developed (Palani and Lalithakumari, 1999). In another case, laboratory mutants of *R. solani*, causing sheath blight of rice, possessing resistance to carbendazim and iprodione were developed by UV irradiation of mycelial discs, adaptation of sclerotia and chemical (EMS) mutagenesis of protoplasts. Both resistant mutants of carbendazim and iprodione obtained through adaptation showed stability in growth on fungicide-amended media after ten generations, while the least stability was observed in UV mutants. The resistant mutants, however, showed delayed growth rate and lost sclerotial production on fungicide-amended medium. All the mutants except iprodione UV mutants produced typical symptoms on sheath blight of rice plants.

17.4 Resistance Cases in the Field

The increase in the use of fungicides, particularly of selective fungicides, on important crops caught the attention of some workers with regards to their likely effects on pathogen populations and, over the past few years, cases of fungicide resistance development have also been reported under field conditions in India (Thind, 2002, 2008). Sensitivity studies through regular monitoring of conidial/sporangial populations of several pathogens have led to the detection of fungicide-resistant strains with low to high resistance levels in some plant pathogens. Some reported field cases of fungicide resistance in India are mentioned in Table 17.2 and are described in the following sections.

Benzimidazoles

Apple scab (Venturia inaequalis)

Apple scab caused by *V. inaequalis* (Cke.) Wint. has become endemic in all the important apple growing belts in India, covering an area of 60,000 ha in Kashmir alone. As most of the commercial apple cultivars are susceptible to scab, orchardists depend mainly on fungicides such as mancozeb, zineb, ziram, carbendazim, benomyl, captan, triazoles, etc., for its control. At least six applications of fungicides are recommended against this disease in Kashmir and seven in Himachal Pradesh

(Gupta and Gupta, 1996) at various phenological stages, starting from silver tip/green tip stage till harvest; but growers usually give 12–15 applications of different fungicides to ensure good disease control. Based on the observations of some growers on the decreased level of disease control after prolonged and exclusive usage of mancozeb and carbendazim in their orchards in the Kashmir valley, sensitivity studies of conidial populations of *V. inaequalis* from 40 affected orchards were carried out. Apparently, one isolate was obtained from each orchard. The results were reported to indicate mancozeb-resistant strains in 12 orchards and carbendazim-resistant strains in 3 orchards (Basu Chaudhary and Puttoo, 1984).

Although apparently unusual, mancozeb-resistant isolates could tolerate 2.5 times higher levels of the fungicide compared to sensitive isolates, and the per cent of disease control ranged from 41 to 77 in these orchards. The isolates proved pathogenic on young apple foliage of cv. Red Delicious during the first and second subculturings only. However, the resistance was found to be unstable, as these isolates lost character after three subculturings on mancozeb-free medium and became non-sporulating. Since the workers themselves found mancozeb resistance in these isolates was not stable, and in any case was at a relatively low level, the reduction in disease control over some years could possibly be attributed to the poorly managed spray schedules of mancozeb. This fungicide

Table 17.2. Reported cases of fungicide resistance under field situations in India.

Fungicide	Pathogen (host)	Reference
Carbendazim	*Venturia inaequalis* (apple)	Basu Chaudhary and Puttoo (1984)
	Gloeosporium ampelophagum (grape)	Kumar and Thind (1992)
	Aspergillus flavus (groundnut)	Gangawane and Reddy (1985)
Edifenphos	*Dreschlera oryzae* (rice)	Annamalai and Lalithakumari (1990)
	Pyricularia oryzae (rice)	Lalithakumari and Kumari (1987)
Metalaxyl	*Plasmopara viticola* (grape)	Rao and Reddy (1988)
	Phytophthora infestans (potato)	Arora *et al.* (1992)
		Thind *et al.* (2001)
	Phytophthora parasitica (citrus)	Thind *et al.* (2009)
	Pseudoperonospora cubensis (cucumber)	Thind *et al.* (2011)
Oxadixyl	*Phytophthora infestans* (potato)	Singh *et al.* (1993)
Triadimefon	*Uncinula necator* (grape)	Thind *et al.* (1998a)

stands low risk of resistance development in the pathogens due to its multi-site mode of action, and cases of practical resistance to mancozeb have not been reported elsewhere, despite its widespread use against many pathogens.

The carbendazim-resistant isolates could tolerate 3–14 times higher levels of the fungicide. These are relatively low resistance factors compared with carbendazim resistance reported elsewhere. In contrast to the mancozeb-resistant isolates, carbendazim resistance was found to be stable. The isolates retained the spore-producing character and were pathogenic on young apple foliage during all subculture inoculations. Strategies for the management of fungicide resistance in *V. inaequalis* involving the need-based application of fungicides and the use of sanitary, physical and cultural practices to control pathogen multiplication have been suggested by Putto and Basu Chaudhary (1986).

Anthracnose of grape (Gloeosporium ampelophagum)

Anthracnose, caused by *G. ampelophagum* (de Bary) Sacc. (syn. *Elsinoe ampelina*), poses a serious threat to grape cultivation in Punjab and other parts of India and requires regular fungicide applications for its control. A number of treatments of benzimidazole and related fungicides like carbendazim, benomyl and thiophanate-methyl, as well as conventional contact fungicides (copper-based compounds, dithiocarbamates, phthalimides, etc.) are applied repeatedly by growers to protect the plants from this disease. Due to excessive and irrational use of benzimidazoles, development of resistance, associated with inferior disease control, has been observed in *G. ampelophagum* and the strains with a high level of resistance to carbendazim have been isolated from vineyards in the Punjab state (Kumar and Thind, 1992). Studies were conducted during 1990–1997 to determine the population structure of *G. ampelophagum* with regard to fungicide sensitivity and strategies for its management.

SCREENING FOR CARBENDAZIM RESISTANCE. In the preliminary screening, a total of 80 isolates of *G. ampelophagum* collected from various regions in the Punjab state during 1990–1997 were studied for their sensitivity to carbendazim (Bavistin 50 WP) using malt agar plates amended with 1 and 5 µg/ml of carbendazim. The majority of isolates showed sensitive or weakly resistant response and were unable to grow beyond 1 µg/ml of carbendazim. However, 36% of the isolates showed growth at 5 µg/ml, indicating resistant response to carbendazim (Thind and Mohan, 1998). Twenty-three isolates found resistant in the preliminary screening were further grown at higher concentration up to 100 µg/ml of carbendazim. Of these, all the isolates showed normal growth up to 50 µg/ml, while 15 were able to grow at an even higher dose of 100 µg/ml of carbendazim, thus exhibiting high resistance factors (Table 17.3). These isolates were obtained from vineyards receiving regular treatments of Bavistin and were mostly from areas near Ludhiana. Resistance to carbendazim was found to be persistent in nature as the resistant isolates were able to grow at 50 µg/ml of carbendazim even after 1 year of subculturing on fungicide-free medium. The isolates of *G. ampelophagum* were categorized into three morphological

Table 17.3. Structuring of *Gloeosporium ampelophagum* isolates from different vineyards for carbendazim sensitivity in Punjab (1990–1997).

Number of isolates tested	ED$_{50}$ (µg/ml)	MIC (µg/ml)	Resistance factor	Sensitivity class
36	0.02–0.04	0.05–0.1	0.0	S
21	0.04–0.16	0.20–0.5	1.8–4.5	WR
8	14–50	40–100	360–900	HR
15	69–100	> 100	> 900	HR

Note: S, sensitive; WR, weakly resistant; HR, highly resistant.
Source: Thind and Mohan (1998).

groups and the majority of the resistant isolates produced reddish-brown to peach-red colonies (Thind *et al.*, 1994).

Further studies conducted from 2000 to 2004 revealed that carbendazim-resistant isolates of *G. ampelophagum* were persistent in natural populations and could be detected frequently in the vineyards around Ludhiana in Punjab state (Mohan *et al.*, 2005).

PATHOGENIC BEHAVIOUR OF RESISTANT ISOLATES. Pathogenic behaviour of two resistant and two sensitive isolates was studied on detached leaves of cv. Perlette treated with different concentrations of Bavistin in the laboratory. While the sensitive isolates did not produce any symptoms above 250 µg/ml, both the resistant isolates Ga 28 and Ga 53 developed normal sporulating lesions at 500 µg/ml and also produced mild symptoms even at 1000 µg/ml of Bavistin, thus confirming their resistant character.

CROSS-RESISTANCE TO OTHER FUNGICIDES. Cross-resistance to other fungicides, viz. Topsin-M (thiophanate-methyl), Captaf (captan), Indofil M-45 (mancozeb), Bordeaux mixture (copper sulfate + calcium hydroxide) and Bayleton-5 (triadimefon), was studied by growth inhibition assay as well as by detached leaf assay by taking one resistant and one sensitive isolate. Observations revealed that resistant isolate Ga 53 possessed cross-resistance to Topsin-M, which had a similar mode of action as Bavistin (Mohan and Thind, 1995). On the other hand, both resistant and sensitive isolates exhibited sensitive response to all other fungicides tested. In another study (Thind *et al.*, 1997) on cross-resistance, three triazole fungicides, viz. Score (difenconazol), Corail (tebuconazol) and Olymp (flusilazole), and one pyridylanine compound, Dirango (fluazinam), were found to possess high inhibitory action against carbendazim-resistant as well as -sensitive isolates, with MIC values of triazoles ranging between 1 and 5 µg/ml for both types. Fluazinam also exhibited good efficacy at 10 µg/ml and above. By detached leaf assay, difenconazole proved most effective and no symptoms developed at 25 µg/ml. Fluazinam arrests disease development completely at 500 µg/ml by both isolates and

holds promise along with difenconazole to check resistance.

MANAGEMENT OF CARBENDAZIM RESISTANCE. Indofil M-45 and Bordeaux mixture to which carbendazim-resistant isolates did not show any cross-resistance were tested in a resistance-affected vineyard near Ludhiana. Bavistin (0.1%) when used alone did not provide the desired control of grape anthracnose. In contrast, when it was applied in alternation with Bordeaux mixture (2:2:250) or Indofil M-45 (0.3%), there was significant reduction in disease severity (Mohan and Thind, 1995). Triazole fungicides such as difenconazole, tebuconazole and flusilazole and fluazinam, a pyridylanine compound, which showed promising efficacy against resistant and sensitive isolates of *G. ampelophagum* in laboratory studies using detached leaf assays (Thind *et al.*, 1997; Thind and Mohan, 1998), are now used as anti-resistance measures in field conditions.

Application of an effective fungicide immediately after the first rain shower in March/April helps in checking the primary infection and multiplication of inoculum for subsequent infections, thereby reducing fungicide sprays and minimizing the risk of resistance development to site-specific fungicides like carbendazim.

Sugarbeet leaf spot (Cercospora beticola)

Leaf spot of sugarbeet, caused by *Cercospora beticola* Sacc., is a serious disease problem of sugarbeet in India. Various fungicides including carbendazim formulations are widely used to control this disease. Some natural populations of the fungus were screened for sensitivity to carbendazim (Bavistin 50 WP).

Organophosphorus

Brown spot of rice (Drechslera oryzae)

Organophosphorus fungicides such as edifenphos and iprobenfos are widely used in the southern parts of India for the control of brown leaf spot of rice caused by *Drechslera oryzae* (Breda de Haan) Subram. and Jain, which causes severe crop losses if not controlled in the early stages. Edifenphos (Hinosan) is used quite regularly in Tamil Nadu and other rice growing states for reducing crop losses due to this disease. Risk of resistance development to edifenphos has been determined in *D. oryzae* under selection pressure of the fungicide in the field during 1984–1988 at a village farm near Chingleput, Madras (Annamalai and Lalithakumari, 1990). Field isolates of the pathogen were collected to study the baseline data on the sensitivity before commencing the application of edifenphos in 1984, and subsequently after the application of edifenphos every year up to 1988. Repeated applications of edifenphos resulted in patches of paddy crop cv. IR-50 with severe disease manifestation. The disease intensity in the treated plots was surprisingly much higher than in the untreated plots. Every year, 400 leaf samples were collected at random, the pathogen was isolated (one lesion/plate) and screened for edifenphos sensitivity by measuring the radial growth on PDA amended with 10, 20, 50, 200 and 300 μg/ml of the fungicide. To characterize the isolates for resistance, these were grouped into four categories based on ED_{50} values, i.e. sensitive, low-level resistance, moderate resistance and high-level resistance, having ED_{50} below 50, between 50 and 100, between 101 and 150 and above 150 μg/ml, respectively.

A shift in the level of sensitivity to edifenphos was noticed from year to year. In 1984, before the application of edifenphos, 96% of the isolates were sensitive at 50 μg/ml, while in 1985, 1986, 1987 and 1988 (i.e. after fungicide application), 74, 64, 50 and 48%, respectively, of isolates showed the same level of sensitivity. The sensitivity data of the field isolates thus showed a clear shift in the level of sensitivity of *D. oryzae* due to frequent applications of edifenphos (Table 17.4). Rate of uptake of edifenphos was less in resistant strains of *D. oryzae* and the reduced membrane permeability was suggested as the mechanism of resistance.

When tested for cross-resistance to other fungicides, mancozeb showed significant inhibitory effect on the growth of edifenphos-resistant isolates compared with the sensitive strain and exhibited ED_{50} value of 45–50 μg.

Table 17.4. Sensitivity range of *Drechslera oryzae* field isolates against edifenphos.

Range of ED_{50} values (μg/ml)	Isolates (%)				
	1984	1985	1986	1987	1988
20–50	96	74	64	50	48
51–100	4	26	16	18	16
101–150	0	0	20	18	10
151–180	0	0	0	14	26

Source: Annamalai and Lalithakumari (1990).

Cross-resistance was observed to iprobenfos, an organophosphorus fungicide with the same mode of action. Other fungicides tested, such as copper oxychloride, benomyl, bitertanol, carbendazim and pyroquilon, were less effective against resistant isolates and inhibited the growth of the fungus at higher concentrations (Annamalai and Lalithakumari, 1992). Cross-resistance studies and field treatments indicated that edifenphos resistance in *D. oryzae* could be counteracted by spraying mancozeb as an alternative (replacement) fungicide. However, edifenphos is still used against *D. oryzae* and is working well in most areas.

Blast of rice (Pyricularia oryzae)

Blast disease of rice caused by *P. oryzae* Cav. (syn. *P. grisea* Cav. and *Magnaporthe oryzae*) is a serious disease in rice growing areas of South Indian states. Edifenphos (Hinosan) is widely used as a foliar application for the effective control of this disease. A preliminary study has been done to estimate the sensitivity of natural populations of *P. oryzae* isolates to edifenphos (Lalithakumari and Kumari, 1987). Diseased leaves of rice cv. IR 50 were collected from fields sprayed regularly with edifenphos and 50 monosporic isolates of the pathogen were obtained. Sensitivity of these isolates to edifenphos was tested at ten concentrations ranging from 10 to 100 μm on oatmeal agar by mycelial growth inhibition technique. The ED_{50} values were compared with a sensitive isolate, unexposed to edifenphos.

The sensitive isolate had an ED_{50} value of 36.3 μg. Seventeen isolates from treated fields showed a shift in their ED_{50} values above

50 µg, and out of these 17 isolates, 9 isolates had ED_{50} values above 60 µg (up to 75.8 µg), thus indicating resistant response to edifenphos. Not much variation was observed in the growth pattern, conidial morphology and pathogenicity among the 50 isolates tested. The resistant isolates were equally pathogenic when inoculated on 1-month-old seedlings of rice cultivar IR 50. In cross-resistance studies with other fungicides, the resistant isolates showed positive cross-resistance to iprobenfos, a related fungicide. Ziram inhibited the growth of all the resistant isolates effectively and its use was suggested as a companion fungicide in mixture with edifenphos or as an alternative spray fungicide (Lalithakumari and Mathivanan, 1990).

Phenylamides

Grape downy mildew (Plasmopara viticola)

Grape downy mildew caused by *Plasmopara viticola* (Burk. & Curt.) Verl. & de Toni causes severe losses in the southern states of India such as Maharashtra, Andhra Pradesh and Karnataka, especially on two commercial grape varieties, Anab-e-Shahi and Thompson Seedless, affecting the production and quality of grapes. When metalaxyl became available in the late 1970s, it caught the attention of Indian farmers, who found it miraculous in controlling grape downy mildew, which earlier had been difficult to control by traditional contact fungicides.

Grape growers around Hyderabad started using metalaxyl (Ridomil 25WP) in 1981 to control downy mildew. Based on reports by some grape growers in 1986 regarding the loss

of effectiveness of metalaxyl in controlling grapevine downy mildew in areas around Hyderabad, monitoring and sensitivity studies were undertaken to ascertain the cause of the reduced efficacy of metalaxyl (Rao and Reddy, 1988). Infected leaves were collected from three affected orchards situated at three villages near Hyderabad which had received metalaxyl applications for 3, 4 and 3 years, respectively. Sporangial populations from these samples were assayed for sensitivity to metalaxyl (Ridomil 25 WP) at 25, 50, 100 and 250 µg a.i./ml, following the detached leaf method of Pappas (1980). The preliminary sensitivity assays with *P. viticola* populations indicated that the loss of efficacy of metalaxyl in these vineyards was attributed to the development of resistance to metalaxyl, which had been used frequently by growers (Rao and Reddy, 1988). Considerable difference in minimal inhibitory concentration was seen among the three populations.

Metalaxyl was almost completely inactive against pathogen populations collected from two of the three villages, which confirmed the reports of grape growers regarding the loss of efficacy of metalaxyl in controlling grape downy mildew (Table 17.5). The continuous and exclusive use of metalaxyl (Ridomil 25 WP) by grape growers in these villages for 3–4 years had led to the development of resistant populations of *P. viticola*. No further work has been done on the problem after this report. Several rounds of metalaxyl-based combination products with mancozeb, viz. Ridomil-MZ (now Ridomil Gold), are used for the control of grape downy mildew in India. Although these combination products are known to minimize the risk of resistance development, regular monitoring to determine the changes

Table 17.5. Sensitivity of *Plasmopara viticola* populations to metalaxyl by detached leaf assay.

Population (village)	Average disease ratings[a] Metalaxyl conc. (µg a.i./ml)					Sensitivity response
	0	25	50	100	250	
Royavalli	7	5	2	2	1	Resistant
Yellampeta	8	4	2	1	1	Resistant
Y. Somavaram	6	0	0	0	0	Sensitive

Note: [a]Based on the 0–10 rating scale of Pappas (1980).
Source: Rao and Reddy (1988).

in sensitivity levels of pathogen populations is necessary where these fungicides are used frequently. A simple laboratory technique based on sporulation on leaf discs has been developed for laboratory testing of fungicides (Thind et al., 1988), which requires less space and can be employed easily for determining fungicide resistance in a large number of P. viticola populations.

Late blight of potato (Phytophthora infestans)

For the management of late blight of potato caused by *Phytophthora infestans* (Mont.) de Bary, traditional fungicides such as mancozeb, zineb, copper oxychloride, chlorothalonil, etc., have been in use for many years in India. However, these fungicides provided poor disease control under heavy disease pressure. The introduction of phenylamide fungicides provided much needed relief to Indian farmers, as these gave excellent control of late blight, even under severe disease conditions. Metalaxyl in combination with mancozeb (Ridomil MZ 72WP) was commercially introduced in India during autumn 1988, and since then has been widely used for the control of late blight in different potato growing areas of the country. Now, Ridomil Gold containing metalaxyl-M (also called mefenoxam) has been introduced in India recently for the control of late blight. Although the mixture fungicides are expected to delay the onset of resistance build-up, their use does not guarantee prevention of resistance development (Gisi and Staehle-Csech, 1989).

Following reports of resistance development to metalaxyl in other countries (Davidse et al., 1981; Davidse, 1987), monitoring for metalaxyl-resistant strains of P. infestans was carried out in the Nilgiri Hills from 1989 to 1991 (Arora et al., 1992). Sporangial populations of P. infestans collected from potato fields sprayed with Ridomil MZ were analysed by the detached leaf method for their response to metalaxyl. Metalaxyl-resistant isolates of the pathogen were absent from early to mid-summer potato crop seasons. These, however, appeared towards the end of the summer season, starting from the last week of July and with a maximum frequency of 13% in the autumn. Variations in tolerance to metalaxyl

from 50 to 700 µg/ml were observed among different isolates resistant to metalaxyl.

Highly tolerant isolates (300–900 µg/ml) were observed during the autumn season only and comprised up to 6% of the total samples examined. Resistant isolates could be obtained in plots with a combined spray of metalaxyl and mancozeb, and also in plots with individual sprays of mancozeb or chlorothalonil and the control plots later during the season. The resistant isolates were found to be more aggressive in traits like short incubation period, quick germination of sporangia to zoospores and the ability to cause larger lesions, as compared to the sensitive isolates (Arora, 1994). In another study of resistance monitoring in the Nilgiri Hills following leaf-disc assay, Gangawane et al. (1995) reported that out of the isolates of P. infestans tested, 82% were sensitive, 5% were moderately resistant (RF 15–40) and 4% were highly resistant (RF 60–70). Use of metalaxyl in mixture with chlorothalonil was highly effective against both sensitive and metalaxyl-resistant isolates. Singh et al. (1993) have also reported resistance to metalaxyl in P. infestans from the Shimla Hills after 5 years of use of Ridomil MZ, but the resistance level reported was quite low. They have also reported isolates of this pathogen developing moderate resistance to oxadixyl under experimental conditions.

In the Punjab state of India, metalaxyl in mixture with mancozeb (Ridomil MZ, Matco 8-64) has been widely used for the control of late blight of potato since 1989. Quite often, farmers also use self-prepared mixtures (tank mixed) of metalaxyl (35% SD) and mancozeb in various proportions. Sensitivity levels of 68 P. infestans populations collected during the 1996–1999 crop seasons from various fields treated with fungicides in Punjab were monitored for their sensitivity to metalaxyl following the detached leaf method (Thind et al., 1989). Thirty-one populations, mostly from the Hoshiarpur district, showed mild to severe infection at 10 µg/ml, while 12 populations, collected mostly during 1998–1999, showed varying levels of infection at 50 µg/ml in the initial screening. When tested at higher concentrations of metalaxyl, three populations were able to produce symptoms at 100 and

200 µg/ml, thus showing a higher resistant response to metalaxyl (Thind *et al.*, 2002). The resistance factors of populations with varying levels of decreased sensitivity to metalaxyl ranged between 2.8 and 28.5. A marked decrease in the efficacy of Ridomil MZ was also observed in the field from where highly resistant populations were collected.

During the 2005–2008 crop seasons, 48 sporangial populations of *P. infestans* collected from different potato growing areas in Punjab were tested for metalaxyl sensitivity. Among these, ten populations showed resistant response, causing infection at 200 µg/ml of metalaxyl with a resistance factor of up to 60 (Kaur *et al.*, 2010). The resistant population exihibited competitive fitness in a mixture with the sensitive population (Table 17.6). RAPD analysis of metalaxyl-resistant populations of *P. infestans* was done with ten oligonucleotide primers. Of 50 primers used initially for amplification, 23 showed polymorphism and 10 were able to distinguish resistant and susceptible populations, producing 2–3 unique bands. Information on banding pattern for all the primers was used to determine the genetic distance between resistant and sensitive isolates and to construct a dendrogram. RAPD data distinguished the test isolates into two groups, thus separating the resistant and susceptible isolates. Using primer P9, a unique band of 100 bp was found in susceptible (S) isolates, indicating that these isolates were different from resistant ones in this 100 bp region (Fig.17.1).

The resistant populations were found to be highly pathogenic when inoculated on leaves of potato cv. Kufri Chandramukhi. Their disease severities (78–90%) were comparable with those of sensitive populations (82–100%). Incubation period varied from 4 to 5 days in both resistant and sensitive populations. Sporulation was also comparable in both types of populations. No cross-resistance was observed to dimethomorph, mandipropamid, cymoxanil, benalaxyl, previcur, fluopicolide, azoxystrobin and multi-site contact fungicides chlorothalonil, fluazinam and mancozeb. Dimethomorph has been reported to be effective in controlling late blight of potato and tomato caused either by metalaxyl-sensitive or -tolerant strains of *P. infestans* (Cohen *et al.*, 1995). Metalaxyl resistance was managed effectively under field conditions through the application of novel action fungicides such as Infinito 68.75 SC (fluopicolide + propamocarb chloride), Amistar 25 SC (azoxystrobin), Acrobat 50 WP (dimethomorph), Mandipropamid 250 SC and Curzate M-8 72 WP (cymoxanil + mancozeb). Combination of fungicides with different modes of action retards development of resistance and ensures sustainable management of late blight. The potential of several of these new fungicides has been documented in a recent review (Stevenson, 2009).

Downy mildew of cucumber
(Pseudoperonospora cubensis)

Phenylamide fungicides are regularly used by farmers in various states of India, including Punjab, to manage downy mildew infection on cucurbits such as cucumber and melons. Sensitivity changes to metalaxyl in *Pseudoperonospora cubensis* populations collected from cucumber and muskmelon fields were monitored during 2007 and 2008.

Table 17.6. Competitive fitness of metalaxyl-resistant and -sensitive populations of *P. infestans*.

Metalaxyl concs. (µg/ml)	PDI with different combinations of R and S populations				
	R	S	R (50): S (50)	R (25): S (75)	R (75): S (25)
0	86.0	85.0	74.8	74.1	75.6
10	65.3	13.4	38.5	29.3	71.1
50	49.2	0.0	31.2	23.3	46.6
100	44.2	0.0	22.1	7.2	33.5

Note: R, resistant strain, PI-24; S, susceptible strain, PI-31; PDI, per cent disease index.
Source: Kaur *et al.* (2010).

The maximum number of sporangial populations (12 out of 25) that exhibited resistant response were collected in the Amritsar district, with ED_{90} values of metalaxyl ranging between 30 and 150 μg/ml and resistance factor (RF) between 6 and 30. Most of the populations from Jalandhar and Kapurthala districts showed normal sensitive response. Resistant populations possessed normal pathogenic potential and exhibited strong competitive fitness when inoculated in mixture with sensitive populations (Table 17.7). Resistant populations did not show cross-resistance to fungicides with different modes of action, such as azoxystrobin, cymoxanil, dimethomorph, fluopicolide, propamocarb, chlorothalonil and mancozeb. These fungicides were also found effective against metal-axyl-resistant populations under field conditions and in practice could form part of the strategy to manage metalaxyl resistance (Thind et al., 2011).

Foot rot of citrus (Phytophthora parasitica)

Metalaxyl-based fungicides are commonly used to manage foot rot of citrus (*P. parasitica*) in different states of India. A significant reduction in fungicide efficacy has been observed in many orchards over the years. In the Punjab state, investigations were carried out to determine changes in the sensitivity levels of *P. parasitica* isolates from different citrus orchards where reduced efficacy had been reported after metalaxyl applications. Of the 56 isolates of the fungus tested, 9 isolates

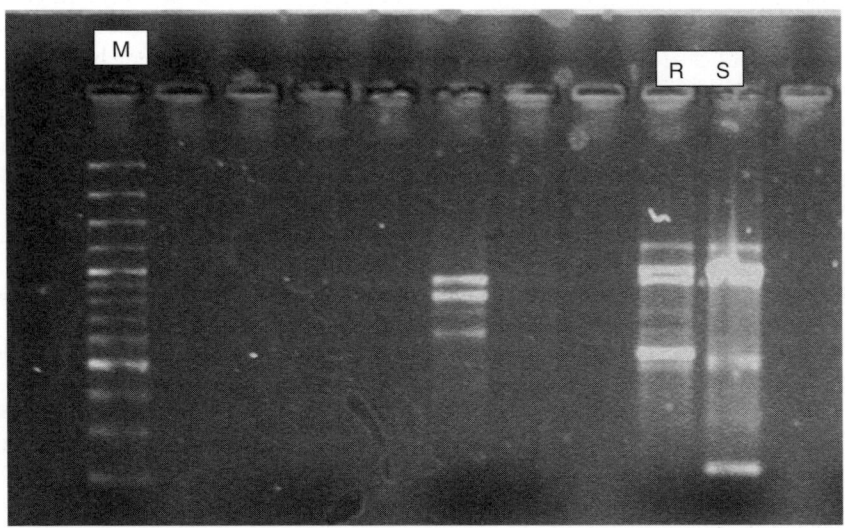

Fig. 17.1. Metalaxyl-resistant and -sensitive strains of *P. infestans* showing amplification with P9. R, resistant strain; S, sensitive strain; M, marker.
Source: Kaur *et al.* (2010).

Table: 17.7. Pathogenic potential of metalaxyl-resistant (R) and -sensitive (S) populations of *P. cubensis*.

P. cubensis population	MIC (μg/ml)	Incubation period (days)	Disease severity (%)	Sporulation (spores/cm²)
DM-12 (R)	100	5	80	75.5×10^3
DM-13 (R)	150	4	85	73.0×10^3
DM-15 (S)	10	4	90	75.0×10^3
DM-16 (S)	10	4	80	72.0×10^3

Note: R, resistant population; S, sensitive population.
Source: Thind *et al.* (2011).

showed resistant response, with ED_{50} values of metalaxyl ranging between 38 and 200 µg/ml. Pathogenic potential, colony growth and sporulation of the resistant isolates were comparable with sensitive isolates (Thind *et al.*, 2009). The resistant isolates did not show cross-resistance to azoxystrobin, cymoxanil, fluopicolide and previcur. Cymoxanil is providing effective control of foot rot in the orchards where metalaxyl resistance has been a problem.

Triazoles

Powdery mildew of grapevine
(Uncinula necator)

Powdery mildew, incited by *Uncinula necator* (Schw.) Burr., is another serious disease of grapevine in India causing more damage to the developing berries. For the past 20 years, various demethylation inhibitor (DMI) fungicides, in addition to the traditional sulfur and dinocap, have been used to control this disease. Apart from triadimefon, which is widely used against this disease in India, other DMI fungicides like penconazol, flusilazole and fenarimol are also applied. In recent years, azoxystrobin has also been introduced for the control of powdery mildew, and also downy mildew in grapevine. Conidial populations of this fungus collected from various regions during 1995–1997 were studied for detection of resistant strains. The first case of development of resistance to triadimefon in *U. necator* was reported in India by Thind *et al.* (1998a).

Fifteen populations of *U. necator*, each obtained from five infected leaves bearing profuse sporulation, were collected from treated vineyards in Punjab, Maharashtra and Karnataka. These were studied to determine their sensitivity levels to triadimefon (Bayleton 25) following criteria of conidial germ tube length and sporulation on leaf discs treated with different concentrations ranging from 0.01 to 10 µg/ml of this fungicide. Criterion of germ tube length of more than 250 µ at 0.3 µg/ml or above was taken to determine resistance to triadimefon (Steva and Clerjeau, 1990; Thind and Mohan, 1995). The majority of the populations showed a

typical sensitive reaction, as their germ tubes measured less than 250 µm at 0.3 µg/ml, which was taken as the discriminatory concentration to distinguish resistant strains in the conidial populations. Germ tubes of such conidia were distorted and deformed at the tips and comparable in their response with the reference sensitive strain Ane-17. However, three populations showed 4–6% conidia, with normal germ tubes measuring more than 250 µm at a higher concentration of 3 µg/ml of triadimefon. Two of these populations, 1a from Bangalore and 7a from Pune, showed 1% conidia producing normal germ tubes at a still higher concentration of 10 µg/ml, thus demonstrating low to moderate levels of resistance development in these populations (Thind *et al.*, 1998a,b). These three populations also developed sporulating colonies on the leaf discs at 3 and 10 µg/ml, thus confirming resistance to triadimefon. However, no apparent decline in the field performance of this fungicide was observed, except in the vineyard at Bangalore where a reduced disease control was recorded.

CROSS-RESISTANCE TO OTHER STEROL BIOSYNTHESIS INHIBITOR (SBI) FUNGICIDES. One isolate each from populations 1a and 7a, showing moderate resistance to triadimefon, was further studied for sensitivity to two other sterol-inhibiting fungicides, triadimenol (Baytan 5) and fenarimol (Rubigan 4), by sporulation test on treated leaf discs. Compared to the sensitive strain Ane-17, which showed negligible sporulation at 0.3 µg/ml of triadimenol, isolate 1a and 7a produced some sporulation even at 1 µg/ml, thus confirming cross-resistance to triadimefon. However, on fenarimol-treated discs, the two isolates were found to be as equally sensitive as the reference strain, thereby showing no cross-resistance to fenarimol (Thind *et al.*, 1998a,b). Fenarimol is now being used where reduced sensitivity has been observed to triazole fungicides.

Triazoles and other sterol-inhibiting fungicides such as bitertanol (Baycor), hexaconazole (Contaf), myclobutanil (Systhane), penconazole (Topas) and fenarimol (Rubigan) are also being used at present to control apple scab in India. Reduced sensitivity in this pathogen to these fungicides has not yet been reported from Indian orchards. Since *V. inae-qualis* has been reported to encounter resistance development to DMI fungicides in other countries (Thind *et al.*, 1986), it is necessary to initiate monitoring programmes to determine changes in the sensitivity levels of *V. inaequalis* populations to these fungicides. A simple and quick method based on spore germination, germ tube length and morphology has been developed which can be used effectively to determine resistance to DMI fungicides (Thind *et al.*, 1987).

17.5   Conclusions and Future Outlook

Several site-specific fungicides are used in Indian agriculture for managing different crop diseases. Reports of resistance build-up to some commonly used fungicides in field populations of certain pathogens indicate the

likely risk these new-generation fungicides may pose in managing plant diseases more effectively. Risk assessment is crucial for the newly developed fungicides before they are introduced for commercial use by farmers. New research initiatives need to be developed to predict the actual risk of resistance. New technologies are required to monitor the performance of resistance management strategies and to help predict problems before they occur. New techniques based on molecular biology may prove useful for the rapid detection of resistance in pathogen populations and may provide necessary information about the performance of an anti-resistance strategy. Apart from using at-risk fungicides in mixture or alternation with compounds from different modes of action, their integration with other control methods may help greatly in resistance management by keeping disease pressure at a low level.

References

Anil Kumar, T.B. and Sastri, M.N.L. (1979) Development of tolerance to fungicides in *Rhizoctonia bataticola*. *Phytopathology Zeitschrift* 94, 126–131.

Annamalai, P. and Lalithakumari, D. (1990) Decreased sensitivity of *Drechslera oryzae* field isolates to edifenphos. *Indian Phytopathology* 43(4), 553–558.

Annamalai, P. and Lalithakumari, D. (1992) Development of resistance in *Drechslera oryzae* to edifenphos – monitoring and control strategies. *Tropical Pest Management* 38(4), 349–353.

Arora, R.K. (1994) Aggressiveness of metalaxyl resistant and sensitive isolates of *Phytophthora infestans*. In: Shekhawat, G.S., Paul Khurana, S.M., Pandey, S.K. and Chandra, V.K. (eds) *Potato: Present and Future*. Indian Potato Association, Shimla, India, pp. 179–183.

Arora, R.K., Kamble, S.S and Gangawane, L.V. (1992) Resistance to metalaxyl in *Phytophthora infestans* in Nilgiri Hills of southern India. *Phytophthora Newsletter* 18, 8–9.

Bains, S.S. and Mohan, C. (1982) Location and comparative behaviour of *Helminthosporium maydis* isolates sensitive and tolerant to Dithane M-45. *Indian Phytopathology* 35, 585–589.

Basu Chaudhary, K.C. and Puttoo, B.L. (1984) Fungicide resistant strains of *Venturia inaequalis* in Kashmir – a prediction. *British Crop Protection Conferance – Pests and Diseases*. British Crop Protection Council, Farnham, UK, pp. 509–514.

Brent, K.J. (1995) *Fungicide resistance in crop pathogens: how can it be managed?* FRAC Monograph No. 1. GCPF, Brussels, 48 pp.

Brent, K.J. and Hollomon, D.W. (2007) *Fungicide Resistance in Crop Pathogens: How Can it be Managed?* FRAC Monograph No. 1 (2nd edn). CropLife International, Brussels, 55 pp.

Cohen, Y., Baider, A. and Cohen, B.H. (1995) Dimethomorph activity against oomycete fungal pathogens. *Phytopathology* 85, 1500–1506.

Davidse, L.C. (1987) Resistance to acylalanines in *Phytophthora infestans* in Netherlands. *EPPO Bulletin* 15, 129.

Davidse, L.C., Looijen, D., Turkensteen, I.J. and Van der Wal, D. (1981) Occurrence of metalaxyl-resistance strains of *Phytophthora infestans* in Dutch potato fields. *Netherland Journal of Plant Pathology* 87, 65–68.

Dekker, J. (1985) The development of resistance to fungicides. *Progress in Pesticide Biochemistry and Toxicology* 4, 165–218.

Gangawane, L.V. (1997) Management of fungicide resistance in plant pathogens. *Indian Phytopathology* 50, 305–315.

Gangawane, L.V. and Reddy, B.R.C. (1985) Resistance of *Aspergillus flavus* to certain fungicides. *ISPP Chemical Control Newsletter* 6, 23.

Gangawane, L.V., Kamble, S.S. and Arora, R.K. (1995) Synergistic effect of other fungicides on metalaxyl resistant isolates of *Phytophthora infestans* from Nilgiri Hills. *Indian Journal of Plant Protection* 23, 159–162.

Gisi, U. and Staehle-Csech, U. (1989) Resistance risk evaluation of new candidates for disease control. In: Delp, C.J. (ed.) *Fungicide Resistance in North America*. APS Press, St Paul, Minnesota, pp. 101–106.

Grover, R.K. and Chopra, B.L. (1971) Adaptation of *Rhizoctonia* species to two oxathiin compounds and manifestations of adapted isolates. *Acta Phytopathology Academy of Science, Hungary* 5, 113–121.

Gupta, P.K.S. and Dass, S.N. (1971) Induced tolerance of spores of *Helminthosporium oryzae* Breda de Haan to fungicides. *Current Science* 40, 168–169.

Gupta, V.K. and Gupta, A.K. (1996) Use of forecasting system in rational use of fungicides for apple scab control. In: Sokhi, S.S. and Thind, T.S. (eds) *Rational Use of Fungicides in Plant Disease Control*. National Agricultural Technology, Information Centre, Ludhiana, India, pp. 121–131.

Kamble, S.V. and Gangawane, L.V. (1994) Effect of passage on the development of benomyl resistance in *Fusarium udum in vitro*. *Indian Phytopathology* 47, 354–356.

Kaur, R., Thind, T.S. and Goswami, S. (2010) Profiling of *Phytophthora infestans* populations for metalaxyl resistance and its management with novel action fungicides. *Journal of Mycology and Plant Pathology* 40(1), 14–21.

Kumar, S. and Thind, T.S. (1992) Reduced sensitivity in *Gloeosporium ampelophagum* to carbendazim in Punjab. *Plant Disease Research* 7, 103–105.

Lalithakumari, D. and Kumari, D.S. (1987) Screening for field resistance in *Pyricularia oryzae* to edifenphos. *Indian Phytopathology* 40(3), 342–347.

Lalithakumari, D. and Mathivanan, K. (1990) Strategies to avoid or delay or break resistance in *Pyricularia oryzae* to edifenphos. *Journal of Ecobiology* 2(2), 155–160.

Mohan, C. and Thind, T.S. (1995) Development of carbendazim resistance in *Gloeosporium ampelophagum* and strategies for its management. *Indian Journal of Mycology and Plant Pathology* 25, 25–33.

Mohan, C., Singh, J. and Thind, T.S. (2005) Prevalance of grape anthracnose and carbendazim resistance in *Gloeosporium ampelophagum* in Punjab. *Plant Disease Research* 20(2), 194–195.

Palani, P.V. and Lalithakumari, D. (1999) Resistance of *Venturia inaequalis* to the sterol biosynthesis inhibiting fungicide, penconazole [1-(2-(2,4-dichlorophenyl) pentyl)-1H-1,2,4-triazole]. *Mycological Research* 103(9), 1157–1164.

Pan, S. and Sen, A. (1980) Relative virulence of parental and fungicide tolerant isolates of *Macrophomina phaseolina* towards soybean. *Indian Phytopathology* 33, 642–643.

Pappas, A.C. (1980) Effectiveness of metalaxyl and phosethyl-Al against *Pseudoperonospora cubensis* (Berk. and Curt.) Rostow isolates from cucumbers. *Proceedings 5th Congress of the Mediterranean Phytopathological Union*, Patras, Greece, 21–27 September 1980. Hellenic Phytopathological Society, Athens, pp. 146–148.

Puttoo, B.L. and Basu Chaudhary, K.C. (1986) Management of apple scab pathogen resistant to fungicides. In: Verma, A. and Verma, J.P. (eds) *Vistas in Plant Pathology*. Malhotra Publishing House, New Delhi, pp. 361–366.

Ramamoorthy, K.K. (1991) Morphological, physiological and biochemical studies on carbendazim resistance in *Fusarium oxysporum* f. sp. *lycopersici* (Sacc.) Snyder and Hansen. PhD thesis, University of Madras, Chennai, India.

Ramaswamy, G.R. and Anilkumar, T.B. (1991) Comparative studies on sporulation and spore germination in mancozeb resistant and sensitive strains of *Alternaria solani*. *Indian Phytopathology* 44 (Supplementary), XCI.

Rao, C.S. and Reddy, M.S. (1988) Resistance to metalaxyl in isolates of grapevine downy mildew fungus (*Plasmopara viticola*) around Hyderabad. *Indian Journal of Plant Protection* 16, 297–299.

Reddy, M.S. and Apparao, A. (1967) *In vivo* and *in vitro* studies on the adaptation of *Gloeosporium ampelophagum* to copper fungicides. In: *Proceedings of First International Symposium of Plant Pathology*. Indian Phytopathological Society, New Delhi, pp. 553–557.

Reddy, M.S., Rama, P. and Appa Rao, A. (1979) Effect of using combinations and alternate use of fungicides on the *in vitro* development of fungicide resistance in fungi. *Indian Phytopathology* 32, 507–510.

Revathi, R., Annamalai, P. and Lalithakumari, D. (1992) Laboratory mutants of *Venturia inaequalis* resistant to sterol inhibitors. *Indian Phytopathology* 45(3), 331–336.

Singh, B.P., Roy, S. and Bhattacharrya, S.K. (1993) Scheduling of systemic fungicides and development of fungicide resistant strains in *Phytophthora infestans*. *Indian Journal of Potato Association* 20, 48–52.

Steva, H. and Clerjeau, M. (1990) Cross resistance to sterol biosynthesis inhibitor fungicides in strains of *Uncinula necator* isolated in France and Portugal. *Mededelingen Faculteit Landbouwwetenschappen Rijksuniversiteit Gent* 55, 983–988.

Stevenson, W.R. (2009) Late blight control strategies in the United States. *Acta Horticulturae (ISHS)* 834, 83–86.

Sullia, S.B. and Maria, R. (1985) Acquired cycloheximide resistance in *Neurospora crassa* and *Sclerotium rolfsii*. *Proceedings of Indian Academy of Science* 95, 417–427.

Thind, T.S. (1995) Fungicide resistance in plant disease control with reference to Indian situation. *Indian Journal of Mycology and Plant Pathology* 25, 121–122.

Thind, T.S. (2002) *Fungicide Resistance in India*. Technical Bulletin, Crop Life India, Mumbai, India, pp. 50.

Thind, T.S. (2008) Fungicide resistance: a perpetual challenge in disease control. *Journal of Mycology and Plant Pathology* 38(3), 407–417.

Thind, T.S. and Jhooty, J.S. (1980) Adaptability of *Colletotrichum capsici* to Dithane M-45 and Blitox. *Indian Phytopathology* 33, 570–573.

Thind, T.S. and Mohan, C. (1995) A laboratory method for evaluating fungicides against *Uncinula necator*. *Indian Phytopathology* 48(2), 207–209.

Thind, T.S. and Mohan, C. (1998) Management of carbendazim resistance in *Gloeosporium ampelophagum* in India. *9th International Congress of Pesticide Chemistry – The Food Environment Challenge*, London, 2–7 August 1998. The Royal Society of Chemistry, London, pp. 4D-022.

Thind, T.S., Clerjeau, M. and Olivier, J.M. (1986) First observations on resistance in *Venturia inaequalis* and *Guignardia bidwelli* to ergosterol biosynthesis inhibitors in France. *Proceedings of British Crop Protection Conference – Pests and Diseases* 4C, 1. British Crop Protection Council, Farnham, UK, pp. 491–498.

Thind, T.S., Clerjeau, M. and Olivier, J.M. (1987) A rapid method for detecting resistance to ergosterol biosynthesis inhibitors in *Venturia inaequalis*. *ISPP Chemical Control Newsletter* 9, 28–29.

Thind, T.S., Munshi, G.D. and Sokhi, S.S. (1988) A method for maintenance of *Plasmopara viticola* and laboratory evaluation of fungicides. *Current Science* 57(13), 734–735.

Thind, T.S., Mohan, C., Sokhi, S.S. and Bedi, J.S. (1989) A detached leaf technique for maintenance and multiplication of *Phytophthora infestans* and evaluation of fungicides. *Current Science* 58, 388–389.

Thind, T.S., Mohan, C., Kumar, S. and Azmi, O.R. (1994) Observations on field isolates of *Gloeosporium ampelophagum* with reduced sensitivity to carbendazim in Punjab. *Indian Journal of Mycology and Plant Pathology* 24(1), 46–50.

Thind, T.S., Mohan, C., Clerjeau, M., Sokhi, S.S and Jailloux, F. (1997) Management of carbendazim resistance in *Gloeosporium ampelophagum* with new fungicides. International Conference on Integrated Plant Disease Management for Sustainable Agriculture, 10–15 November 1997, New Delhi. Abstract p. 324.

Thind, T.S., Clerjeau, M., Sokhi, S.S., Mohan, C. and Jailloux, F. (1998a) Observations on reduced sensitivity in *Uncinula necator* to triadimefon in India. *Indian Phytopathology* 51(1), 97–99.

Thind, T.S., Mohan, C., Clerjeau, M., Sokhi, S.S. and Jailloux, F. (1998b) Resistance to triazole fungicides in *Uncinula necator* in India. *7th International Congress of Plant Pathology*, Edinburgh, Scotland, 9–16 August 1998. British Society for Plant Pathology, Birmingham, UK, p. 5.5.24.

Thind, T.S., Singh, L., Mohan, C. and Pal, J. (2001) Monitoring for metalaxyl resistance in populations of *Phytophthora infestans* from 1996 to 1999 in Punjab and their characteristics. *Indian Phytopathology* 54(1), 68–74.

Thind, T.S., Mohan, C. and Raj, P. (2002) Competitive fitness of metalaxyl resistant population of *Phytophthora infestans* and its cross resistance to strobilurins and other fungicides. *Journal of Mycology and Plant Pathology* 32(1), 122–124.

Thind, T.S., Goswami, S., Thind, S.K. and Mohan, C. (2009) Resistance in *Phytophthora parasitica* against metalaxyl in citrus orchards. *Indian Phytopathology* 62(4), 536–538.

Thind, T.S., Goswami, S., Kaur, R., Raheja, S. and Mohan, S. (2011) Development of metalaxyl resistance in *Pseudoperonospora cubensis* and its management with novel action fungicides. *Indian Phytopathology* 63(4), 387–391.

18 Resistance to QoI and SDHI Fungicides in Japan

Hideo Ishii

National Institute for Agro-Environmental Sciences,
Tsukuba, Ibaraki, Japan

18.1 Introduction

In Japan, two quinone outside inhibitor (QoI) fungicides (inhibitors of mitochondrial respiration at Qo site of cytochrome bc_1 enzyme complex), kresoxim-methyl and azoxystrobin, were officially registered in December 1997 and April 1998, respectively. Most cucumber growers followed the manufacturers' usage recommendation and applied their products only a couple of times per crop in alternation with other fungicides, which possessed different modes of action. However, between 1998 and 1999, shortly after the introduction of QoIs, control failures of powdery mildew and downy mildew diseases were reported on cucumber. To determine the cause of the reported rapid decline in fungicide efficacy, QoI sensitivities of powdery and downy mildew pathogens were tested using isolates collected from cucumber greenhouses or fields where control failure was observed. Results from inoculation tests carried out on intact cucumber plants and leaf disks clearly showed the distribution of pathogen isolates highly resistant to azoxystrobin and kresoxim-methyl (Heaney *et al.*, 2000; Ishii *et al.*, 2001; Fig. 18.1).

Similarly, the development of many succinate dehydrogenase inhibitors (SDHIs, so-called complex II inhibitors in the mitochondrial respiration chain) is now in progress worldwide. Boscalid, a novel SDHI fungicide was registered for the control of grey mould and Sclerotinia rot diseases on vegetables in January 2005 in Japan. Subsequently, registration was expanded to control Corynespora leaf spot on cucumber and some diseases on other crops in July 2006. On cucumber, boscalid was sprayed only a couple of times per crop according to the manufacturer's recommendation. Despite that, boscalid-resistant strains of *Corynespora cassiicola*, the cause of Corynespora leaf spot disease, appeared rapidly in 2007 and are now increasing their populations widely, representing a serious problem in disease control (Miyamoto *et al.*, 2009, 2010a).

18.2 History and Current Situation of QoI Resistance

QoI fungicides have been widely used for the control of many diseases on various crops during the past decade, as they inherently have a broad spectrum of control. But it is well known now that QoI fungicides carry very high risk for resistance development in target pathogens. Actually, QoI resistance has also occurred in aubergine leaf mould (Yano and Kawada, 2003; Ishii *et al.*, 2007), cucumber Corynespora leaf spot

Powdery mildew

Downy mildew

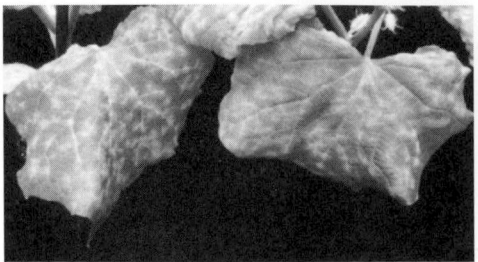 

Distilled water azoxystrobin 100 mg/l   Distilled water azoxystrobin 100 mg/l

Fig. 18.1. Resistance of cucumber powdery and downy mildew pathogens to azoxystrobin.

(Date *et al.*, 2004; Ishii *et al.*, 2007), strawberry anthracnose (Inada *et al.*, 2008) and many other diseases (Table 18. 1).

18.3 Molecular Mechanism of QoI Resistance

To understand the molecular mechanism of resistance, fragments of the mitochondrial cytochrome *b* gene encoding fungicide-targeted proteins were amplified by polymerase chain reaction (PCR) using QoI-resistant and -sensitive isolates of both pathogens, *Podosphaera xanthii* (= *P. fusca*) and *Pseudoperonospora cubensis*. In the sequence analysis, a single point mutation (GGT to GCT) in the cytochrome *b* gene, resulting in substitution of glycine (G) by alanine (A) at amino acid position 143 on a possible fungicide-binding site, was found in resistant isolates of downy mildew. The same mutation was found in some, but not all, resistant isolates of powdery mildew (Ishii *et al.*, 2001).

Subsequently, the G143A mutation was also detected from resistant isolates of several other pathogens such as *C. cassiicola*, *Mycovellosiella nattrassii*, *Colletotrichum gloeosporioides* and others, all collected in Japan (Ishii *et al.*, 2007, 2009; Inada *et al.*, 2008), suggesting that high QoI resistance was based mainly on a single point mutation from GGN to GCN at position 143 of the cytochrome *b* gene, which caused an amino acid change G143A (Fig. 18.2; Table 18.2).

Another mutation of F129L, resulting in the substitution of phenylalanine (F) by leucine (L), was found in the cytochrome *b* gene, PCR-amplified from QoI-resistant isolates of *Passalora fulva* (= *Fulvia fulva*), the causal fungus of tomato leaf mould (Watanabe, 2009). This mutation has been known to confer a moderate level of QoI resistance in general. In *P. fulva*, however, F129L seems to be involved in resistance enough to decrease fungicide efficacy on tomato in the field.

QoI-resistant isolates of *Botrytis cinerea* were first detected incidentally from commercial citrus orchards and subsequently from vegetables in Japan (Kansako *et al.*, 2005; Banno *et al.*, 2009; Ishii *et al.*, 2009). Banno *et al.* (2009) mentioned that all QoI-resistant isolates had the same mutation (GGT to GCT) in the cytochrome *b* gene that led to the substitution of glycine by alanine at position 143 of cytochrome *b* protein. They also found an intron after codon 143 in the gene of all azoxystrobin-sensitive isolates, suggesting that the QoI-resistant mutation at codon 143 prevented self-splicing of the cytochrome *b* gene, as proposed in some other plant pathogens.

18.4 Molecular Diagnostic Methods for Identifying QoI Resistance

Based on the elucidation of possible resistance mechanisms, molecular methods for identifying QoI resistance have been developed. PCR-RFLP (restriction fragment-length polymorphism) has been the most frequently

Table 18.1. Field occurrence of QoI resistance in Japan.

Disease	Pathogen
Wheat powdery mildew	*Erysiphe* (*Blumeria*) *graminis* f. sp. *tritici*
Cucurbit powdery mildew	*Podosphaera xanthii* (= *Podosphaera fusca*)
Cucumber downy mildew	*Pseudoperonospora cubensis*
Cucumber Corynespora leaf spot	*Corynespora cassiicola*
Cucurbit gummy stem blight	*Didymella bryoniae*
Eggplant leaf mould	*Mycovellosiella nattrassii*
Strawberry anthracnose	*Colletotrichum gloeosporioides*
Strawberry powdery mildew	*Sphaerotheca aphanis* var. *aphanis*
Grapevine leaf blight	*Pseudocercospora vitis*
Grapevine downy mildew	*Plasmopara viticola*
Grapevine ripe rot	*Colletotrichum gloeosporioides*
Apple Alternaria leaf blotch	*Alternaria alternata* apple pathotype
European pear black spot	*Alternaria alternata* apple pathotype
Grey mould	*Botrytis cinerea*
Tomato leaf mould	*Passalora fulva* (= *Fulvia fulva*)
Tea grey blight	*Pestalotiopsis longiseta*
Turf grass anthracnose[a]	*Colletotrichum graminicola*

Note: [a]Strobilurins not registered in Japan.

Table 18.2. Deduced amino acid sequences of cytochrome *b* in field isolates of plant pathogenic fungi.

Species	Isolate	Response to QoI[a]	Sequence Codon 129	143
Podosphaera xanthii	K-7-2	S	••FLGYGLPYGQMSLWGAT••	
	R-2	R	••FMGYGLPWGGMSFWAAT••	
Pseudoperonospora cubensis	S	S	••FMGYVLPWGQMSFWGAT••	
	R	R	••FMGYVLPWGQMSFWAAT••	
Corynespora cassiicola	C6-2	S	••FTGYVLPYGQMSLWGAT••	
	ST-20S-1	R	••FTGYVLPYGQMSLWAAT••	
Mycovellosiella nattrassii	T-1	S	••FLGYVLPYGQMSLWGAT••	
	K-1	R	••FLGYGLPYGQMSLWAAT••	

Note: [a]S, sensitive; R, resistant. Amino acid sequences corresponding to the positions 129 and 143 of cytochrome *b* protein are highlighted by shaded boxes.
Source: Reprinted from Ishii *et al.* (2001), *Phytopathology* 91, 1166–1171 and Ishii *et al.* (2007), *Phytopathology* 97, 1458–1466.

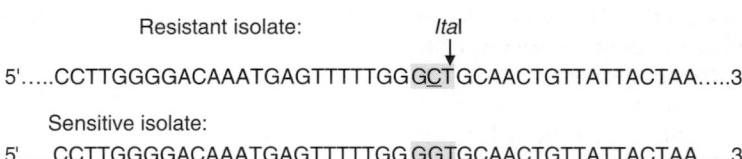

Resistant isolate: *Ita*I
                                                  ↓
5'.....CCTTGGGGACAAATGAGTTTTTGGGCTGCAACTGTTATTACTAA.....3'

Sensitive isolate:
5'.....CCTTGGGGACAAATGAGTTTTTGGGGTGCAACTGTTATTACTAA.....3'

Fig. 18.2. Partial nucleotide sequences of the mitochondrial cytochrome *b* gene from QoI-resistant and -sensitive isolates of *Pseudoperonospora cubensis*. Nucleotide sequences corresponding to position 143 of cytochrome *b* protein are shaded. The site of point mutations is underlined. The arrow indicates the restriction site of the enzyme *Ita*I. Reprinted from Ishii *et al.* (2002) in *Modern Fungicides and Antifungal Compounds III*, pp. 149–159.

used method. A partial fragment of the cyto-chrome *b* gene was PCR amplified and products were then treated with a mutation-specific restriction enzyme, *Ita*I or *Fnu*4HI (Fig. 18.2, Ishii *et al.*, 2002, 2007). The PCR products from resistant isolates carrying the mutated sequence GCN at position 143 could be digested by these restriction enzymes, resulting in the appearance of two smaller bands on a gel in electrophoresis, whereas those from sensitive isolates remained undi-gested (Fig. 18.3).

The method of PCR-RFLP has been employed for the detection of QoI-resistant isolates in *P. xanthii*, *P. cubensis*, *C. cassiicola*, *M. nattrassii*, *C. gloeosporioides* and others. How-ever, the cytochrome *b* gene is encoded by multi-copy mitochondrial DNA and the problem of heteroplasmy (concomitance of mutated DNA with wild-type DNA) in the cytochrome *b* gene of some fungi has raised a question about the reliability of this molecular diagnostic method. In contrast, this method has been used routinely for identifying QoI resistance successfully in other pathogens such as *Plasmopara viticola*, the cause of grapevine downy mildew disease, where heteroplasmy of the cytochrome *b* gene has not been reported so far (Furuya *et al.*, 2009, 2010).

When G143A mutant alleles are present at a low level (below approximately 10%) in the cytochrome *b* gene, it is generally difficult

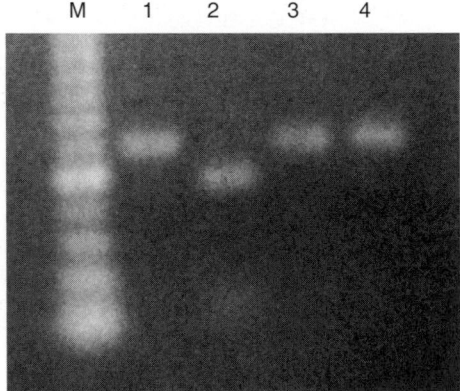

Fig. 18.3. Digestion of the cytochrome *b* gene fragment from *Podosphaera xanthii* by *Ita*I. M, 50 bp ladder; 1, untreated (resistant); 2, *Ita*I-treated (resistant); 3, untreated (sensitive); 4, *Ita*I-treated (sensitive).

to detect it using conventional ethidium bromide staining of an electrophoresis agar-ose gel in the PCR-RFLP method. However, the protective efficacy of the full dose of azoxystrobin decreased when the popula-tions of powdery and downy mildew patho-gens contained resistant isolates at 10% in model experiments. So, a fluorescence bioim-aging analyser (FMBIO) was tested for more sensitive quantification of mutant alleles in the cytochrome *b* gene (Ishii *et al.*, 2007). Using the FMBIO, the mutant alleles from QoI-resistant isolates of *P. cubensis* could be detected at a 1% level. Detection sensitivity of ethidium bromide staining was thus ≈ 10 times lower, indicating that the PCR-RFLP method might overlook the existence of QoI-resistant isolates in pathogen populations.

To detect G143A mutation, a hybridiza-tion probe assay based on real-time PCR amplification and melting curve analysis was further developed (Banno *et al.*, 2009). This assay could detect individual cytochrome *b* genes from a mixture of DNAs from both QoI-resistant and -sensitive *B. cinerea* isolates, suggesting that the assay might be used to assess the population ratio of mutant and wild-type isolates in the field.

18.5 Decline of Resistance After Withdrawal of QoI Fungicides

In cucumber greenhouses where the efficacy of QoI fungicides had been lost, resistance moni-toring was conducted every 6 months. Resistant strains were still widely distributed in patho-gen populations of powdery and downy mil-dew diseases 2 years after stopping use of the products (Ishii *et al.*, 2002). However, resistant powdery mildew populations declined 3 years after QoI withdrawal, although they recovered rapidly when azoxystrobin was applied again for the control of Corynespora leaf spot disease (Ishii *et al.*, 2007).

In the laboratory, resistant isolates of cucumber powdery mildew fungus derived from bulk conidial mass of lesions were main-tained individually through sub-inoculation on detached healthy cucumber leaves for 3.5 years. High azoxystrobin resistance in these isolates still remained during the whole period. However, results from the restriction enzyme *Ita*I digestion experiment exhibited a decrease of the proportion of mutated cytochrome *b* gene and an increase of wild-type DNA following repeated sub-inoculation in the absence of the selection pressure by QoI fungicides (Fig.18.4). Sequencing of the cytochrome *b* gene fragment also showed the reversion of the mutated-type sequence GCT to that of the wild-type GGT at position 143. Eventually, the resistant bulk isolates became sensitive 8 years after the onset of sub-inoculations. It also might have been possible that resistant isolates used in the experiment contained small amounts of sensitive isolates. If so, it was also likely that resistant isolates were replaced gradually by sensitive ones due to a fitness penalty, giving rise to wild-type DNA that increased in the fungal mass. Fuji *et al.* (2000) reported that the proportion of resistant strains of *P. cubensis* sometimes decreased over successive generations in the mixture with sensitive strains.

Monoconidial isolates of cucumber powdery mildew fungus which were resistant to azoxystrobin also became QoI-sensitive after 2 years of sub-inoculations in the absence of fungicide selection pressure. The involvement

of lower competitiveness of resistant isolates was ruled out in this case because all of these monoconidial isolates were sub-inoculated separately.

18.6 Heteroplasmy in Mitochondrial Cytochrome *b* Gene

To examine the instability of QoI resistance further and to confirm the existence of heteroplasmy of the cytochrome *b* gene, QoI-resistant and -sensitive isolates of *C. cassiicola* and *M. nattrassii*, together with an isolate of *C. gloeosporioides* (all monoconidial isolates) carrying the mutated sequence in cytochrome *b* gene, were subcultured on azoxystrobin-amended or -unamended media (Ishii *et al.*, 2007). The mutated sequence found in the resistant isolates of *C. cassiicola* was lost earlier in the absence of fungicide than in the presence of fungicide. Furthermore, in *C. gloeosporioides*, the mutated sequence reverted rapidly to the wild-type sequence after subculturing in both the presence and absence of fungicide. This phenomenon strongly suggested that heteroplasmy was occurring in the cytochrome *b* gene and the proportion of mutated sequences in cytochrome *b* gene decreased over time in the pathogen individuals and their populations (Fig. 18.5). Contrasts with the fast-growing fungi were seen when compared with the slow-growing fungus *M. nattrassii*, which did not lose its mutated sequence after continuous subculturing. Differences in the speed of the reversion to wild-type sequence may indicate the differential stability of QoI resistance in the field.

In 2004 and 2005, a number of single-spore isolates of cucumber powdery mildew fungus were obtained in Japan and many of them were found to be highly resistant to azoxystrobin using a leaf-disk assay. However, the G143A mutation of cytochrome *b* gene was seldom detected in these isolates by gel staining PCR-RFLP methods or by direct sequencing of the PCR products. Other researchers also reported a similar phenomenon in the same pathogen (Fernández-Ortuño *et al.*, 2008). Therefore, the status of heteroplasmy in the mitochondrial genome which contains the

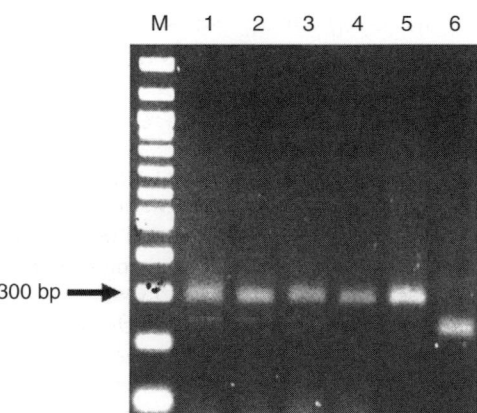

Fig. 18.4. *Ita*I digestion pattern of cytochrome *b* gene of cucumber powdery mildew isolates after successive inoculation on untreated leaves for 3 years. M, 100-bp ladder; 1, R-2 (R); 2, S-1 (R); 3, S-3 (R); 4, K-7-2 (S); 5, negative control; 6, positive control. Reprinted from Ishii *et al.* (2007), *Phytopathology* 97, 1458–1466.

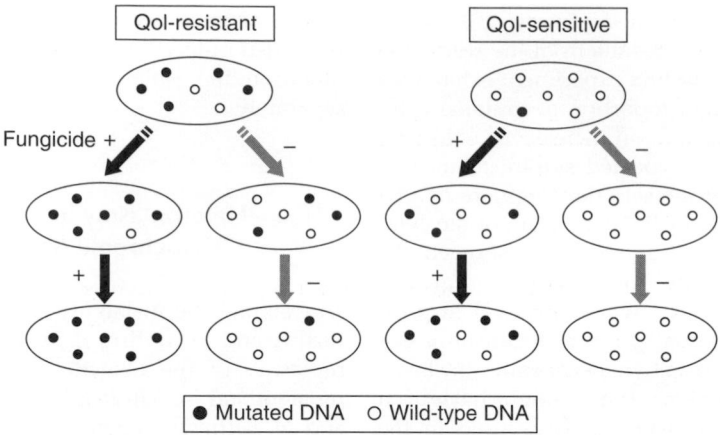

Fig. 18.5. Hypothetical scheme of effects of presence (+) and absence (–) of QoI fungicide selection pressure on heteroplasmic status of mitochondrial cytochrome *b* gene in some fungi. Reprinted from Ishii *et al.* (2007), *Phytopathology* 97, 1458–1466.

cytochrome *b* gene can cause instability over time, making it difficult to monitor QoI resistance precisely by DNA-based methods in some pathogens (Ishii *et al.*, 2009; Ishii, 2010).

Insufficiency of the conventional RFLP and sequence analyses of PCR-amplified cytochrome *b* gene was also reported from grey mould (Ishii *et al.*, 2009). In resistant isolates, the mutation at amino acid position 143 of the cytochrome *b* gene, known to cause high QoI resistance in various fungal pathogens, was found, but only occasionally. The heteroplasmy of cytochrome *b* gene was confirmed by cloning of this gene, and the wild-type sequence was often present in the majority of resistant isolates, indicating that the proportion of mutated cytochrome *b* gene was very low. In contrast, when those resistant isolates were grown under the selective pressure exerted by QoI fungicides, clear G143A mutation was again detected by PCR-RFLP (F. Faretra, personal communication, 2009). This may further strengthen the hypothetical scheme shown in Fig. 18.5.

18.7 *In vitro* Methods and Bioassay for Identifying QoI Resistance

Cyanide-insensitive alternative oxidase (AOX) is induced by reactive oxygen species accumulating in mitochondria when normal mitochondrial electron transfer is blocked by QoI fungicides. QoI-resistant isolates of *B. cinerea* grew well on potato dextrose agar (PDA) plates containing kresoxim-methyl or azoxystrobin at 1 mg/l, supplemented with 1 mM of *n*-propyl gallate, an inhibitor of AOX, whereas the growth of sensitive isolates was strongly suppressed (Table 18.3). When QoI-resistant isolates collected in Japan were tested, the activity of azoxystrobin was elevated by coapplications of 2,3-dihydroxybenzaldehyde, suggesting that disruption of the fungal oxidative stress defence system through chemosensitization could be an effective way to control fungal pathogens, as already mentioned by Kim *et al.* (2008).

Results from this *in vitro* test coincided well with those of inoculation tests performed on cucumber cotyledons (Table 18.4). Salicylhydroxamic acid (SHAM), another AOX inhibitor, is also commonly used for QoI resistance monitoring of *C. cassiicola*, *C. gloeosporioides* and other culturable fungi on PDA medium. It is most likely that AOX shares an important part in the expression of QoI resistance in field isolates of *B. cinerea* and the role of AOX in phytopathogenic fungi has been studied earlier (Yukioka *et al.*, 1998; Tamura *et al.*, 1999). Expression analysis of the *AOX* gene and measurement of AOX activity after treatment with QoI fungicides will be the next subjects.

Table 18.3. Sensitivity of *Botrytis cinerea* isolates to azoxystrobin (AZ) and *n*-propyl gallate (PG) on potato dextrose agar medium.

Isolate	Mycelial growth inhibition (%) at:		
	AZ 1 mg/l	AZ 1 mg/l + 1 mM PG	1 mM PG
R-1-2	1.9	32.5	38.5
R-1-4	6.0	27.9	−8.3
Arimi 4-3	3.4	23.5	26.0
Arimi 4-5	−1.5	26.0	29.4
Arimi 7-1	5.8	21.7	27.4
Arimi 7-5	−1.2	24.5	21.1
Himi 8-1	−6.6	17.1	22.2
Himi 8-4	−0.4	25.5	28.5
Nishimi 8-3	7.3	16.0	−0.9
Nishimi 8-5	2.8	0.0	0.8
Aichi 4-12	41.0	87.5	45.1
Aichi 4-17	43.9	72.2	36.7
S-1-1	30.4	100	35.6
S-1-3	31.5	100	37.1
S-1-5	54.6	100	26.9
S-1-6	37.4	100	24.4
A03-5-2	19.4	100	0.7

Source: Reprinted from Ishii *et al.* (2009), *Pest Management Science* 65, 916–922.

Table 18.4. Effect of kresoxim-methyl[a] and azoxystrobin[b] on suppression of grey mould development (cucumber cotyledon paper disk methods).

Botrytis cinerea isolate	Suppression of lesion development (%)	
	Kresoxim-methyl	Azoxystrobin
R-1-2	0	11.0
R-1-4	−13.6	−0.1
Arimi 4-3	−20.2	22.7
Arimi 4-5	4.7	2.2
Arimi 7-1	8.0	−41.0
Arimi 7-5	−58.6	−114.1
Himi 8-1	−16.3	30.7
Himi 8-4	15.9	22.1
Nishimi 8-3	7.4	2.8
Nishimi 8-5	40.8	13.4
Aichi 4-12	−[c]	100
Aichi 4-16	100	100
Aichi 4-17	−	100
S-1-1	100	100
S-1-3	100	100
S-1-5	100	100
S-1-6	100	100
A03-5-2	100	100

Note: [a]Kresoxim-methyl was treated at 235 mg (a.i.)/l; [b]azoxystrobin was treated at 133 mg (a.i.)/l; [c]not determined.
Source: Reprinted from Ishii *et al.* (2009), *Pest Management Science* 65, 916–922.

It is well known that mitochondrial DNA mutates in higher frequency than nuclear DNA in general. In the presence of QoIs, mutation rates in mitochondrial cytochrome *b* gene might be increased indirectly owing to elevated levels of reactive oxygen species. Further research will be needed to demonstrate this possibility.

18.8 Strategies Against QoI Resistance on Rice

It is of great concern whether QoI-resistant isolates appear in rice blast fungus (*Magnaporthe oryzae*) in paddy fields. In QoI-resistant isolates of *Pyricularia grisea* found on turf in the USA, two types of point mutation, G143A and F129L, were detected in the cytochrome *b* gene and the mutation was associated closely with high and moderate levels of resistance, respectively (Kim *et al.*, 2003). Based on these mutations, PCR-RFLP methods were developed.

Metominostrobin sensitivity of rice blast fungus was monitored between 2001 and 2003 in Japan using PCR-RFLP analysis of the cytochrome *b* gene, but neither G143A nor F129L mutation was observed in the field isolates tested (Araki *et al.*, 2005). Subsequently, sensitivity to orysastrobin was monitored based on genetic assays by pyrosequencing, but no samples with reduced sensitivity could be detected in extensive studies carried out in 2004 and 2005 (Stammler *et al.*, 2007).

However, the formulations of orysastrobin developed for seedling box treatments provide long-lasting excellent fungicidal efficacy against leaf and panicle blast and against sheath blight in rice. Therefore, when orysastrobin was marketed on rice, the Research Committee on Fungicide Resistance, the Phytopathological Society of Japan, proposed a guideline indicating how to use orysastrobin and other QoI fungicides such as azoxystrobin and metominostrobin, which were already on the market. In this guideline, it was recommended to use QoIs only once per year on rice plants if necessary (So and Yamaguchi, 2008). It was also recommended to use QoIs as seedling box treatment in alternation with other unrelated fungicides such as MBI-R fungicides (inhibitors of polyhydroxynaphthalene reductase in melanin biosynthesis) or disease resistance inducers, e.g. probenazole, every 2–3 years since neither MBI-R fungicides nor disease resistance inducers caused resistance development in rice blast fungus in the field. According to the guideline, applications of QoIs were not recommended in paddy fields where seeds were produced for commercial use so that long-distance movement of resistant strains through purchased seeds could be stopped.

As QoI-resistant isolates of M. oryzae were not detected from rice fields, site-directed mutagenesis was employed and the G143A point mutation was introduced into a plasmid containing the cytochrome b gene sequence of rice blast fungus (Wei et al., 2009). Subsequently, molecular diagnostic methods including PCR-Luminex were developed for identifying QoI resistance in rice blast fungus. PCR-Luminex analysis was employed originally for identifying resistance to MBI-fungicides (inhibitors of scytalone dehydratase in melanin biosynthesis) of this fungus and species identification of fungi causing Fusarium head blight disease on wheat (Ishii et al., 2008). The novel system PCR-Luminex is based on the hybridization between PCR-amplified products and sequence-specific oligonucleotide probes coupled with fluorescent beads. Several laborious and time-consuming procedures such as restriction enzyme treatment, electrophoresis and gel staining are not required for this method and are suitable for rapid high throughput analysis of single nucleotide polymorphisms (SNPs).

To avoid the development of QoI resistance in rice blast fungus, a mixture of orysastrobin with probenazole has also been marketed. It will be necessary to conduct intensive resistance monitoring because less QoI-sensitive isolates of blast fungus have been found in pathogen inoculation tests on rice very recently (Takeda et al., unpublished).

18.9 Development of Novel QoI Fungicides

Pyribencarb, methyl {2-chloro-5-[(E)-1-(6-methyl-2-pyridylmethoxyimino)ethyl]benzyl} carbamate, developed by the Kumiai Chemical Industry Co Ltd, is a novel fungicide having excellent activity against a wide range of plant pathogenic fungi, especially grey mould and Sclerotinia rot fungi (Takagaki et al., 2010). The primary target of this fungicide is the Qo site on the cytochrome b in the electron transfer system of the respiratory chain.

Interestingly, however, pyribencarb showed differential patterns of cross-resistance to pre-existing QoI fungicides (Table 18.5) and exhibited higher control efficacy against QoI-resistant isolates of grey mould than other QoI fungicides, indicating that pyribencarb is a new type of fungicide belonging to the QoI fungicide group (Kida et al., 2010). Based on the relationship of biological activities and amino acid sequence of QoI-binding sites on cytochrome b protein, it was suggested that pyribencarb might differ slightly

Table 18.5. Sensitivity of grey mould isolates to two QoI fungicides.

Isolate	Suppression of lesion development (%)	
	Kresoxim-methyl 235 mg/l	Pyribencarb 200 mg/l
Kresoxim-methyl-resistant:		
R-1-2	0	95.0
R-1-4	−13.6	100
Arimi 4-3	−20.2	80.4
Kresoxim-methyl-sensitive:		
S-1-1	100	91.4
S-1-3	100	65.0
A03-5-2	100	100

Source: Reprinted from Ishii (2008) in Modern Fungicides and Antifungal Compounds V, pp. 11–17.

in the binding sites from other QoI fungicides (Kataoka *et al.*, 2010).

18.10 Resistance to SDHI Fungicides

Occurrence of SDHI resistance

Isolates of *C. cassiicola* (Corynespora leaf spot of cucumber) with resistance to boscalid have been observed in Japan. These isolates were divided into four groups based on their sensitivity to boscalid, as sensitive (S, complete inhibition of mycelial growth on YBA agar amended with 10 mg/l), moderately resistant (MR, EC_{50}: 2.0–5.9 mg/l), highly resistant (HR, EC_{50}: 8.9–10.7 mg/l) and very highly resistant (VHR, EC_{50}: > 30 mg/l) (Miyamoto *et al.*, 2010a).

Boscalid-resistant isolates have also been found recently in the cucumber powdery mildew fungus *P. xanthii*, although this fungicide is not registered for the control of this disease in Japan (Miyamoto *et al.*, 2010b). Penthiopyrad, in the same cross-resistance group as boscalid, was commercialized in early 2010 and used for the control of cucumber powdery mildew. It is very likely that the resistant strains of *P. xanthii* that pre-existed at a low frequency in natural populations received selection pressure by boscalid, as this fungicide was applied for Corynespora leaf spot disease control. At present, isolates of *B. cinerea* resistant to boscalid are also emerging from vegetable crops.

18.11 Unique Characteristics of Fluopyram

Fluopyram (Fig. 18.6), a new broad-spectrum SDHI fungicide developed by Bayer CropScience, is now under development and this

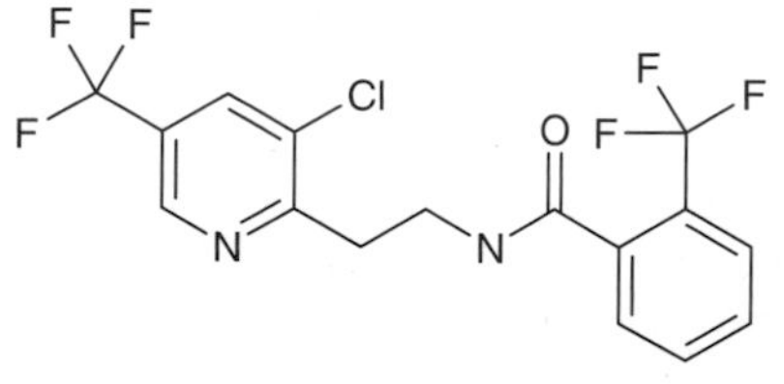

Fig. 18.6. The chemical structure of fluopyram.

fungicide is also thought to belong to the same cross-resistance group as boscalid. Very interestingly, however, no cross-resistance has been observed between boscalid and fluopyram when very highly (VHR) and highly (HR) boscalid-resistant isolates as well as sensitive (S) isolates of *C. cassiicola* were used, not only in mycelial growth tests on YBA agar medium but also in inoculation tests on potted cucumber plants (Table 18.6; Ishii *et al.*, 2011). The lack of cross-resistance to fluopyram was further confirmed in VHR isolates of *P. xanthii* in cucumber pot experiments and leaf-disk tests (Table 18.7). In contrast, when moderately boscalid-resistant (MR) isolates of *C. cassiicola* and *P. xanthii* were served for sensitivity tests, slight positive cross-resistance was present between boscalid and fluopyram both *in vitro* and *in planta*.

18.12 Molecular Mechanism of SDHI Resistance

Molecular interactions of boscalid and fluopyram with target SDH proteins are unknown in detail yet, but it is possible that various mutations found in *SDH* genes influence the binding affinity of both fungicides with target proteins, resulting in differential sensitivity of boscalid-resistant isolates to fluopyram. Such differential sensitivity of two pathogens, *C. cassiicola* and *P. xanthii*, may indicate the involvement of a slightly distinct active molecule of fluopyram from boscalid. This finding may also lead to the discovery of unique SDHIs contributing to resistance management in the future.

SDH (succinate: ubiquinone oxidoreductase) is an iron-sulfur flavoprotein that resides in the mitochondrial inner membrane and functions to oxidize succinate to fumarate and reduce ubiquinone to ubiquinol. The *SDHB* and *SDHC/SDHD* genes encode the iron sulfur and two membrane-anchored subunits of SDH, respectively, that constitute the boscalid binding molecular targets. In *C. cassiicola*, all of the VHR isolates tested carried H278Y in the *SDHB*

Table 18.6. Control efficacy of boscalid and fluopyram against boscalid-resistant and -sensitive isolates of the cucumber Corynespora leaf spot fungus (*Corynespora cassiicola*) on potted plants.

Boscalid sensitivity[a]	Isolate	Fungicide[b]	Control (%)[c] Mean	SD
VHR	070508C3	Boscalid	11.6	16.40
		Fluopyram	97.7	1.70
HR	IbCor3009	Boscalid	39.4	26.16
		Fluopyram	91.4	10.18
MR	IbCor1482	Boscalid	55.0	22.91
		Fluopyram	64.6	7.64
S	070508C1	Boscalid	99.7	0.49
		Fluopyram	99.7	0.49

Note: [a]S, boscalid-sensitive; MR, moderately boscalid-resistant; HR, highly boscalid-resistant; and VHR, very highly boscalid-resistant, based on Miyamoto *et al.* (2010a). [b]Boscalid and fluopyram were applied at 334 mg/l (a.i.) for each. [c]Mean and SD (standard deviation) of control (%) were calculated from the results of two experiments. Control (%) was calculated according to the methods of Ishii *et al.* (2007).
Source: Reprinted from Ishii *et al.* (2011), *Pest Management Science* 67, 474–482.

Table 18.7. Control efficacy of boscalid, penthiopyrad and fluopyram against very highly boscalid-resistant and -sensitive isolates of cucumber powdery mildew fungus (*Podosphaera xanthii*) on potted plants.

Boscalid sensitivity	Isolate	Fungicide	Concentration (mg/l in a.i.)	Control (%)
Very highly boscalid-resistant	Chikusei	Boscalid	66.8	−44.2
			334	−74.1
		Penthiopyrad	66.8	14.5
			334	28.1
		Fluopyram	334	100
Boscalid-sensitive	BCS	Boscalid	66.8	94.2
			334	100
		Penthiopyrad	66.8	100
			334	100
		Fluopyram	334	100

Source: Reprinted from Ishii *et al.* (2011), *Pest Management Science* 67, 474–482.

gene (Miyamoto *et al.*, 2010a). Furthermore, the substitution of H278R in the *SDHB* gene was detected in all HR isolates. In contrast, there were no common mutations in the *SDH* genes of all MR isolates. A part of the MR isolates possessed mutations in either the *SDHC* or *SDHD* gene, with no mutations in the *SDHB* gene. It was thus indicated that very high and high levels of resistance to boscalid were probably due to distinct mutations in the *SDH* genes, leading to conformational change of fungicide-target protein molecules. Most recently, the mutation of the *SDHB* gene has also been detected in VHR isolates of *P. xanthii*

(Miyamoto *et al.*, 2010b). Substitution of H278Y was deduced from the gene sequence in VHR isolates but not in MR isolates of this fungus.

18.13 Strategies Against SDHI Resistance

To lower the risk for boscalid resistance development, two mixed formulations of this fungicide with pyraclostrobin have been commercialized for fruit trees and vegetable crops. However, as pyraclostrobin is also a QoI fungicide, it is still necessary to spare applications of these mixtures and to supervise disease occurrence carefully. How to inform farmers precisely about fungicide resistance including cross-resistance is important, but difficult to achieve at present. Establishment of an effective network of information is thus urgently required. The SDHI and QoI Working Groups of the Fungicide Resistance Action Committee (FRAC) regularly report the monitoring results of their resistance and use recommendations on their home page (http://www.frac.info/frac/index.htm). Unfortunately, however, the proposal of other guidelines is needed which include the stricter regulation of fungicide use under warm and humid climatic conditions, favourable for disease occurrence in Japan.

References

Araki, Y., Sugihara, M., Sawada, H., Fujimoto, H. and Masuko, M. (2005) Monitoring of the sensitivity of *Magnaporthe grisea* to metominostrobin 2001–2003: no emergence of resistant strains and no mutations at codon 143 or 129 of the cytochrome *b* gene. *Journal of Pesticide Science* 30, 203–208.

Banno, S., Yamashita, K., Fukumori, F., Okada, K., Uekusa, H., Takagaki, M., *et al.* (2009) Characterization of QoI resistance in *Botrytis cinerea* and identification of two types of mitochondrial cytochrome *b* gene. *Plant Pathology* 58, 120–129.

Date, H., Kataoka, E., Tanina, K., Sasaki, S., Inoue, K., Nasu, H., *et al.* (2004) Sensitivity of *Corynespora cassiicola*, causal agent of Corynespora leaf spot of cucumber, to thiophanate-methyl, diethofencarb and azoxystrobin. *Japanese Journal of Phytopathology* 70, 10–13.

Fernández-Ortuño, D., Torés, J.A., de Vicente, A. and Pérez-García, A. (2008) Field resistance to QoI fungicides in *Podosphaera fusca* is not supported by typical mutations in the mitochondrial cytochrome *b* gene. *Pest Management Science* 64, 694–702.

Fuji, M., Takeda, T., Uchida, K. and Amano, T. (2000) The latest status of resistance to strobilurin type action fungicides in Japan. In: *Proceedings of BCPC Conference – Pests and Diseases*. BCPC, Farnham, UK, pp. 421–426.

Furuya, S., Suzuki, S., Kobayashi, H., Saito, S. and Takayanagi, T. (2009) Rapid method for detecting resistance to a QoI fungicide in *Plasmopara viticola* populations. *Pest Management Science* 65, 840–843.

Furuya, S., Mochizuki, M., Saito, S., Kobayashi, H., Takayanagi, T. and Suzuki, S. (2010) Monitoring of QoI fungicide resistance in *Plasmopara viticola* populations in Japan. *Pest Management Science* 66, 1268–1272.

Heaney, S.P., Hall, A.A., Davies, S.A. and Olaya, G. (2000) Resistance to fungicides in the QoI-STAR cross-resistance group: current perspectives. In: *Proceedings of BCPC Conference – Pests and Diseases*. BCPC, Farnham, UK, pp. 755–762.

Inada, M., Ishii, H., Chung, W.H., Yamada, T., Yamaguchi, J. and Furuta, A. (2008) Occurrence of strobilurin resistant strains of *Colletotrichum gloeosporioides* (*Glomerella cingulata*), the causal fungus of strawberry anthracnose. *Japanese Journal of Phytopathology* 74, 114–117.

Ishii, H. (2008) Fungicide research in Japan – an overview. In: Dehne, H.W., Deising, H.B., Gisi, U., Kuck, K.H., Russell, P.E. and Lyr, H. (eds) *Modern Fungicides and Antifungal Compounds V*. DPG Selbstverlag, Braunschweig, Germany, pp. 11–17.

Ishii, H. (2010) QoI fungicide resistance: current status and the problems associated with DNA-based monitoring. In: Gisi, U., Chet, I. and Gullino, M.L. (eds) *Recent Developments in Management of Plant Diseases, Plant Pathology in the 21st Century 1*. Springer, Dordrecht, The Netherlands, pp. 37–45.

Ishii, H., Fraaije, B.A., Sugiyama, T., Noguchi, K., Nishimura, K., Takeda, T., *et al.* (2001) Occurrence and molecular characterization of strobilurin resistance in cucumber powdery mildew and downy mildew. *Phytopathology* 91, 1166–1171.

Ishii, H., Sugiyama, T., Nishimura, K. and Ishikawa, Y. (2002) Strobilurin resistance in cucumber pathogens: persistence and molecular diagnosis of resistance. In: Dehne, H.W., Gisi, U. and Kuck, K.H. (eds) *Modern Fungicides and Antifungal Compounds III*. AgroConcept, Bonn, Germany, pp. 149–159.

Ishii, H., Yano, K., Date, H., Furuta, A., Sagehashi, Y., Yamaguchi, T., *et al.* (2007) Molecular characterization and diagnosis of QoI resistance in cucumber and eggplant pathogens. *Phytopathology* 97, 1458–1466.

Ishii, H., Tanoue, J., Oshima, M., Chung, W.-H., Nishimura, K., Yamaguchi, J., *et al.* (2008) First application of PCR-Luminex system for molecular diagnosis of fungicide resistance and species identification of fungal pathogens. *Journal of General Plant Pathology* 74, 409–416.

Ishii, H., Fountaine, J., Chung, W.-H., Kansako, M., Nishimura, K., Takahashi, K., *et al.* (2009) Characterisation of QoI-resistant field isolates of *Botrytis cinerea* from citrus and strawberry. *Pest Management Science* 65, 916–922.

Ishii, H., Miyamoto, T., Ushio, S. and Kakishima, M. (2011) Lack of cross resistance to a novel succinate dehydrogenase inhibitor fluopyram in highly boscalid-resistant isolates of *Corynespora cassiicola* and *Podosphaera xanthii*. *Pest Management Science* 67, 474–482.

Kansako, M., Yoneda, Y., Shimadu, K. and Ishii, H. (2005) Occurrence of strobilurin-resistant *Botrytis cinerea*, pathogen of citrus grey mold (abstract in Japanese). *Japanese Journal of Phytopathology* 71, 249.

Kataoka, S., Takagaki, M., Kaku, K. and Shimuzu, T. (2010) Mechanism of action and selectivity of a novel fungicide, pyribencarb. *Journal of Pesticide Science* 35, 99–106.

Kida, K., Takagaki, M., Kansako, M., Watanabe, H., Ishii, H. and Fujimura, M. (2010) The efficacy of pyribencarb to the several QoI resistant (F129L, G143A) fungal strains. In: *Abstracts of the 12th IUPAC International Congress of Pesticide Chemistry*, Melbourne, Australia, 424 pp.

Kim, J.H., Mahoney, N., Chan, K.L., Molyneux, R., May, G.S. and Campbell, B.C. (2008) Chemosensitization of fungal pathogens to antimicrobial agents using benzo analogs. *FEMS Microbiology Letters* 281, 64–72.

Kim, Y.S., Dixon, E.W., Vincelli, P. and Farman, M.L. (2003) Field resistance to strobilurin (QoI) fungicides in *Pyricularia grisea* caused by mutations in the mitochondrial cytochrome *b* gene. *Phytopathology* 93, 891–900.

Miyamoto, T., Ishii, H., Seko, T., Kobori, S. and Tomita, Y. (2009) Occurrence of *Corynespora cassiicola* isolates resistant to boscalid on cucumber in Ibaraki Prefecture, Japan. *Plant Pathology* 58, 1144–1151.

Miyamoto, T., Ishii, H., Stammler, G., Koch, A., Ogawara, T., Tomita, Y., *et al.* (2010a) Distribution and molecular characterization of *Corynespora cassiicola* isolates resistant to boscalid. *Plant Pathology* 59, 873–881.

Miyamoto, T., Ishii, H. and Tomita, Y. (2010b) Occurrence of boscalid resistance in cucumber powdery mildew disease in Japan and the molecular characterization of iron-sulfur protein of succinate dehydrogenase of the causal fungus. *Journal of General Plant Pathology* 76, 261–267.

So, K. and Yamaguchi, J. (2008) Management of MBI-D and QoI fungicide resistance on rice blast caused by *Magnaporthe grisea*. In: *Abstracts of the 18th Symposium of Research Committee on Fungicide Resistance*, Matsue, Japan, pp. 70–80.

Stammler, G., Itoh, M., Hino, I., Watanabe, A., Kojima, K., Motoyoshi, M., *et al.* (2007) Efficacy of orysastrobin against blast and sheath blight in transplanted rice. *Journal of Pesticide Science* 32, 10–15.

Takagaki, M., Kataoka, S., Kida, K., Miura, I., Fukumoto, S. and Tamai, R. (2010) Disease-controlling effect of a novel fungicide pyribencarb against *Botrytis cinerea*. *Journal of Pesticide Science* 35, 10–14.

Tamura, H., Mizutani, A., Yukioka, H., Miki, N., Ohba, K. and Masuko, M. (1999) Effect of the methoxyimi-noacetamide fungicide, SSF129, on respiratory activity in *Botrytis cinerea*. *Pesticide Science* 55, 681–686.

Watanabe, H. (2009) Reduced sensitivity of tomato leaf mold fungus (*Passalora fulva*) to azoxystrobin in Gifu Prefecture, Japan. In: *Abstracts of the 19th Symposium of Research Committee on Fungicide Resistance*, Yamagata, Japan, pp. 42–49.

Wei, C.-Z., Katoh, H., Nishimura, K. and Ishii, H. (2009) Site-directed mutagenesis of the cytochrome *b* gene and development of diagnostic methods for identifying QoI resistance of rice blast fungus. *Pest Management Science* 65, 1344–1351.

Yano, K. and Kawada, Y. (2003) Occurrence of strobilurin-resistant strains of *Mycovellosiella nattrassii*, causal fungus of leaf mold of eggplants. *Japanese Journal of Phytopathology* 69, 220–223.

Yukioka, H., Inagaki, S., Tanaka, R., Katoh, K., Miki, N. and Mizutani, A. (1998) Transcriptional activation of the alternative oxidase gene of the fungus *Magnaporthe grisea* by a respiratory-inhibiting fungicide and hydrogen peroxide. *Biochimica et Biophysica Acta* 1442, 161–169.

Part IV

Genetics and Multi-Drug Resistance

19 Genetics of Fungicide Resistance in *Botryotinia fuckeliana* (*Botrytis cinerea*)

Rita Milvia De Miccolis Angelini, Stefania Pollastro and Franco Faretra
Department of Environmental and Agro-Forestry Biology and Chemistry,
University of Bari, Italy

19.1 Introduction

Botryotinia fuckeliana (de Bary) Whetz., teleomorph of *Botrytis cinerea* Pers., is the causal agent of grey mould. The fungus is a cosmopolitan, necrotrophic and polyphagous pathogen causing heavy yield losses on numerous crops, such as grapevine, horticultural and ornamental crops, glasshouse crops, etc., in all worldwide temperate areas. Moreover, it induces one of the most important postharvest diseases on fruit and vegetables. The main features of *B. fuckeliana* and grey mould have been reviewed exhaustively on several occasions (i.e. Verhoeff *et al.*, 1992; Elad *et al.*, 2004; Williamson *et al.*, 2007).

The control of grey mould on various crops is still based largely on the use of fungicides and is difficult because the disease is particularly severe close to harvest time, when fungicide sprays may leave residues on edible products. The broad variability and adaptability of *B. fuckeliana* often underlie acquired resistance to fungicides under the selection pressure exerted by spray schedules.

Resistance to plant protection products (PPPs) is a key challenge in modern crop protection (Gullino *et al.*, 2000; Clark and Yamaguchi, 2002; Ishii, 2006; Brent, 2007; Brent and Hollomon, 2007; van den Bosch and Gilligan, 2008). Agrochemical companies are well aware of the need for common actions to prevent resistance. The FRAC website is a useful and updated source of information on several aspects of resistance to the main groups of fungicides and guidelines for their appropriate usage (http://www.frac.info).

Acquired resistance to fungicides in fungal pathogens is due to genetic modifications transmissible to the progeny. Resistant isolates can be detected in the field or can be obtained under laboratory conditions. The variability of field isolates often hampers the understanding of genetic bases of resistance, but the resistance of laboratory mutants is sometimes due to genetic and physiological mechanisms that are not necessarily the same as those occurring in the field (Grindle, 1987; Grindle and Faretra, 1993). The genetic and molecular aspects of fungicide resistance have been reviewed recently (i.e. Leroux, 2004; Ma and Michailides, 2005; Gisi and Sierotzki, 2008).

B. fuckeliana is recognized by FRAC among pathogens at high risk of resistance. Fungicide resistance in practice has indeed been experienced with almost all fungicides used against grey mould on various crops and in numerous countries (www.frac.info).

B. fuckeliana is a heterothallic bipolar Ascomycete. Studies on its sexual behaviour and mating system and on experimental conditions inducive to carpogenesis (Faretra and Antonacci, 1987; Faretra *et al.*, 1988a,b) brought about setting up a method to obtain meiotic progeny (ascospores) of sexual crosses. This made classical genetic analysis feasible in *B. fuckeliana*, which has been exploited to investigate the mode of inheritance of resistance phenotypes in ascospore progeny of sexual crosses and to clarify the genetic bases of resistance to various fungicides. In the last decades, due to the progress in molecular biology, research efforts have been focused more and more on the molecular characterization of genes potentially responsible for resistance. Hence, it has been possible to identify specific single nucleotide polymorphisms (SNPs) responsible for or associated with resistance to several groups of fungicides and to develop molecular techniques for the detection and quantification of resistant isolates in the field.

This chapter, with no claims to exhaustiveness, reviews the most significant progress in the knowledge on genetic mechanisms underlying acquired resistance in *B. fuckeliana*

to the main groups of fungicides used against grey mould. A synthesis of the current knowledge of genes and putative mutations responsible for resistance to various fungicides is reported in Table 19.1.

19.2 Benzimidazoles

Benzimidazole fungicides are inhibitors of the mitotic division by binding to β-tubulin, a globular protein that is the basic structural constituent of microtubules of the mitotic spindle. In several fungi, resistance to benzimidazoles is the result of a reduced binding affinity between fungicides and β-tubulin. The usage of benzimidazoles against grey mould lost its importance a long time ago due to the widespread occurrence in the field of benzimidazole-resistant *B. fuckeliana* isolates.

The genetic bases of resistance have been investigated in numerous phytopathogenic

Table 19.1. Nucleotide substitutions (in bold) in specific target genes and amino acid changes associated with different phenotypes of resistance to fungicides in *B. fuckeliana*.

Gene	Codon substitution	Amino acid substitution	Resistance phenotype[*]	Occurrence	References
			Benzimidazoles		
Mbc1	GA**G** to G**C**G	E198A	Ben^{HR}NPC^S	Field isolates	Yarden and Katan (1993);
	GAG to **A**AG	E198K	Ben^{HR}NPC^R	and	Leroux et al. (2002)
	TTC to TA**C**	F200Y	Ben^{MR}NPC^R	laboratory	
	GA**G** to G**T**G	E198V	Ben^{HR}NPC^{LR}	mutants	Banno et al. (2008)
			Dicarboximides		
Daf1	AT**C** to A**G**C	I365S	Daf^{LR}	Field isolates	Leroux et al. (2002);
				and	Oshima et al. (2002);
				laboratory	Cui et al. (2004)
	AT**C** to A**A**C	I365N	Daf^{LR}	mutants	Leroux et al. (2002);
	AT**C** to **CG**C	I365R	Daf^{LR}		Cui et al. (2004)
	AT**C** to A**G**C+	I365S	Daf^{LR}		Cui et al. (2004)
	to C**C**G	Q369P			
	GTC to **TT**C	V368F+Q369H+	Daf^{LR}		Oshima et al. (2006);
	CA**G** to CA**C** ?	T447S			Banno et al. (2008)
	CA**G** to C**C**G +	Q369P	Daf^{MR}		Oshima et al. (2006);
	AA**C** to A**G**C	N373S			De Miccolis Angelini et al. (2007);
					Ma et al. (2007);
					Banno et al. (2008)
	G**G**T to G**A**T	G357D	Daf^{HR}	Laboratory	De Miccolis Angelini
	G**G**T to **A**GT	G446S	Daf^{HR}	mutants	et al. (2007);
	CCT to **A**CT	P742T	Daf^{HR}		Rotolo (2010)

Continued

Table 19.1. Continued.

Gene	Codon substitution	Amino acid substitution	Resistance phenotype[*]	Occurrence	References
		Fenhexamid (SBI fungicides)			
Erg27	TTC to ATC	F412I	Fen^{HR} (HydR3$^+$)	Field isolates	Albertini and Leroux (2004); Fillinger *et al.* (2008); De Miccolis Angelini *et al.* (2010a)
	TTC to GTC	F412V	Fen^{HR} (HydR3$^+$)		Fillinger *et al.* (2008);
	TTC to TCC	F412S	Fen^{HR} (HydR3$^+$)		De Miccolis Angelini *et al.* (2010a)
	?	R496T	Fen^{HR} (HydR3$^+$)		Albertini and Leroux (2004)
	?	N369D+F412I	Fen^{HR} (HydR3$^+$)		Fillinger *et al.* (2008)
	?	L195F	Fen^{LR} (HydR3$^-$)		
	?	V309M	Fen^{LR} (HydR3$^-$)		
	?	A314V	Fen^{LR} (HydR3$^-$)		
	?	S336C	Fen^{LR} (HydR3$^-$)		
	?	N369D	Fen^{LR} (HydR3$^-$)		
	?	L400F	Fen^{LR} (HydR3$^-$)		
	?	L400S	Fen^{LR} (HydR3$^-$)		
	?	S336C+N369D	Fen^{LR} (HydR3$^-$)		
	TTT to TCT	F26S	Fen^{LR}		De Miccolis Angelini *et al.* (2010a)
	AAC to GAC	N369D			
	GGC to AGC	G23S	Fen^{HR}	Laboratory	De Miccolis Angelini *et al.* (2007); Rotolo (2010)
	GGC to GTC	G23V	Fen^{LR}	mutants	
	ACA to ATA	T63I	Fen^{LR}		
	CCT to CTT	P427L	Fen^{HR}		
		Boscalid (SDHI fungicides)			
SdhB	CCC to CTC	P225L	Bos^{HR}	Field isolates	Stammler *et al.* (2008); De Miccolis Angelini *et al.* (2010a,b)
	CCC to CTT	P225L	Bos^{HR}	and	
	CCC to TTC	P225F	Bos^{HR}	laboratory	
	CCC to ACC	P225T	Bos^{HR}	mutants	
	CAC to TAC	H272Y	Bos^{HR}		
	CAC to CGC	H272R	Bos^{HR}		
	AAC to ATC	N230I	Bos^{MR}		
		QoI fungicides			
cytB	GGT to GCT	G143A	QoI^{HR}	Field isolates	Banno *et al.* (2009); Ishii *et al.* (2009); Jiang *et al.* (2009); De Miccolis Angelini *et al.* (2010a)

Note: ?, mutations not reported.

fungi, yeasts and other organisms. In most species, it is caused by mutations in single chromosomal genes. In *B. fuckeliana*, genetic analysis of meiotic progeny of suitable sexual crosses showed that resistance was caused by a single major gene, or closely linked genes, which was designated, according to nomenclature rules proposed by Yoder *et al.* (1986), *Mbc1* (response to methyl benzimidazol-2-yl carbamate, the common derivative of benzimidazoles) (Faretra and Pollastro, 1991). The *Mbc1* gene showed at least three main allelic variants, *Mbc1S*, *Mbc1LR* and *Mbc1HR*, conferring sensitivity, low and high resistance to benzimidazoles, respectively. The *Mbc1HR* alleles generally cause, as pleiotropic effect, hypersensitivity to *N*-phenylcarbamates. Additional allelic classes of the *Mbc1* gene cause phenotypes combining low or high resistance to benzimidazoles with wild-type or moderate insensitivity to N-phenylcarbamates (Faretra and Pollastro, 1991, 1993a; Pollastro and Faretra, 1992). These phenotypes were found at very low frequency in fungal populations exposed only to benzimidazoles, but spread in the field when the mixture carbendazim (benzimidazole) plus diethofencarb (N-phenylcarbamate) was used (i.e. Beever and O'Flaherty, 1985; Faretra et al., 1989; Elad et al., 1992). Observations on near-isogenic strains carrying different Mbc1 alleles revealed that these isolates showed a slight decrease of vegetative vigour and sporulation, but not of aggressiveness, and this could explain their dynamics in the field (Pollastro et al., 1996a). Tetrad analysis showed that the Mbc1 gene was weakly linked to its centromere and was linked (41 map units) to the Daf1 gene responsible for resistance to dicarboximides (Faretra and Pollastro, 1996).

Yarden and Katan (1993) cloned and sequenced the β-tubulin gene in B. fuckeliana. They found that high resistance to benzimidazoles was associated with point mutations in the codon 198 coding for glutamic acid in wild-type isolates; the amino acid was substituted by alanine (E198A) or lysine (E198K) in resistant isolates showing or not hypersensitivity to N-phenylcarbamates, respectively. A further substitution with valine (E198V) confers high resistance to benzimidazoles and intermediate hypersensitivity to N-phenylcarbamates (Nakazawa and Yamada, 1997; Banno et al., 2008). Low resistance to benzimidazoles is associated with change of phenylalanine to tyrosine at position 200 (F200Y) (Yarden and Katan, 1993). Field-resistant isolates show generally the mutations described above, although other resistance mutations have been selected under laboratory conditions. It is likely that these do not occur in the field because they reduce the fitness of mutants (Davidse and Ishii, 1995).

The knowledge on the molecular bases of resistance made diagnostic tools available for monitoring benzimidazole resistance, such as AS-PCR (allele specific-PCR) or PCR-RFLP (PCR-restriction fragment length polymorphism) (Luck and Gillings, 1995) or real-time PCR (Banno et al., 2008).

19.3 Dicarboximides and Phenylpyrroles

Dicarboximides and phenylpyrroles interfere with the same process of osmotic regulation, although they probably have different sites of action (Pillonel and Meyer, 1997). Resistance to dicarboximides is quite common on numerous crops. Generally, resistant field isolates show low to moderate levels of resistance (LR phenotype), while laboratory mutants show high resistance (HR). Classical genetic analysis showed that the response to dicarboximides was due to a single major gene, named Daf1 (dicarboximides and aromatic hydrocarbon fungicides which are related by cross-resistance), or closely linked genes. Three allelic classes, Daf1S, Daf1LR and Daf1HR, confer wild-type sensitivity, low and high resistance (Faretra and Pollastro, 1991, 1993a; Hilber et al., 1993). An additional allelic class, Daf1UR, conferring an ultra-low level of resistance, has been distinguished by Beever and Parkes (1993). Most Daf1HR alleles cause, as a pleiotropic effect, a reduced tolerance to high osmotic pressure, although a rare allele did not show such effect (Beever, 1983; Faretra and Pollastro, 1991, 1993a). The pleiotropic effects of alleles of the Daf1 gene have been investigated in near-isogenic strains, showing that Daf1HR alleles and, to a less extent, Daf1LR alleles cause reduction of vegetative vigour and osmotolerance; Daf1HR but not Daf1LR alleles even cause low sporulation and aggressiveness (Pollastro et al., 1996a).

B. fuckeliana is multi-nucleate in all its vegetative stages. The coexistence of Daf1HR nuclei with Daf1S or Daf1LR nuclei in heterokaryotic state can yield a phenotype combining high resistance with almost normal osmotolerance. Heterokaryon can indeed mitigate the negative pleiotropic effects of resistance mutations because high resistance is dominant (or partially dominant) on sensitivity, and hypersensitivity to high osmolarity is recessive (or partially recessive) on normal osmotolerance (Faretra and Pollastro, 1993a).

Dicarboximide-resistant laboratory mutants, but not field isolates, often show cross-resistance to phenylpyrroles (Faretra and Pollastro, 1993b; Leroux et al., 1999). At least six phenotypic classes have been discriminated on the grounds of their response to dicarbo-ximides and phenylpyrroles: S/S, LR/S, HR/S; HR/LR, HR/MR and HR/HR (where: S = sensitivity; LR = low resistance; MR = moderate resistance; HR = high resistance) (Faretra and Pollastro, 1993b; Hilber *et al.*, 1995). An additional gene, *Daf2*, not linked to the *Daf1* gene, was found responsible for high resistance to both groups of fungicides in few laboratory mutants (Faretra and Pollastro, 1993b). Vignutelli *et al.* (2002) showed that field resistance to phenylpyr-roles was due to mutations in a gene other than the *Daf1* and hypothesized that an addi-tional gene closely linked to the *Daf1* was responsible for laboratory resistance to these fungicides.

The *Daf1* gene is the same gene named *BcOS1* (Oshima *et al.*, 2002) or *Bos1* (Cui *et al.*, 2002; Liu *et al.*, 2008), homologous of the *os1* gene of *Neurospora crassa*, coding for a group III hybrid histidine kinase which acts as a virulence factor in *B. fuckeliana* (Viaud *et al.*, 2006). The gene, cloned and sequenced in numerous *B. fuckeliana* isolates, showed a broad polymorphism in nucleotide sequence (Fig. 19.1). Most dicarboximide low-resistant field isolates (*Daf1LR*) carry one of three possible mutations at codon 365, leading to substitution of isoleucine (wild type) with

serine (I365S) or, less frequently, with arginine (I365R) or asparagine (I365N) (Leroux *et al.*, 2002; Oshima *et al.*, 2002; Cui *et al.*, 2004; De Miccolis Angelini *et al.*, 2007; Ma *et al.*, 2007). Two further mutations, Q369P and N373S, have been detected in moderately resistant field isolates from Japan and Lebanon (Oshima *et al.*, 2006; De Miccolis Angelini *et al.*, 2007; Ma *et al.*, 2007). The mutations G357N or G446S were detected in *Daf1HR* mutants showing normal osmotolerance or hypersensitivity to high osmolarity, respectively (De Miccolis Angelini *et al.*, 2007).

Sequence analysis of the *Daf1* gene in phenylpyrrole-sensitive and -resistant iso-lates did not reveal any SNPs associated with resistance, although a substitution of proline with threonine was found in codon 742 (P742T) in two UV-induced laboratory mutants displaying high resistance (Rotolo, 2010).

Molecular methods have been devel-oped for detection of mutation I365S, which is most often responsible for field resistance to dicarboximides based on cleaved ampli-fied polymorphic sequences (CAPS) with the restriction enzyme *Taq* I (Sierotzki and Gisi, 2003), or on PCR or real-time AS-PCR (Oshima *et al.*, 2006; De Miccolis Angelini *et al.*, 2007; Rotolo, 2010).

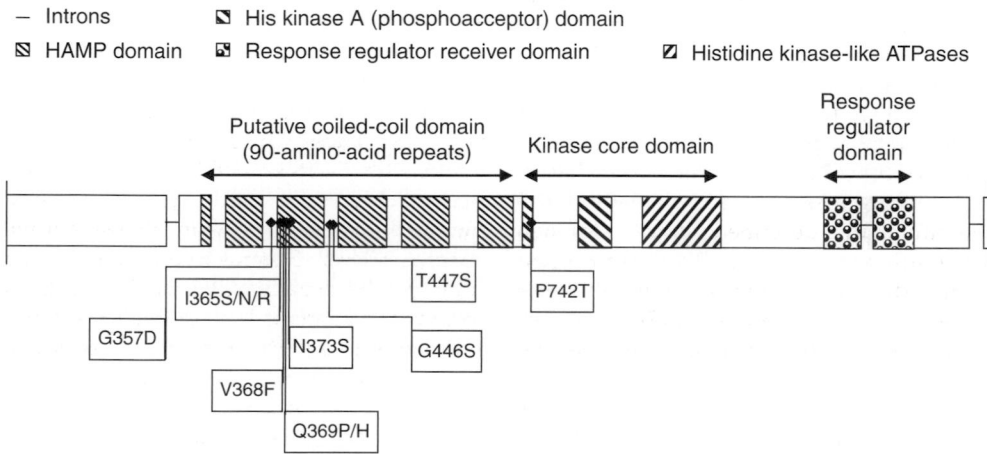

Fig. 19.1. Schematic representation of the *B. fuckeliana Daf1* gene (4386 bp) encoding a group III histidine kinase (accession number AF435964). The intron position, the conserved domains (Conserved Domains Database; http://www.ncbi.nlm.nih.gov/cdd) and the changed amino acid residues putatively involved in the resistance to dicarboximides are shown.

19.4   Dichlofluanide

Dichlofluanide is a multi-site fungicide reacting with thiols, sulfhydryl and aminic groups yielding thiophosgene and sulfhydric acid. Although multi-site fungicides generally show a low resistance risk, resistance to dichlofluanide has been reported in several countries. Observations on field isolates identified three phenotypic classes: sensitivity, moderate and high resistance. Classical genetic analysis showed that two major genes, *Dic1* (with its alleles: *Dic1S*, sensitivity; *Dic1LR*, moderate resistance; *Dic1HR*, high resistance) and *Dic2* (with its alleles: *Dic2S*, sensitivity; *Dic2LR*, moderate resistance) were responsible for resistance (Pollastro *et al.*, 1996b).

19.5   Anilinopyrimidines and 'Multi-Drug Resistance' (MDR)

Although numerous studies have been carried out, the primary mode of action of anilinopyrimidines has not yet been clarified. Inhibition of methionine biosynthesis has been supposed in *B. fuckeliana*. The target enzyme was believed to be cystathionine β-lyase, but it was shown that the inhibitor activity of anilinopyrimidines on the enzyme was negligible and no mutations in the *Cbl* or *metC* gene coding for the enzyme were found associated with resistance (Sierotzki *et al.*, 2002; Fritz *et al.*, 2003). Other key enzymes in the methionine biosynthesis could be involved in resistance since increased cystathionine and reduced methionine and homocysteine concentrations were observed in *B. fuckeliana* mycelium exposed to pyrimethanil (Fritz *et al.*, 1997).

Field isolates and laboratory mutants with low resistance to anilinopyrimidines showed a marked phenotypic instability due to their heterokaryotic state. Resistance mutations showed such strong negative pleiotropic effects to be lethal in the homokaryotic state in ascospores. This finding confirms that heterokaryon confers to *B. fuckeliana* a broad genetic adaptability keeping mutated nuclei in more or less latent state, lessening their negative effects on the phenotype (Santomauro *et al.*, 2000; De Miccolis Angelini *et al.*, 2002).

Hilber and Hilber-Bodmer (1998) reported a Mendelian mode of inheritance of resistance traits in ascospore progeny of sexual crosses. Afterwards, classical genetic analysis showed that at least three major genes were associated to anilinopyrimidine-resistance phenotypes, Ani[R1], Ani[R2] and Ani[R3]. Ani[R1] isolates show moderate to high levels of resistance to anilinopyrimidines not associated with decreased sensitivity to other fungicides. Ani[R2] and Ani[R3] isolates, named also MDR1 and MDR2, show multi-drug resistance to anilinopyrimidines, dicarboximides, phenylpyrroles and several sterol biosynthesis inhibitors (SBIs), which is caused by an active efflux of fungicides from mycelium due to ATP-dependent membrane transporters ('ABC' or 'MFS') (Chapeland *et al.*, 1999; Vermeulen *et al.*, 2001; Leroux *et al.*, 2002; Kretschmer *et al.*, 2009).

Recently, the molecular bases of MDR phenotypes in *B. fuckeliana* have been clarified. Mutations activating a transcription factor (*mrr1*) regulating the expression of the *atrB* gene coding for an ABC transporter are responsible for the MDR1 phenotype. A specific rearrangement in the promoter of the *mfsM2* gene with the insertion of a 1,326-bp sequence, probably an LTR-type retrotransposon, causes the MDR2 phenotype. An additional phenotype, named MDR3, showing resistance to more numerous fungicides, derives from meiotic recombination of the *mrr1* and *mfsM2* genes carried by MDR1 and MDR2 isolates, respectively (Kretschmer *et al.*, 2009).

Molecular characterization of genes coding for key enzymes in the methionine biosynthesis, cystathionine β-lyase, (*Cbl*), cystathionine γ-lyase (*Cgl*), cystathionine β-synthase (*Cbs*) and cystathionine γ-synthase (*Cgs*), in anilinopyrimidine-resistant field isolates evidenced several single nucleotide polymorphisms, some leading to amino acidic changes, but none associated with anilinopyrimidine resistance (De Miccolis Angelini *et al.*, 2010a).

19.6   Hydroxianilides (Fenhexamid)

Fenhexamid is a sterol biosynthesis inhibitor (SBI: class III) that acts by interfering with the enzyme 3-keto-reductase operating C-4 demethylation.

Resistant isolates were detected on grapevine in France even before the introduction of the fungicide in agricultural practice. Such isolates, named HydR1, showed a high level of resistance of mycelial growth *in vitro*, but the fungicide was still effective in practice because it inhibited their germ tube growth (Leroux *et al.*, 1999; Suty *et al.*, 1999; Baroffio *et al.*, 2003). The HydR1 isolates show increased sensitivity to edifenphos and some SBIs. Such a phenotype seems a distinct genetic entity (group I) or even a sympatric species (*Botrytis pseudocinerea*) naturally resistant to fenhexamid (Leroux *et al.*, 2002; Leroux, 2004). HydR1 and non-HydR1 isolates show morphological differences, vegetative incompatibility and apparently sexual incompatibility (Leroux *et al.*, 2002; Albertini *et al.*, 2002; Fournier *et al.*, 2003).

The high polymorphism observed comparing the sequences of the *Erg27* and *CYP51* genes, encoding respectively 3-keto-reductase and eburicol 14-α-demethylase, in HydR1 and non-HydR1 isolates seems to confirm the genetic diversity between the two groups of isolates. Expressed mutations in the *CYP51* gene carried by HydR1 isolates could explain their higher sensitivity to SBIs, while the specific sequence of the *Erg27* gene could explain their natural resistance to fenhexamid (Albertini *et al.*, 2002; Albertini and Leroux, 2004).

Two other resistance phenotypes have been detected in *B. fuckeliana* populations exposed to the fungicide. HydR2 isolates show a moderate level of resistance in the stage of mycelium growth; HydR3 isolates show a high level of resistance in both stages of germ tube elongation and mycelium growth (Albertini and Leroux, 2004). Unlike HydR1 isolates, fenhexamid resistance in HydR2 and HydR3 isolates is not associated with increased sensitivity to SBIs and edifenphos (Leroux *et al.*, 2002). More recently, Fillinger *et al.* (2008) discriminated two levels of resistance among HydR3 isolates and distinguished them as HydR3+ and HydR3-.

Mutants displaying various levels of resistance to fenhexamid were obtained under laboratory conditions, and classical genetic analysis showed that they carried mutations in a single major gene, or closely linked genes (De Guido *et al.*, 2007). Laboratory mutants showing different levels of resistance carried mutations in the *Erg27* gene, leading to one of the following amino acid substitutions, G23S, G23V, T63I or P427L, in 3-keto-reductase (Fig. 19.2) (De Miccolis Angelini *et al.*, 2007; Rotolo, 2010).

In field isolates from France and Germany, HydR3+ resistance was associated with mutations in the *Erg27* gene, leading to the substitution of a phenylalanine at the carboxylic end of the putative transmembrane domain of the 3-keto reductase (position 412) with serine, isoleucine or valine. Many different amino acid substitutions have been detected in HydR3- isolates showing a moderate resistance and none in HydR2 isolates (Fig. 19.2) (Albertini and Leroux, 2004; Fillinger *et al.*, 2008). These findings

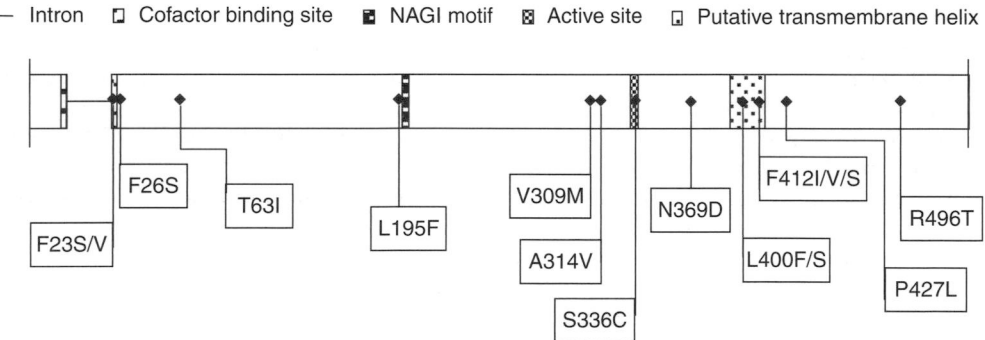

Fig. 19.2. Schematic representation of the *B. fuckeliana Erg27* gene (1687 bp) (accession number AY220532) coding for 3-keto reductase. The intron position, the conserved regions of the encoded protein and the changed amino acid residues associated to different phenotypes of resistance to hydroxyanilides are shown.

were confirmed by De Miccolis Angelini *et al.* (2010a), who characterized Italian field isolates with different levels of resistance. The significance of mutations in resistance was studied through genetic transformation of a sensitive strain with a replicative plasmid. *Erg27*[HydR3+] alleles proved to confer high resistance, while *Erg27*[HydR3–] alleles did not cause any phenotypic variation in response to fenhexamid, probably because of weak expression of the transgene. Nevertheless, classical genetic analysis showed a strict co-segregation of such mutations and the resistance phenotype in ascospore progeny of suitable crosses (Fillinger *et al.*, 2008).

It has been hypothesized that mono-oxigenases (P_{450} cytochromes) are involved in fenhexamid detoxification and edifenphos activation in HydR1 and HydR2 isolates (Leroux *et al.*, 2002).

19.7 SDHI (Succinate Dehydrogenase Inhibitor) Fungicides

The target enzyme of SDH inhibitors is succinate dehydrogenase (SDH, so-called complex II in the mitochondrial respiration chain), which is a functional part of the tricarboxylic cycle and linked to the mitochondrial electron transport chain. SDH consists of four subunits (A, B, C and D), and the binding site of ubiquinone (and of SDHIs) is formed by the subunits B, C and D. Resistance to SDHIs in various fungi has been associated with amino acidic substitutions in cysteine-rich, highly conserved regions of the Fe-S protein (subunit B of succinate dehydrogenase) (Avenot and Michailides, 2010).

Boscalid-resistant mutants have been obtained in *B. fuckeliana* under laboratory conditions. Analysis of meiotic progeny showed that resistance was caused by a single major gene, or closely linked genes (De Miccolis Angelini *et al.*, 2010b). Their molecular characterization identified specific mutations in the *SdhB* gene, causing amino acid substitutions in conserved regions of the Fe-S protein (Fig. 19.3). In particular, a replacement of proline in position 225 with leucine (P225L), phenylalanine (P225F) or threonine (P225T), or the replacement of histidine at position 272 with tyrosine (H272Y) or arginine (H272R), have been found in laboratory mutants and field isolates (Stammler, 2008; Stammler *et al.*, 2008; De Miccolis Angelini *et al.*, 2010a,b). Both substitutions occur in the cysteine-rich regions of the protein associated with Fe-S (S2 and S3) centres involved in electron transfer to ubiquinone. Therefore, boscalid resistance could be due to conformational changes of the protein, suggesting that SDHs can interfere with the functionality of Fe-S centres, preventing their reoxidation. An additional N230I substitution was found in some field isolates displaying a moderate level of resistance to boscalid (De Miccolis Angelini *et al.*, 2010a; Leroux *et al.*, 2010). The close association between mutations and resistance phenotype was confirmed by their co-segregation in ascospore progeny of appropriate sexual crosses.

— Intron   ▨ Cys-rich clusters

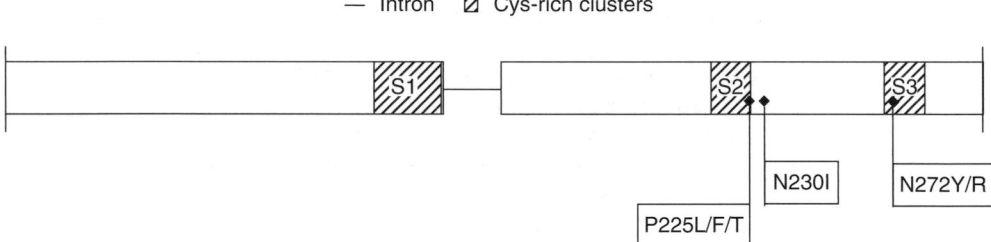

Fig. 19.3. Schematic representation of the *B. fuckeliana SdhB* gene (959 bp) coding for the iron-sulfur protein of the succinate dehydrogenase (SDH) complex (*Botrytis cinerea* Sequencing Project database, accession number BC1G_13286.1). The unique intron, the highly conserved Cys-rich clusters of the protein associated with the iron-sulfur centres (S1, S2 and S3) and the amino acid substitutions responsible for the resistance to boscalid are represented.

Classical analysis showed a genetic linkage among the genes responsible for resistance to boscalid (*SdhB*), dicarboximides (*Daf1*) and benzimidazoles (*Mbc1*) (De Miccolis Angelini *et al.*, 2010b). The presence of the three genes on the same chromosome was confirmed by S. Fillinger and A.S. Walker (personal communication) on the grounds of the complete genome sequence available at the Broad Institute of Harvard and MIT, USA (*Botrytis cinerea* Sequencing Project; http://www.broad.mit.edu).

19.8 Quinone Outside Inhibitors (QoI)

QoI fungicides are inhibitors of respiration acting at the quinol outer (Qo) binding site of the cytochrome bc_1 complex in the electron chain. At present, QoIs are not used against *B. fuckeliana*, probably because an alternative oxidase (AOX) is present in its mitochondria (Tamura *et al.*, 1999; Leroux, 2004). However, *B. fuckeliana* populations are often exposed to these fungicides applied against other diseases (especially downy and powdery mildews).

In many fungal species, a high level of resistance to QoIs is often caused by a single mutation in codon 143 of the mitochondrial gene *cytb* coding for cytochrome *b*, leading to the substitution of glycine with alanine (G143A) (Gisi *et al.*, 2002). The same mutation has been detected in *B. fuckeliana* field isolates resistant to QoIs, although often in conditions of heteroplasmy (coexistence of resistant and sensitive mitochondria in the same cell) (Fig. 19.4; Banno *et al.*, 2009; Ishii *et al.*, 2009; Jiang *et al.*, 2009; De Miccolis Angelini *et al.*, 2010b,c). Classical genetic analysis of progeny from sexual crosses between resistant and sensitive isolates confirmed that resistance was due to cytoplasmic determinants showing maternal (i.e. sclerotial) inheritance (De Miccolis Angelini *et al.*, 2010b,c).

Spontaneous or UV-induced laboratory mutants selected for their high resistance to trifloxystrobin did not display any mutations in the *cytb* gene. No mutations were found even in the genes coding for alternative oxidase (AOX) and the Rieske protein, which have been involved in resistance to QoIs in some phytopathogenic fungi. It is possible that laboratory mutants were heteroplasmic and their condition could have hampered the detection of the G143A mutation in the *cytb* gene (De Miccolis Angelini *et al.*, 2008). However, different genes or mechanisms of resistance, other than the decreased affinity with the target site, could be responsible for their resistance phenotype. It has been hypothesized that resistance to QoIs in some fungi could be caused by active efflux due to the family of ATP-binding cassette (ABC) or the major facilitator superfamily (MFS) transporters (Fernández-Ortuño *et al.*, 2008).

Sequence analysis of the *cytb* gene in *B. fuckeliana* showed polymorphism in its genetic structure due to the presence of a sequence (about 1200 bp) flanking codon

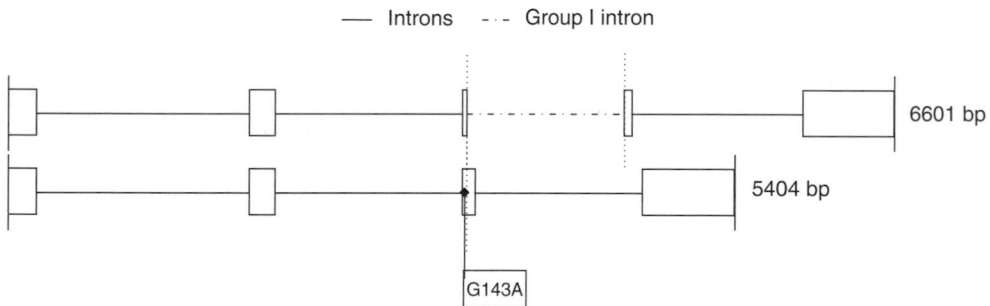

Fig. 19.4. Schematic representation of the polymorphic structure of the mitochondrial cytochrome *b* gene (*cytb*) (5404 bp) in *B. fuckeliana* (accession number AB262969) showing the introns and the G143A mutation responsible for the resistance to QoI fungicides. Dotted line indicates the 1.197-bp intron (group I) flanking downstream codon 143 of the gene in some of the QoI-sensitive isolates (AB428335).

143 showing high homology to conserved regions of introns of group I characterized by the capability of 'autosplicing' (Fig. 19.4) (De Miccolis Angelini *et al.*, 2008; Banno *et al.*, 2009; Jiang *et al.*, 2009). It has been hypothesized that the presence of such introns prevents the occurrence of mutation in codon 143 since it would be lethal, impeding the correct excision of the intronic sequence from the gene transcript (Grasso *et al.*, 2006; Sierotzki *et al.*, 2007). This seems to be confirmed by the finding that in *B. fuckeliana* the intron has been detected in 15–43% of sensitive isolates and never in resistant field isolates (Jiang *et al.*, 2009; De Miccolis Angelini *et al.*, 2010c).

Molecular detection of the G143A mutation in the *cytb* gene in *B. fuckeliana* isolates can be based on AS-PCR assay (Jiang *et al.*, 2009), or on CAPS analysis with the restriction enzymes *Alu*I (De Miccolis Angelini *et al.*, 2008; Banno *et al.*, 2009), *Fnu*4HI or *Ita*I (Ishii *et al.*, 2009).

19.9 Conclusions

B. fuckeliana is a pathogen well known for its broad genetic variation and great adaptability, which likely derive from its particular genetic characteristics (haploidy, probable heteroploidy, heterokaryosis, parasexuality (?), transposons, mycoviruses, etc.). As a result, acquired resistance has been experienced in *B. fuckeliana* to all the fungicides used specifically to control grey mould. In the last decades, the knowledge on the genetic bases of fungicide resistance has improved notably due to classical genetic analysis based on the mode of inheritance of resistance phenotypes in ascospore progeny of suitable sexual crosses, and especially to the progress of molecular biology. Nowadays, the complete genome sequence of *B. fuckeliana* is available at the Broad Institute of Harvard and MIT, USA (*Botrytis cinerea* Sequencing Project; http://www.broad.mit.edu), and this will certainly speed up all future research on this issue.

The knowledge on the genetic bases of fungicide resistance has crucial outcomes, even for the prevention and management of resistance in practice. The knowledge on the genes and mutations responsible for resistance enables the development of molecular techniques for the detection (even quantitative) of fungicide-resistant isolates in field samples. These will be helpful tools to clarify in deeper detail the possible role of fungal variants (or subpopulations, biotypes) in the occurrence, spread and stability of resistance, and the dynamics of resistant subpopulations and resistance alleles under the selective pressure exerted by fungicide sprays. It is hoped that more joint efforts between private and public research, combining complementary skills, can bring significant advances on the topic.

References

Albertini, C. and Leroux, P. (2004) A *Botrytis cinerea* putative 3-keto reductase gene (*ERG27*) that is homologous to the mammalian 17 β-hydroxysteroid dehydrogenase type 7 gene (*17β-HSD7*). *European Journal of Plant Pathology* 110, 723–733.
Albertini, C., Thebaud, G., Fournier, E. and Leroux, P. (2002) Eburicol 14*a*-demethylase gene (*CYP51*) polymorphism and speciation in *Botrytis cinerea*. *Mycological Research* 106, 1171–1178.
Avenot, H.F. and Michailides, T.J. (2010) Progress in understanding molecular mechanisms and evolution of resistance to succinate dehydrogenase inhibiting (SDHI) fungicides in phytopathogenic fungi. *Crop Protection* 29, 643–651.
Banno, S., Fukumori, F., Ichiishi, A., Okada, K., Uekusa, H., Kimura, M., *et al.* (2008) Genotyping of benzimidazole-resistant and dicarboximide-resistant mutations in *Botrytis cinerea* using real-time polymerase chain reaction assays. *Phytopathology* 98, 397–404.
Banno, S., Yamashita, K., Fukumori, F., Okada, K., Uekusa, H., Takagaki, M., *et al.* (2009) Characterization of QoI resistance in *Botrytis cinerea* and identification of two types of mitochondrial cytochrome b gene. *Plant Pathology* 58, 120–129.

Baroffio, C.A., Siegfried, W. and Hilber, V.W. (2003) Long-term monitoring for resistance of *Botryotinia fuckeliana* to anilinopyrimidine, phenylpyrrole and hydroxyanilide fungicides in Switzerland. *Plant Disease* 87, 662–666.

Beever, R.H. (1983) Osmotic sensitivity of fungal variants resistant to dicarboximide fungicides. *Transactions of the British Mycological Society* 80, 327–331.

Beever, R.H. and O'Flaherty, B.F. (1985) Low-level benzimidazole resistance in *Botrytis cinerea* in New Zealand. *New Zealand Journal of Agricultural Research* 28, 289–292.

Beever, R.H. and Parkes, S.L. (1993) Mating behaviour and genetics of fungicide resistance of *Botrytis cinerea* in New Zealand. *New Zealand Journal of Crop and Horticultural Science* 21, 303–310.

Brent, K.J. (2007) *Fungicide Resistance in Crop Pathogens: How Can it be Managed?* FRAC Monograph No. 1. Global Crop Protection Federation, Brussels.

Brent, K.J. and Hollomon, D.W. (2007) *Fungicide Resistance: The Assessment of Risk.* FRAC Monograph No. 2. Global Crop Protection Federation, Brussels.

Chapeland, F., Fritz, R., Lanen, C., Gredt, M. and Leroux, P. (1999) Inheritance and mechanisms of resistance to anilinopyrimidine fungicides in *Botrytis cinerea* (*Botryotinia fuckeliana*). *Pesticide Biochemistry and Physiology* 64, 85–100.

Clark, J.M. and Yamaguchi, I. (2002) *Agrochemical Resistance: Extent, Mechanism, and Detection.* American Chemical Society Symposium Series. American Chemical Society, Washington, DC.

Cui, W., Beever, R.E., Parkes, S.L., Weeds, P.L. and Templeton, M.D. (2002) An osmosensing histidine kinase mediates dicarboximide fungicide resistance in *Botryotinia fuckeliana* (*Botrytis cinerea*). *Fungal Genetics and Biology* 36, 187–198.

Cui, W., Beever, R.E., Parkes, S.L. and Templeton, M.D. (2004) Evolution of an osmosensing histidine kinase in field strains of *Botryotinia fuckeliana* (*Botrytis cinerea*) in response to dicarboximide fungicide usage. *Phytopathology* 94, 1129–1135.

Davidse, L.C. and Ishii, T. (1995) Biochemical and molecular aspects of benzimidazoles, N-phenylcarbamates and N-phenylformamidoxines and the mechanisms of resistance to these compounds in fungi. In: Lyr, H. (ed.) *Modern Selective Fungicides*. Gustav Fisher, Jena, Germany, pp. 305–322.

De Guido, M.A., De Miccolis Angelini, R.M., Pollastro, S., Santomauro, A. and Faretra, F. (2007) Selection and genetic analysis of laboratory mutants of *Botryotinia fuckeliana* resistant to fenhexamid. *Journal of Plant Pathology* 89, 203–210.

De Miccolis Angelini, R.M., Santomauro, A., De Guido, M.A., Pollastro, S. and Faretra, F. (2002) Genetics of anilinopyrimidine-resistance in *Botryotinia fuckeliana* (*Botrytis cinerea*). In: Vannacci, G. and Sarrocco, S. (eds) *Proceedings of 6th European Conference on Fungal Genetics*. Edizioni Plus, University of Pisa, Italy, 434 pp.

De Miccolis Angelini, R.M., Rotolo, C., Habib, W., Pollastro, S. and Faretra, F. (2007) Single nucleotide polymorphisms (SNPs) in *Botryotinia fuckeliana* genes involved in fungicide resistance. In: *Abstract Book of the 14th International Botrytis Symposium*. Cape Town, South Africa, 64 pp.

De Miccolis Angelini, R.M., Rotolo, C., Pollastro, S. and Faretra, F. (2008) Characterization of the cytochrome b gene in laboratory mutants of *Botryotinia fuckeliana* resistant to QoI fungicides. In: *Abstract Book of the 3rd Botrytis Genome Workshop*. Tenerife, Canary Islands, Spain, 51 pp.

De Miccolis Angelini, R.M., Rotolo, C., Pollastro, S. and Faretra, F. (2010a) Phenotypic and molecular characterization of fungicide-resistant field isolates of *Botryotinia fuckeliana* (*Botrytis cinerea*). In: *Abstract Book of the XV International Botrytis Symposium*. Càdiz, Spain, 46 pp.

De Miccolis Angelini, R.M., Habib, W., Rotolo, C., Pollastro, S. and Faretra, F. (2010b) Selection, characterization and genetic analysis of laboratory mutants of *Botryotinia fuckeliana* (*Botrytis cinerea*) resistant to the fungicide boscalid. *European Journal of Plant Pathology* 128, 185–199.

De Miccolis Angelini, R.M., Rotolo, C., Pollastro, S., Santomauro, A., Masiello, M. and Faretra, F. (2010c) Characterization of resistance to QoI fungicides in *Botryotinia fuckeliana* (*Botrytis cinerea*). *Journal of Plant Pathology* 92(S4), 79–80.

Elad, Y., Yunis, H. and Katan T. (1992) Multiple fungicide resistance to benzimidazoles, dicarboximides and diethofencarb in field isolates of *Botrytis cinerea* in Israel. *Plant Pathology* 41, 41–46.

Elad, Y, Williamson, B., Tudzynski, P. and Delen, N. (2004) *Botrytis: Biology, Pathology and Control.* Kluwer Academic Publishers, Dordrecht, The Netherlands.

Faretra, F. and Antonacci, E. (1987) Production of apothecia of *Botryotinia fuckeliana* (de Bary) Whetz. under controlled environmental conditions. *Phytopathologia Mediterranea* 26, 29–35.

Faretra, F. and Pollastro, S. (1991) Genetic basis of resistance to benzimidazole and dicarboximide fungicides in *Botryotinia fuckeliana* (*Botrytis cinerea*). *Mycological Research* 95, 943–951.

Faretra, F. and Pollastro, S. (1993a) Genetics of sexual compatibility and resistance to benzimidazole and dicarboximide fungicides in isolates of *Botryotinia fuckeliana* (*Botrytis cinerea*) from nine countries. *Plant Pathology* 42, 48–57.

Faretra, F. and Pollastro, S. (1993b) Isolation, characterization and genetic analysis of laboratory mutants of *Botryotinia fuckeliana* (*Botrytis cinerea*) resistant to the phenylpyrrole fungicide CGA 173506. *Mycological Research* 97, 620–624.

Faretra, F. and Pollastro, S. (1996) Genetic studies of the phytopathogenic fungus *Botryotinia fuckeliana* (*Botrytis cinerea*) by analysis of ordered tetrads. *Mycological Research* 100, 620–624.

Faretra, F., Antonacci, E. and Pollastro, S. (1988a) Improvement of the technique used for obtaining apothecia of *Botryotinia fuckeliana* (*Botrytis cinerea*) under controlled conditions. *Annali di Microbiologia* 38, 29–40.

Faretra, F., Antonacci, E. and Pollastro, S. (1988b) Sexual behaviour and mating system of *Botryotinia fuckeliana*, teleomorph of *Botrytis cinerea*. *Journal of General Microbiology* 134, 2543–2550.

Faretra, F., Pollastro, S. and Di Tonno, A.P. (1989) New natural variants of *Botryotinia fuckeliana* (*Botrytis cinerea*) coupling resistance to benzimidazoles to insensitivity towards the *N*-phenylcarbamate diethofencarb. *Phytopathologia Mediterranea* 28, 98–104.

Fernández-Ortuño, D., Tores, J.A., De Vicente, A. and Perez-Garcia, A. (2008) Mechanisms of resistance to Qol fungicides in phytopathogenic fungi. *International Microbiology* 11, 1–9.

Fillinger, S., Leroux, P., Auclair, C., Barreau, C., Al Hajj, C. and Debieu, D. (2008) Genetic analysis of fenhexamid-resistant field isolates of the phytopathogenic fungus *Botrytis cinerea*. *Antimicrobial Agents and Chemotherapy* 52, 3933–3940.

Fournier, E., Levis, C., Fortini, D., Giraud, T., Leroux, P. and Brygoo, Y. (2003) Characterization of Bc-*hch*, the *Botrytis cinerea* homolog of the *Neurospora crassa het-c* vegetative incompatibility locus, and its use as a population marker. *Mycologia* 95, 251–261.

Fritz, R., Lanen, C., Colas, V. and Leroux, P. (1997) Inhibition of methionine biosynthesis in *Botrytis cinerea* by the anilinopirimidine fungicide pyrimethanil. *Pesticide Science* 49, 40–46.

Fritz, R., Lanen, C., Chapeland-Leclerc, F. and Leroux, P. (2003) Effect of the anilinopyrimidine fungicide pyrimethanil on the cystathionine β-lyase of *Botrytis cinerea*. *Pesticide Biochemistry and Physiology* 77, 54–65.

Gisi, U. and Sierotzki, H. (2008) Molecular and genetic aspects of fungicide resistance in plant pathogens. In: Dehne, H.W., Gisi, U., Kuck, K.H., Russell, P.E. and Lyr, H. (eds) *Modern Fungicides and Antifungal Compounds V*, Proceedings of the 15th International Reinhardsbrunn Symposium, Friedrichroda, Germany. BCPC, DPG, Braunschweig, Germany, pp. 52–61.

Gisi, U., Sierotzki, H., Cook, A. and McCaffery, A. (2002) Mechanisms influencing the evolution of resistance to Qo inhibitor fungicides. *Pest Management Science* 58, 859–867.

Grasso, V., Sierotzki, H., Garibaldi, A. and Gisi, U. (2006) Characterization of the cytochrome b gene fragment of *Puccinia* species responsible for the binding site of QoI fungicides. *Pesticide Biochemistry and Physiology* 84, 72–82.

Grindle, M. (1987) Genetics of fungicide resistance. In: Ford, M.G., Hollomon, D.H., Khambay, B.P.S. and Sawiki, R.M. (eds) *Combating Resistance to Xenobiotics – Biological and Chemical Approaches*. Ellis Horwood, Chichester, UK, pp. 75–93.

Grindle, M. and Faretra, F. (1993) Genetic aspects of fungicide resistance. In: Lyr, H. and Polter, C. (eds) *Proceedings of the 10th International Symposium 'Modern Fungicides and Antifungal Compounds'*. Ulmer, Wollgrasweg, Germany, pp. 33–43.

Gullino, M.L., Leroux, P. and Smith, C.M. (2000) Uses and challenges of novel compounds for plant disease control. *Crop Protection* 19, 1–11.

Hilber, U.W. and Hilber-Bodmer, M. (1998) Genetic basis and monitoring of resistance of *Botryotinia fuckeliana* to anilinopyrimidines. *Plant Disease* 82, 496–500.

Hilber, U.W., Schüepp, H. and Scwinn, F.J. (1993) Resistance of *Botryotinia fuckeliana* (de Bary) Whetzel to phenylpyrrole fungicides as compared to dicarboximides. In: Lyr, H. and Polter, C. (eds) *Modern Fungicides and Antifungal Compounds*. Ulmer, Wollgrasweg, Germany, pp. 105–111.

Hilber, U.W., Schwinn, F.J. and Schüepp, H. (1995) Comparative resistance patterns of fludioxonil and vinclozolin in *Botryotinia fuckeliana*. *Journal of Phytopathology* 143, 423–428.

Ishii, H. (2006) Impact of fungicide resistance in plant pathogens on crop disease control and agricultural environment. *Japan Agricultural Research Quarterly* 40, 205–211.

Ishii, H., Fountaine, J., Chung, W.H., Kansago, M., Nishimura, K., Takahashi, K., *et al.* (2009) Characterization of QoI resistant field isolates of *Botrytis cinerea* from citrus and strawberry. *Pest Management Science* 65, 916–922.

Jiang, J., Ding, L., Michailides, T.J., Li, H. and Ma, Z. (2009) Molecular characterization of field azoxystrobin-resistant isolates of *Botrytis cinerea*. *Pesticide Biochemistry and Physiology* 93, 72–76.

Kretschmer, M., Leroch, M., Mosbach, A., Walker, A.S., Fillinger, S., Mernke, D., *et al.* (2009) Fungicide-driven evolution and molecular basis of multidrug resistance in field populations of the grey mould fungus *Botrytis cinerea*. *PLoS Pathogen* 5(12), e1000696.

Leroux, P. (2004) Chemical control of *Botrytis* and its resistance to chemical fungicides. In: Elad, Y., Williamson, B., Tudzynski, P. and Delen, N. (eds) *Botrytis: Biology, Pathology and Control*. Kluwer Academic Publishers, Dordrecht, The Netherlands, pp. 195–222.

Leroux, P., Chapeland, F., Debrosses, D. and Gredt, M. (1999) Patterns of cross-resistance in *Botryotinia fuckeliana* (*Botrytis cinerea*) isolates from French vineyards. *Crop Protection* 18, 687–697.

Leroux, P., Fritz, R., Debieu, D., Albertini, C., Lanen, C., Bach, J., *et al.* (2002) Mechanisms of resistance to fungicides in field strains of *Botrytis cinerea*. *Pest Management Science* 58, 876–888.

Leroux, P., Gredt, M., Leroch, M. and Walker, A.S. (2010) Exploring mechanisms of resistance to respiratory inhibitors in field strains of *Botrytis cinerea*, the causal agent of gray mold. *Applied and Environmental Microbiology* 76, 6615–6630.

Liu, W., Leroux, P. and Fillinger, S. (2008) The HOG1-like MAP kinase *Sak1* of *Botrytis cinerea* is negatively regulated by the upstream histidine kinase *Bos1* and is not involved in dicarboximide- and phenylpyrrole-resistance. *Fungal Genetics and Biology* 45, 1062–1074.

Luck, J.E. and Gillings, M.R. (1995) Rapid identification of benomyl resistant strains of *Botrytis cinerea* using the polymerase chain reaction. *Mycological Research* 99, 1483–1488.

Ma, Z. and Michailides, T.J. (2005) Advances in understanding molecular mechanisms of fungicide resistance and molecular detection of resistant genotypes in phytopathogenic fungi. *Crop Protection* 24, 853–863.

Ma, Z., Yan, L., Luo, Y. and Michailides, T.J. (2007) Sequence variation in the two-component histidine kinase gene of *Botrytis cinerea* associated with resistance to dicarboximide fungicides. *Pesticide Biochemistry and Physiology* 88, 300–306.

Nakazawa, Y. and Yamada, M. (1997) Chemical control of grey mould in Japan. A history of combating resistance. *Agrochemicals Japan* 71, 2–6.

Oshima, M., Fujimura, M., Banno, S., Hashimoto, C., Motoyama, T., Ichiishi, A., *et al.* (2002) A point mutation in the two-component histidine kinase BcOS-1 gene confers dicarboximide resistance in field isolates of *Botrytis cinerea*. *Phytopathology* 92, 75–80.

Oshima, M., Banno, S., Okada, K., Takeuchi, T., Kimura, M., Ichiishi, A., *et al.* (2006) Survey of mutations of a histidine kinase gene BcOS1 in dicarboximide resistant field isolates of *Botrytis cinerea*. *Journal of General Plant Pathology* 72, 65–73.

Pillonel, C. and Meyer, T. (1997) Effect of phenylpyrroles on glycerol accumulation and protein kinase activity of *Neurospora crassa*. *Pesticide Science* 49, 229–236.

Pollastro, S. and Faretra, F. (1992) Genetic characterization of *Botryotinia fuckeliana* (*Botrytis cinerea*) field isolates coupling high resistance to benzimidazoles to insensitivity toward the *N*-phenylcarbamate diethofencarb. *Phytopathologia Mediterranea* 31, 148–153.

Pollastro, S., Faretra, F., Santomauro, A., Miazzi, M. and Natale, P. (1996a) Studies on pleiotropic effects of mating type, benzimidazole-resistance and dicarboximide-resistance genes in near-isogenic strains of *Botryotinia fuckeliana* (*Botrytis cinerea*). *Phytopathologia Mediterranea* 35, 48–57.

Pollastro, S., Faretra, F., Di Canio, V. and De Guido, M.A. (1996b) Characterization and genetic analysis of field isolates of *Botryotinia fuckeliana* (*Botrytis cinerea*) resistant to dichlofluanid. *European Journal of Plant Pathology* 102, 607–613.

Rotolo, C. (2010) Molecular characterization of fungicide resistance and mating type in *Botryotinia fuckeliana*. PhD Thesis, University of Bari, Italy, pp. 198 [in Italian].

Santomauro, A., Pollastro, S., De Guido, M.A., De Miccolis Angelini, R.M., Natale, P. and Faretra, F. (2000) A long-term trial on the effectiveness of new fungicides against grey mould on grapevine and on their influence on the pathogen's population. In: *Abstract Book of the XII International Botrytis Symposium*. Reims, France, 75 pp.

Sierotzki, H. and Gisi, U. (2003) Molecular diagnostics for fungicide resistance in plant pathogens. In: Voss, G. and Ramos, G. (eds) *Chemistry of Crop Protection*. Wiley-VCH, Weinheim, Germany, pp. 71–88.

Sierotzki, H., Wullschleger, J., Alt, M., Bruyère, T., Pillonel, C., Parisi, S., *et al.* (2002) Potential mode of resistance to anilinopyrimidine fungicides. In: Dehne, H.W., Gisi, U., Juck, K.H., Russel, P.E. and Lyr, H. (eds) *Modern Fungicides and Antifungal Compounds III*. Agro Concept GmbH, Bonn, Germany, pp. 141–148.

Sierotzki, H., Frey, R. and Wullschleger, J. (2007) Cytochrome b gene sequence and structure of *Pyrenophora teres* and *P. tritici-repentis* and implications for QoI resistance. *Pest Management Science* 63, 225–233.

Stammler, G. (2008) Mode of action, biological performance and latest monitoring results of boscalid sensitivity. In: *Book of Abstracts of the 18th Symposium of Research Committee on Fungicide Resistance.* Matsue, Japan, pp. 30–43.

Stammler, G., Brix, H.D., Nave, B., Gold, R. and Schoefl, U. (2008) Studies on the biological performance of boscalid and its mode of action. In: Dehne H.W., Gisi, U., Kuck, K.H., Russell, P.E. and Lyr, H. (eds) *Modern Fungicides and Antifungal Compounds V,* Proceedings of the 15th International Reinhardsbrunn Symposium, Friedrichroda, Germany. BCPC, DPG, Braunschweig, Germany, pp. 45–51.

Suty, A., Pontzen, R. and Stenzel, K. (1999) Fenhexamid-sensitivity of *Botrytis cinerea*: determination of baseline sensitivity and assessment of the risk of resistance. *Pflanzenschutz-Nachrichten Bayer* 52, 145–157.

Tamura, H., Mizutani, A., Yukioka, H., Miki, N., Ohba, K. and Masuko, M. (1999) Effect of the methoxyiminoacetamide fungicide, SSF129, on respiratory activity in *Botrytis cinerea. Pesticide Science* 55, 681–686.

van den Bosch, F. and Gilligan, C.A. (2008) Models of fungicide resistance dynamics. *Annual Review of Phytopathology* 46, 123–147.

Verhoeff, K., Malathrakis, N.E. and Williamson, B. (1992) *Recent Advances in Botrytis Research.* Pudoc Scientific Publisher, Wageningen, The Netherlands.

Vermeulen, T., Schoonbeek, H. and De Waard, M.A. (2001) The ABC transporter BcatrB from *Botrytis cinerea* is a determinant of the activity of the phenylpyrrole fungicide fludioxonil. *Pest Management Science* 57, 393–402.

Viaud, M., Fillinger, S., Liu, W., Polepalli, J.S., Le Pêcheur, P., Kunduru, A.R., *et al.* (2006) A class III histidine kinase acts as a novel virulence factor in *Botrytis cinerea. Molecular Plant–Microbe Interactions* 19, 1042–1050.

Vignutelli, A., Hilber-Bodmer, M. and Hilber, U.W. (2002) Genetic analysis of resistance to the phenylpyrrole fludioxonil and the dicarboximide vinclozolin in *Botryotinia fuckeliana* (*Botrytis cinerea*). *Mycological Research* 106, 329–335.

Williamson, B., Tudzynski, B., Tudzynski, P. and Van Kan, J.A.L. (2007) *Botrytis cinerea*: the cause of grey mould disease. *Molecular Plant Pathology* 8, 561–580.

Yarden, O. and Katan, T. (1993) Mutations leading to substitutions at amino acids 198 and 200 of beta-tubulin that correlate with benomyl-resistance phenotypes of field strains of *Botrytis cinerea. Phytopathology* 83, 1478–1483.

Yoder, O.C., Valent, B. and Chumley, F. (1986). Genetic nomenclature and practice for plant pathogenic fungi. *Phytopathology* 76, 383–385.

20 Emergence of Multi-Drug Resistance in Fungal Pathogens: A Potential Threat to Fungicide Performance in Agriculture

Matthias Kretschmer

Michael Smith Laboratories, University of British Columbia, Vancouver, British Columbia, Canada

20.1 Introduction

Infections of crops with a fungal plant pathogen can have severe impacts on crop quality and/or crop yield (Oerke, 2006). As a result, whole harvests can be destroyed under pathogen-favouring conditions. Losses can occur during different production stages, respectively during plant growth, harvest, or during crop storage. Depending on the crops, average losses per year to plant pathogens range from 9.2% (cotton) to 29.4% (potato; Oerke, 2007). In commercial wine production, losses of up to US$2 billion/year are estimated for *Botrytis cinerea* alone (Vivier and Pretorius, 2002). Thus, it is of major interest to reduce the impact of these microorganisms during crop production. Biological control mechanisms such as the use of antagonistic microorganisms, creation of pathogen-unfavourable conditions, or breeding of resistant plant varieties are difficult and not always successful. Thus, plant pathogen control is still dependent on the use of chemical compounds with different modes of action.

Antifungal compounds had already been developed in the 19th century and were widely introduced into the market in the second half of the 20th century (Oerke, 2006). Because of the frequent use of compounds with similar modes of action, fungal pathogens easily developed a resistance against these compounds (Leroux *et al.*, 2002). This so-called 'target site resistance' is a modification in the amino acid sequence of the protein which interacts directly with the fungicide and leads to its insensitivity. For example, a modification in the ß-tubulin amino acid sequence at codons A198G or F200Y leads to an almost total resistance of *B. cinerea* against carbendazim (Leroux and Clerjeau, 1985; Park *et al.*, 1997). Similar modifications are responsible for the resistance of other fungi against benzidimazole fungicides (Ma and Michailides, 2005).

Because of fungicide resistance development against the older fungicides and the introduction of new fungicide classes, it was recommended to limit the use of one fungicide class to 1–2 treatments per vegetation season. Further, it was recommended to alternate fungicides with different modes of action (Staub, 1991). This should reduce the development of strains which are resistant against one specific fungicide, because they can be further controlled by other fungicides. The accumulation of multiple mutations leading to a build-up of several target site resistances in one strain was considered as unlikely, because mutations could lead to disturbed protein function, and thus to reduced fitness of resistant strains.

Cancer cells, bacteria such as *Mycobacterium tuberculosis* or fungi such as *Candida albicans* can develop a broad-range resistance against chemically unrelated compounds (Gulshan and Moye-Rowley, 2007; Zhang and Yew, 2009). This resistance mechanism is correlated with the increased export of toxic compounds by energy-dependent and low substrate-specific plasma membrane transporters. Such a resistance against multiple compounds is called multi-drug resistance (MDR). It was shown that in *C. albicans*, mutations in transcription factors were responsible for the overexpression of these efflux transporters (Morschhäuser *et al.*, 2007). The increased export of toxic compounds results in lower cytosolic concentrations and leads to higher tolerance of the cells against all exported compounds. In fungi, two exporter classes are mainly responsible for MDR phenotypes. ATP binding cassette transporters (ABCs) are energized by the hydrolysis of ATP, while major facilitator superfamily transporters (MFSs) use an ion gradient, often H+, for the export of toxic compounds. In *C. albicans*, it was shown that prolonged fluconazole treatment of patients leads to the selection of strains with reduced sensitivity, mainly against azole fungicides. This MDR phenotype is correlated with the constitutive overexpression of the ABC transporters CDR1 and CDR2, or the MFS transporter MDR1 (Coste *et al.*, 2004; Morschhäuser *et al.*, 2007).

Until recently, such a fungicide resistance mechanism was unknown for greenhouse or field populations of fungal plant pathogens. Today, the best-documented MDR in field populations is known from *B. cinerea*, a necrotrophic pathogen infecting more than 200 plant species, including important crops like strawberries, grapes or tomatoes (Kretschmer *et al.*, 2009). Other fungi, such as *Mycosphaerella graminicola* and *Penicillium digitatum*, also show evidence of the occurrence of MDR in field populations or during crop storage, respectively (Nakaune *et al.*, 1998; Roohparvar *et al.*, 2008). In this chapter, the involvement of ABC and MFS transporters in the export of toxic compounds and their involvement in MDR in different fungal plant pathogens are discussed. However, as MDR is best documented for *B. cinerea*, this chapter will focus mainly on this important pathogen.

20.2 Analysis of Efflux Transporters in Laboratory Strains

ABC transporters are important for basal fungicide sensitivity

It was first shown in the yeast *Saccharomyces cerevisiae* that the ABC transporter ScPDR5 was responsible for sensitivity against DMI (ergosterol demethylation inhibitor) fungicides, herbicides, antibiotics and other chemical compounds (Kolaczkowski *et al.*, 1998). In filamentous fungi, ABC transporters play a role in the defence against microbial toxins, plant defence compounds, chemicals and fungicides.

In *Aspergillus nidulans*, the ABC transporters AnAtrA, AnAtrB, AnAtrC and AnAtrD are well described. Expression of these genes was upregulated in response to the plant defence metabolites pisatin and reserpine, as well as azole fungicides or antibiotics like cycloheximid. Deletion mutants in *Aspergillus* of *anAtrB* and heterologous expression in *S. cerevisiae* indicated that this transporter was able to export cycloheximid, anilinopyrimidines, benzimidazoles, phenylpyrroles, azoles, strobilurins and the plant compounds camptothecin and resveratrol (Table 20.1). In addition, the disruption of *anAtrD* led to hypersensitivity to cycloheximid, cyclosporine, nigericin and valinomycin (De Waard *et al.*, 2006).

In *M. graminicola*, five ABC transporters have been described so far and analysed in laboratory strains. The induction of these transporters showed a distinct pattern for the yeast-like and the mycelial form of *M. graminicola*. Gene expression of *mgAtr1* was induced mainly after a treatment with cycloheximid, eugenol, fatty acids, cyclosporine, nystatin and cyproconazole in the yeast form, while it was induced only by eugenol and psoralen during mycelial growth. The expression of *mgAtr2*, *mgAtr4* and *mgAtr5* was induced by some of the compounds which also induced the expression of *mgAtr1*. However,

Table 20.1. Substrate spectra of ABC and MFS transporters identified in laboratory strains of *Aspergillus nidulans* and of the plant pathogens *Mycosphaerella graminicola, Magnaporthe grisea, Penicillium digitatum* and *Botrytis cinerea.*

Organism	Transporter	Fungicides	Antibiotics	Plant defence compounds	Other chemicals
ABC transporter					
A. nidulans[a]	AtrA (De Waard et al., 2006)				
	AtrB (De Waard et al., 2006)	Azole Anilinopyrimidin Benzimidazole Phenylpyrrole Phenylpyridylamine Strobilurin		Camptothecin Resveratrol	Acriflavine 4-NQO
	AtrC (De Waard et al., 2006)				
	ArtD (De Waard et al., 2006)		Cycloheximid Nigericin Valinomycin		Cyclosporin
M. graminicola[a]	Atr1 (De Waard et al., 2006)	Azole	Cycloheximid	Berberine Camptothecin	Sterols/fatty acids Diacetoxyscirpenol
	Atr2 (De Waard et al., 2006)	Azole			Sterols/fatty acids RhodaminG
	Atr3 (De Waard et al., 2006)				
	Atr4 (De Waard et al., 2006)	Azole		Berberine Camptothecin	Sterols/fatty acids Diacetoxyscirpenol RhodaminG
	Atr 5 (De Waard et al., 2006)			Berberine Camptothecin Resveratrol	
	Atr7 (Zwiers et al., 2007)				Iron homeostasis

Continued

Table 20.1. Continued.

Organism	Transporter	Fungicides	Antibiotics	Plant defence compounds	Other chemicals
M. grisea	ABC1 (Urban et al., 1999)			Camptothecin	
	ABC2 (Lee et al., 2005) ABC3 (Sun et al., 2006)	Azole	Cycloheximid Valinomycin ActinomycinD		
	ABC4 (Gupta and Chattoo, 2008)	Azole	Cycloheximid	Resveratrol	
P. digitatum	PMR5 (Nakaune et al., 2002)	Benzimidazole Azole Quinone		Resveratrol Camptothecin	
B. cinerea[a]	AtrB (De Waard et al., 2006)	Phenylpyrrole		Eugenol Resveratrol	
	AtrD (De Waard et al., 2006)	Azole			
	BMR1 (De Waard et al., 2006)	Iprobenfos	Polyoxin		
MFS transporter					
B. cinerea	Mfs1 (Hayashi et al., 2002)	Azole Anilinopyrimidins Dicarboximide Phthalimide Phenylpyridylamine		Camptothecin	Cercosprorin
M. graminicola	Mfs1 (Roohparvar et al., 2007a)	Azole Allylamine Dicarboximide Phenylpyrrole Strobilurin	Cycloheximid	Berberine Camptothecin	Cercosprorin Sterols/fatty acids RhodaminG Diacetoxyscirpenol

Note: [a]Reviewed in De Waard et al. (2006).

distinguished differences were found for all transporters and for the yeast and the mycelial form of *M. graminicola*. In contrast to the other transporters, the expression of *mgAtr3* was not induced by any tested compound. Heterologous expression of MgAtr1, MgAtr2, MgAtr4 and MgAtr5 in *S. cerevisiae* showed distinct, although somewhat overlapping, export spectra for these transporters (Table 20.1). MgAtr1 showed the broadest transport spectrum, including fungicides, steroids, plant metabolites, antibiotics and other chemicals, followed by MgAtr2 and MgAtr4, while MgAtr5 only showed export activity for some plant metabolites like berberin or camptothecin (De Waard *et al.*, 2006). Recently, MgAtr7 was identified and characterized. It is a rather unique ABC transporter, which most likely is not involved in MDR but is connected to iron homeostasis (Zwiers *et al.*, 2007).

In *Magnaporthe grisea*, several ABC transporters like MgABC1, MgABC3 and MgABC4 were identified as important during the infection of rice plants. *MgABC1* showed an induction of gene expression after treatment with azole fungicides, the phytoalexin sakuranetin and hygromycinB. However, its deletion had no effect on sensitivity against these compounds (Urban *et al.*, 1999). *MgABC3* played a major role in defence against oxidative stress and an increased sensitivity of the deletion mutants was observed in the presence of valinomycin and actinomycinD (Sun *et al.*, 2006). The ABC transporters MgABC2 and MgABC4 showed a distinct MDR phenotype. The deletion mutants of *mgABC4* were more sensitive against resveratrol, miconazole and cycloheximid (Gupta and Chattoo, 2008). MgABC2 was not essential to the infection process of *M. grisea*, but it played a role in the export of different toxic compounds (Table 20.1). Its expression was upregulated by fungicides, mainly DMIs, antibiotics and plant defence compounds. Deletion mutants showed increased sensitivity to some DMI fungicides, camptothecin and cycloheximid (Lee *et al.*, 2005).

The ABC transporter *pdPMR5* from *P. digitatum* was induced after treatment with plant defence compounds like resveratrol, phloretin, camptothecin or fungicides such as dithianone and benzimidazoles. Deletion mutants in sensitive laboratory and DMI-resistant field strains showed that PdPMR5 played no role in DMI resistance, but had a function as baseline defence against the toxic compounds camptothecin, resveratrol, dithianone and benzimidazoles (Table 20.1; Nakaune *et al.*, 2002).

The best-characterized ABC transporters of *B. cinerea* are BcatrB, BcatrD and BcBMR1. *BcatrB* is induced after treatment with antibiotics, phytoalexins, phenylpyrroles, anilinopyrimidines, dicarboximide, DMIs and fenhexamid. Laboratory deletion mutants showed an increased sensitivity against fludioxonil, fenpicolin, resveratrol and eugenol (Schoonbeek *et al.*, 2001, 2003; Vermeulen *et al.*, 2001). *BcatrD* is induced mainly by azole fungicides, but also dicarboximides and benzimidazoles have led to its expression. Deletion mutants showed reduced sensitivity against azole fungicides (Table 20.1). BcBMR1 is involved in the export of the antibiotic polyoxin and the organophosphorus fungicide iprobenfos (De Waard *et al.*, 2006).

These studies indicate that in filamentous fungi, ABC transporters are responsible for a basal defence against toxic compounds, even in sensitive laboratory strains.

MFS transporters are important for basal fungicide sensitivity

MFS transporters are the largest group of transport proteins known so far. They play important roles for the uptake of nutrients, minerals and ions, but they are also responsible for the export of toxins, produced either by the fungus itself or by other competing microorganisms. However, there are only a few examples where MFS transporters in plant pathogens have been shown to be involved in the export of fungicides in an MDR-dependent manner.

The first described fungal MFS transporter with MDR properties was BcMFS1 from *B. cinerea*. Expression of *bcMFS1* was induced by azole fungicides. Deletion mutants of *bcMFS1* in laboratory strains showed an increased sensitivity to camptothecin and cercosporin, while no difference was observed between the mutant and the parental strain in the presence of fungicides. However, when *bcMFS1* was overexpressed in a laboratory strain, it was shown that fungicides of the DMI, anilinopyrimidine, dicarboximide, phthalimide and phenylpyridylamine classes were exported (Table 20.1). Thus, BcMFS1 is the first MFS MDR related transporter of a fungal plant pathogen (Hayashi *et al.*, 2002).

Recently, another MFS transporter from *M. graminicola*, MgMfs1, was identified and characterized as an MDR transporter. Heterologous expression of *mgMfs1* in *S. cerevisiae* showed export activities for azole, strobilurin, phenylpyrrole and dicarboximide fungicides. Antibiotics, toxins, plant alkaloids and sterols were also exported (Table 20.1). Disruption of *mgMfs1* in *M. graminicola* laboratory strains resulted in an increased sensitivity against strobilurin fungicides and cercosporin, showing that this transporter had a function as a baseline defence against these compounds (Roohparvar *et al.*, 2007a).

So far, these two MFS transporters are the only identified transporters which play a role in baseline defence against fungicides in filamentous plant pathogenic fungi.

20.3 Multi-Drug Resistance in Field Strains of Filamentous Plant Pathogenic Fungi

MDR in different plant pathogenic fungi

Nakaune *et al.* (1998) first reported MDR in *P. digitatum* strains isolated from citrus fruit which were resistant against multiple toxic compounds, like azole fungicides, antibiotics and other chemicals. But, only three strains were isolated and analysed and no further study was conducted to clarify the occurrence and the frequency of these strains during production or storage of citrus fruit. Thus, it is unclear if MDR is common in *P. digitatum*, if it has a major effect on fungicide performance and if it is relevant for crop production.

Similar to the situation in *P. digitatum*, it was reported that an MDR mechanism could play a role in strobilurin fungicide resistance in *M. graminicola* field isolates. However, in *M. graminicola* a target site resistance against strobilurins is far more common; thus, it is unknown how often the MDR phenotype occurs, which can only be determined by gene expression or fungicide efflux analysis, or if it has any effect on fungicide performance (Roohparvar *et al.*, 2008). Recently, *M. graminicola* MDR field strains from France and the UK were identified with an increased resistance against mainly azole fungicides, but also sensitivity against tolnaftate, terbinafin and boscalid was affected. In France, 13% of the collected strains showed this MDR phenotype (Leroux and Walker, 2011). The involved efflux transporter has not been identified but, because mainly azole fungicides are affected, the PDR5 homologue known from *S. cerevisiae* could be responsible.

The occurrence, frequency and migration of MDR field strains with different MDR phenotypes is best documented for the necrotrophic plant pathogen *B. cinerea* during grape production in France (Champagne) and Germany (Freiburg i. Br. and Palatine). In a long-term fungicide resistance monitoring programme conducted in Champagne since the early 1990s, three different MDR phenotypes were identified. Initially, two of them were named AniR2 and AniR3 because of their resistance against anilinopyrimidines (Chapeland *et al.*, 1999). But, they also showed increased resistance against other fungicides, antibiotics and plant defence compounds (Table 20.2). Thus, their new designation is MDR1 and MDR2. MDR3 strains occurred later and represented a new MDR phenotype. At the beginning of the study, only low frequencies of MDR1 and MDR2 strains were observed. But, with the introduction of new antifungal compounds like phenylpyrroles, anilinopyrimidines and fenhexamid, the frequency of these strains started to increase and, several years later, MDR3 strains were also observed. Today, around 60% of the total *B. cinerea* population in Champagne shows one of the three MDR phenotypes (Fig. 20.1). Further evidence was provided that the MDR2 phenotype originated in Champagne and migrated within a few years to Germany (Kretschmer *et al.*, 2009).

In 2004, in a fungicide resistance monitoring programme conducted in vineyards in

Table 20.2. Drug sensitivity of *B. cinerea* MDR field strains, respectively their transporter and transcription factor deletion mutants (Kretschmer *et al*, 2009; modified).

	Fludioxonil	Fenhexamid	Cyprodinil	Carbendazim	Boscalid	Iprodione	Tebuconazole	Tolnaftate[a]	Cycloheximid	Camalexin[b]	Eugenol
Sensitive strains (EC$_{50}$; mg/l)	0.03±0.01	0.05±0.01	0.006±0.001	0.04±0.01	0.08±0.01	1.03±0.08	0.66±0.31	0.65±0.08	3.2±0.68	8.7±0.34	81.6±4.15
Resistance factors (× fold of sensitivity)											
MDR1 strains	8.1×±1.3**	1.6×±0.5 n.s.	18.2×±4.5*	2.8×±0.2***	1.6×±0.2 n.s.	1.4×±0.3 n.s.	0.7×±0.4 n.s.	20.4×±2.3***	0.7×±0.1 n.s.	1.3×±0.1*	1.1×±0.0*
ΔBcatrB	0.4×±0.0***	0.5×±0.2**	0.8×±0.0**	1.1×±0.2***	1.8×±0.3 n.s.	0.5×±0.1*	0.7×±0.1 n.s.	0.8×±0.0***	0.9×±0.1 n.s.	0.7×±0.0*	0.8×±0.0*
ΔBcmrr1	0.5×±0.1***	0.9×±0.1***	0.7×±0.2**	n.t.	n.t.	n.t.	n.t.	1.0×±0.2***	n.t.	n.t.	n.t.
MDR2 strains	2.6×±0.3***	9.8×±0.6***	6.2×±2.8**	1.1×±0.1 n.s.	2.0×±0.8*	5.4×±1.0***	1.8×±0.2*	>25×***	13.7×±1.5**	n.t.	1.0×±0.0 n.s.
ΔBcmfsM2	1.1×±0.1***	1.3×±0.3***	1.4×±1.3***	n.t.	1.0×±0.2***	1.5×±0.4***	1.1×±0.6***	2.2×±0.9***	1.4×±0.4***	n.t.	n.t.
MDR3 strains	11.4×±1.9***	14.7×±3.2**	25.7×±5.2**	3.1×±0.5**	3.5×±0.9*	6.4×±1.0**	2.3×±0.6*	>25×***	14.1×±0.3***	n.t.	1.1×±0.0*

Notes: Drug sensitivity values from sensitive strains, MDR1 strains, MDR2 strains, MDR3 strains are shown. Further, the values from MDR1 transporter (BcatrB), transcription factor (Bcmrr1) and MDR2-derived transporter (BcmfsM2) deletion mutants are shown. For sensitive strains, EC$_{50}$ values, for MDR strains and mutants, resistance factors relative to the corresponding values of sensitive strains are shown. [a]Due to limited solubility of tolnaftate, accurate values above 25-fold could not be determined. For MDR field strain, significant differences to mean values of sensitive strains are indicated. In the case of deletion mutants, the significance corresponds to values of the parental strains: n.s., not significant; *p < 0.05; **p < 0.01; ***p < 0.001; n.t., not tested. [b]Stefanato *et al*. (2009).

(a)

(b)

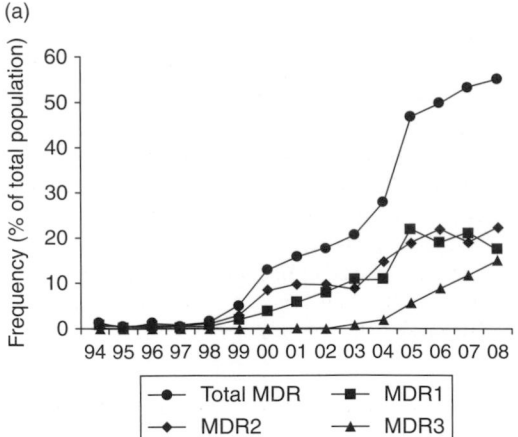

	MDR1 (%)	MDR2 (%)	MDR3 (%)	Total MDR (%)
2006	17	4	2	23
2007	31	1	1	33
2008	28	8	2	38
2009	20	4	7	32

Fig. 20.1. Isolation frequencies of *B. cinerea* MDR strains from French and German wine growing regions (Kretschmer *et al.*, 2009; modified). (a) Appearance of MDR strains in Champagne, France. (b) Frequency of MDR strains in the German Wine Route region.

Freiburg i. Br. (Germany), MDR1 strains were identified, with a frequency of 1% (Kretschmer and Hahn, 2008). Following up this study between 2006 and 2009, a similar investigation was conducted in the Palatine area along the German Wine Route. MDR1 strains were found frequently, in around 20% of the population in 2006 and 2009, reaching a peak of 31% in 2007 (Fig. 20.1). In Champagne, both MDR1 and MDR2 phenotypes could be found with similar frequencies. In contrast, only low frequencies of MDR2 strains were found in Germany, ranging from 0.7% (2007) to 8.0% (2008). MDR3 strains were also present in Germany, but only with low frequency, ranging from 0.7% (2007) to 7.3% (2009). However, these two MDR phenotypes have showed a trend of increased frequencies lately (Kretschmer *et al.*, 2009).

Taken together, MDR phenotypes have evolved in at least three different plant pathogenic filamentous fungi in response to fungicide treatment under field conditions.

Fungicide resistance level and spectra of MDR strains

In *P. digitatum*, the ABC transporter PdPMR1 was identified as a homologue of ScPDR5. It was shown that in DMI-resistant strains, this transporter was upregulated constitutively compared to sensitive strains. DMI-resistant *P. digitatum* strains showed resistance factors ranging from 11.5× for fenarimol to over 85× for pyrifenox based on EC_{50} values. However, increased resistance levels were also observed for the antibiotic cycloheximid (3.1×), oligomycin (4.5×), the plant toxin phloretin (1.5×) and the chemical compounds 4NQO (4.6×) and acriflavin (1.9×; Nakaune *et al.*, 1998, 2002). This indicated that an MDR mechanism was responsible for the resistance against these toxic compounds. Deletion of *pdPMR1* in DMI-resistant strains resulted in increased sensitivity of the mutants for all DMI fungicides tested. However, fungicide resistance was only completely abolished for fenarimol and bitertanol, resulting in EC_{50} values similar to those of sensitive strains. For trifluminzole and pyrifenox, fungicide sensitivity was increased, but did not reach the levels of sensitive strains, thus indicating, at least for these compounds, further resistance mechanisms (Nakaune *et al.*, 1998). Export via other ABC or MFS transporters which could be co-regulated with *pdPMR1* could be one explanation for these results. This would indicate a mutation in a regulatory protein, leading to upregulation of several transporters, which was further supported by the finding that neither the amino acid sequence of PdPMR1 nor its promoter was altered in DMI-resistant isolates compared to sensitive strains (Hamamoto *et al.*, 2001). This study showed for the first time that, also in filamentous plant pathogenic fungi, a DMI resistance of field strains could be caused by an ABC transporter.

In *M. graminicola*, the MFS transporter MgMfs1 was identified as a broad-range exporter of toxic compounds of all major classes, including fungicides (including strobilurin fungicides), antibiotics, toxins and plant alkaloids. This study was conducted with sensitive laboratory strains (Roohparvar *et al.*, 2007a). In addition, this transporter was strongly upregulated in field strains resistant to the strobilurin fungicide trifloxistrobin (Roohparvar *et al.*, 2008). Further, resistant strains showed increased export of this fungicide compared to sensitive strains. The main resistance mechanism against strobilurin fungicides in these field strains, however, is a target site mutation in the cytochrome *b* gene, leading to the amino acid substitution G143A. All tested resistant strains showed this target site mutation and a replacement of the target site-resistant allele with a sensitive allele of cytochrome *b* was not undertaken. Thus, it is impossible to determine to which degree MgMfs1 is responsible for the resistance against strobilurin fungicides.

As described above, three different MDR phenotypes with overlapping fungicide-resistance spectra were identified in *B. cinerea* field strains. MDR1 strains showed increased resistance against fludioxonil, cyprodinil, carbendazim and tolnaftate, but also against the plant compounds eugenol and camalexin (Kretschmer *et al.*, 2009; Stefanato *et al.*, 2009; Kretschmer, unpublished). The resistance factors for fungicides range from 2.8× for carbendazim to 20.4× for tolnaftate (Table 20.2).

MDR2 strains are more resistant against fludioxonil, fenhexamid, cyprodinil, iprodion, tolnaftate, boscalid, tebuconazole and the antibiotic cycloheximid. Resistance factors range from 1.8× for tebuconazole to over 25× for tolnaftate. MDR3 strains showed the combined fungicide resistance of MDR1 and MDR2 strains and thus the broadest resistance spectra. However, for the fungicides fludioxonil, fenhexamid and cyprodinil, these

strains showed higher resistance factors than known from MDR1 and MDR2 strains. Overall, the resistance factors for MDR3 strains range from 2.3× for tebuconazole to over 25× for tolnaftate (Table 20.2).

Depending on the organism and the fungicide classes involved in the MDR phenotype, the resistance spectra can be narrow to broad and the resistance level can be considered as low to high. Thus, fungicide performance in the field could be affected negatively by MDR phenomena.

BcatrB and BcmfsM2 are responsible for the three MDR phenotypes in *B. cinerea*

It was found that MDR1 and MDR3 strains showed a reduced uptake of C-14 labelled fludioxonil compared to sensitive or MDR2 strains. Sensitive and MDR2 strains showed a transient uptake of the fungicide, reaching a peak after 30 min. Afterwards, the cytosolic fungicide concentration decreased, because of activation of efflux systems, and reached a basal level after 60 min. In contrast, MDR1 and MDR3 strains showed no initial uptake and remained at a basal level. With the addition of an uncoupler of ATP synthesis, it was shown that this reduced uptake in the case of MDR1 and MDR3 strains and the transient uptake in the case of sensitive strains were correlated with fungicide export. Similar results were obtained for MDR2 and MDR3 strains tested with C-14 triadimenol and C-14 bitertanol compared to sensitive and MDR1 strains (Chapeland *et al.*, 1999; Kretschmer *et al.*, 2009). Taken together, these results are evidence of an involvement of constitutively expressed and energy-dependent export systems in all three MDR phenotypes.

Two approaches were followed to identify the responsible efflux system. A macroarray analysis identified the ABC transporter *bcatrB* as constitutively highly expressed in MDR1 and MDR3 compared to sensitive strains. The overexpression of *bcatrB* was 50-fold for MDR1 and 150-fold for MDR3 strains, respectively. However, sensitive strains also showed an induction of *bcatrB* after treatment with fludioxonil (Kretschmer *et al.*, 2009). In sensitive strains, the expression of *bcatrB* can be induced by a variety of toxic compounds like fludioxonil, cyprodinil, iprodion, fenhexamid, tebuconazole and cycloheximid (Schoonbeek *et al.*, 2001; Vermeulen *et al.*, 2001; Kretschmer, unpublished). It was also found that other ABC transporters (*bcatrK* and *bcBMR3*) were co-regulated with *bcatrB*. However, the upregulation of these was only 2.5- to 5-fold. Interestingly, these genes were also induced by a fludioxonil treatment in sensitive strains. The induction was 10-fold (*bcatrK*) and 100-fold (*bcBMR3*). This indicates a common transcription activation network for these three ABC transporters.

With a microarray analysis, a single gene encoding a novel MFS transporter (BcmfsM2) was found to be 70.5-fold overexpressed in MDR2 strains. *BcmfsM2* was also constitutively highly expressed in MDR3 strains (Kretschmer *et al.*, 2009). The expression of *bcmfsM2* in sensitive strains was not inducible by fenhexamid, cyprodinil, iprodion, cycloheximid, boscalid, tebuconazole, carbendazim or hydrogen peroxide. Thus, compounds which can induce *bcmfsM2* expression and its biological function in sensitive strains are still unknown.

Taken together, MDR1 strains show a strong constitutive overexpression of *bcatrB*, while MDR2 strains show a strong constitutive overexpression of *bcmfsM2*. MDR3 strains show the combined fungicide resistance spectra of MDR1 and MDR2 strains occurred in the field several years after MDR1 and MDR2 strains were found with increasing frequency and showed the overexpression of both efflux transporters *bcatrB* and *bcmfsM2*. They are the results of natural crosses between MDR1 and MDR2 strains or mutations leading to the MDR1 and MDR2 phenotypes occurring in the same isolate.

To establish the causal relationship between overexpression of efflux transporters and the MDR phenotypes, the ABC transporter *bcatrB* was deleted in MDR1 strains and the MFS transporter *bcmfsM2* in MDR2 strains, respectively. MDR1 deletion mutants showed a strong uptake of C-14 fludioxonil compared to the parental strains and this uptake was not transient, as shown for sensitive strains. Even after 2 h, fungicide uptake remained on a high level. Thus,

the export of fludioxonil was abolished completely in the MDR1 deletion mutants. Deletion mutants of *bcmfsM2* in MDR2 strains showed a high initial uptake of C-14 bitertanol compared to the parental strain but, in contrast to MDR1 mutants tested for fludioxonil uptake, this was only transient, probably because of the induction of *bcatrD* by the DMI fungicide bitertanol (Kretschmer *et al.*, 2009).

The fungicide resistance levels of the MDR1 *bcatrB* deletion mutants showed sensitivity levels similar to sensitive strains (Table 20.2). However, the mutants even showed hypersensitivity against the fungicides fludioxonil and iprodion and against the plant compounds eugenol and camalexin. Thus, BcatrB is the main reason for the MDR1 phenotype.

MDR2 *bcmfsM2* deletion and overexpression mutants confirmed the correlation of this MFS transporter with the MDR2 phenotype. The deletion mutants lost the MDR2 phenotype and reached the fungicide sensitivity levels of sensitive strains and no hypersensitivity for any of the compounds tested was observed (Table 20.2). In contrast, overexpression of *bcmfsM2* in a sensitive strain led to the MDR2 phenotype with similar resistance factors known from natural occurring MDR2 strains. Thus, BcmfsM2 is the single cause for the MDR2 phenotype (Kretschmer *et al.*, 2009).

Several possible mutations could cause a constitutive gene overexpression. This includes gene duplication, mutations in the promoter region or mutations in genes controlling the expression of the overexpressed genes. With southern blot hybridizations, gene duplication was excluded as the cause for the overexpression of *bcatrB* and *bcmfsM2*. Sequencing of the promoter region of *bcatrB* of sensitive, MDR1 and MDR3 strains also did not reveal mutations which were correlated with the MDR phenotypes. Further, because of the upregulation of other ABC transporters in MDR1 and MDR3 strains, it was concluded that mutations in a regulator of gene expression were responsible for the MDR1 and MDR3 phenotype. A map-based cloning approach identified the transcription factor Bcmrr1 as the main cause for the upregulation of *bcatrB* expression, and thus for the MDR1 and MDR3 phenotype. *Bcmrr1* expression was not elevated in MDR1 or MDR3 strains compared to sensitive strains. Thus, the cause for the overexpression of *bcatrB* had to be found in the protein itself. Sequencing of *bcmrr1* of sensitive, MDR1 and MDR3 strains revealed several point mutations in MDR1 and MDR3 strains, leading to amino acid exchanges in Bcmrr1. None of these modifications was found in sensitive strains. This indicates that structural modifications in Bcmrr1 are responsible for the *bcatrB* overexpression. Deletion of *bcmrr1* in MDR1 strains resulted in reduced expression of *bcatrB, bcatrK* and *bcBMR3* compared to MDR1 strains, and also in reduced induction of these genes in the presence of fungicides. In contrast, the transfer of a mutated version of a *bcmrr1* allele into a sensitive strain resulted in constitutive expression of these transporters. The fungicide resistance of MDR1 *bcmrr1* deletion mutants was abolished for all tested fungicides (Table 20.2). Further sensitive strains harbouring a mutated version of *bcmrr1* showed the MDR1 phenotype. Thus, mutations in Bcmrr1 are the cause for the MDR1 and MDR3 phenotype (Kretschmer *et al.*, 2009).

Already, the approach to amplify the promoter region of *bcmfsM2* from sensitive, MDR2 and MDR3 strains revealed a variation in the fragment size of sensitive compared to MDR2 and MDR3 strains, indicating a variation in the promoter of MDR2 and MDR3 strains. Sequencing of promoter fragments from sensitive, MDR2 and MDR3 strains identified transposable element sequences which were integrated in the promoter regions of *bcmfsM2* in MDR2 and MDR3 strains. MDR2-promoter Gus fusions showed that modified promoters of MDR2 and MDR3 strains harbouring the transposable element were able to promote a strong gene expression compared to the natural promoter of sensitive isolates (Kretschmer *et al.*, 2009). Thus, the promoter rearrangement caused by the integration of transposable element sequences in MDR2 and MDR3 strains led to a high expression of *bcmfsM2* and to the MDR2 and MDR3 phenotype.

20.4 MDR Strains are Competitive and are Selected by Fungicide Treatment

It is known that fungicide resistance can lead to reduced pathogen fitness (Anderson, 2005). Thus, it is important to determine the fitness of fungicide-resistant pathogens to estimate the risk of further distribution and persistence of these strains in the natural population, even under non-selective conditions. If fungicide resistance is not correlated with reduced fitness, these strains will spread within the population and result in reduced fungicide performance when the fungicides are applied. However, nothing is known about the fitness of MDR strains in *P. digitatum, M. graminicola* or *B. cinerea*.

Under laboratory conditions, none of the three MDR phenotypes of *B. cinerea* showed reduced fitness compared to sensitive strains, when growth was determined on different artificial media (complete or minimal). Also, temperature sensitivity, osmotic stress, oxidative stress, biomass production and sporulation of the MDR strains were not affected negatively by MDR compared to sensitive strains. When virulence was tested on different host plants or tissues, no major differences were found between MDR and sensitive strains (Kretschmer and Hahn, 2008; Kretschmer, unpublished). Taken together, the different MDR phenotypes did not show a reduced fitness compared to sensitive strains under any tested laboratory condition. Thus, it was suggested that even under non-selective conditions, these strains would persist in the natural population in the field and could further be selected under fungicide selection pressure.

In a field competitiveness experiment with a sensitive and an MDR3 strain, it was found that a single fungicide application with the fungicide switch®, a mixture of fludioxonil and cyprodinil, at early grape berry development led to a higher recovery rate of the MDR3 strain compared to the sensitive strain at harvest. But also under non-selective conditions, the MDR3 strains were highly competitive compared to the sensitive strains. When the recovery rate of the MDR3 strain was compared with the isolation rate of the endogenous population, it was found that the MDR3 strain was competitive even compared to the endogenous population that was already established in the vineyard under both selective and non-selective conditions (Kretschmer *et al.*, 2009). Thus, MDR3 strains as a combination of MDR1 and MDR2 phenotypes are highly competitive under non-selective conditions compared to introduced sensitive strains, or even when compared to the endogenous population. Further, one single fungicide application is sufficient to create a selection pressure to select for MDR3 strains under field conditions. It was found that even after a winter period without fungicide selection pressure and harsh climatic conditions, the MDR3 strain was frequently found compared to the introduced sensitive strain or the endogenous population in fungicide treated, but also in untreated, vineyards. Thus, MDR strains in general show a high degree of competitiveness without fungicide selection pressure. In the case of a fungicide application, these strains are strongly selected. This explains also the drastic increased frequency of strains with MDR phenotypes observed in the Champagne and Palatine areas after the recent introduction of fludioxonil, cyprodinil and fenhexamid in the market and their altered use.

20.5   Increased Fungicide Performance to Control MDR Strains

It is unknown if further evolutionary processes will lead to mutations resulting in higher expression of the efflux transporters *bcatrB* and *bcmfsM2* in *B. cinerea* field strains, and thus leading to higher fungicide resistance of the MDR strains causing problems with fungicide performance. From human medicine, it is known that modulators of transporter activity are able to reduce fungicide export and thus increase in a synergistic way the toxicity of fungicides. Transporter activity modulators are mainly effective against the ABC transporter class. Roohparvar *et al.* (2007b) showed that different ABC transporter modulators like amitriptyline, loperamide and promazine were able to protect wheat seedlings against *M. graminicola*, even without fungicide application. It was suggested that these modulators were interfering with ABC transporters important to the infection process. Thus, these inhibitors alone could already be used as a new class of fungicides.

Hayashi *et al.* (2003) showed that a laboratory strain of *B. cinerea* overexpressing the ABC transporter *bcatrD* could be sensitized against the DMI fungicide oxpoconazole by adding chlorpromazine or tacrolimus. The synergy effect was determined as 3.1-fold for the combination of chlorpromazine and the fungicide oxpoconazole. Further, it was shown that chlorpromazine was able to inhibit the export of C-14 oxpoconazole, leading to higher intracellular fungicide concentrations.

The MDR1 phenotype of *B. cinerea* is caused by the overexpression of *bcatrB*; thus, 11 different putative ABC transporter modulators have been tested for their ability to reduce the MDR phenotype. *In vitro*, the modulator chlorpromazine showed the highest synergistic effect and reduced the resistance against fludioxonil of the tested MDR1 strains by a factor of 4.9 (Kretschmer, unpublished). It was also observed that chlorpromazine was able to inhibit the export of C-14 fludioxonil, leading to higher intracellular fungicide concentrations (Kretschmer, unpublished).

However, most of these inhibitors are unspecific, affect many organisms and have to be used in high concentrations to reduce fungicide resistance. This would result in high environmental pollution. Further, they are used mainly in human medicine and thus cannot be used for agricultural purposes.

20.6   Perspective

The emergence of MDR in field populations of filamentous plant pathogenic fungi is a new and threatening development for fungicide performance. Until now, at least three different plant pathogenic fungi have shown evidence of MDR development against different classes of fungicides. Hence, it is most likely that other fungi are also able to develop adequate fungicide resistance mechanisms under the right selection pressure. Instead of a high-level resistance against one fungicide class, the altered use of fungicides with different modes of action favours the selection of MDR strains in *B. cinerea* vineyard populations (Fig. 20.2; Kretschmer *et al.*, 2009). However, the resistance factors normally are low to medium and do not reach the levels known from target site resistant strains (Table 20.2). Thus, it was considered that MDR strains could still be controlled by a regular fungicide treatment and that they did not affect fungicide performance. None the less, it has been shown recently that MDR1 strains of *B. cinerea* lead to partial reduction of fungicide performance and to increased disease development under field conditions (Leroch, unpublished). In the last years, a further disturbing development has been observed. By far, MDR strains more often show additional target site resistances against other frequently used fungicides compared to non-MDR strains with multiple target site resistances. For example, *B. cinerea* MDR strains often show an additional target site resistance against anilinopyrimidine fungicides (Leroch *et al.*, 2011). This development leads to a higher basal resistance against fungicides, for example for phenylpyrroles, where no target site resistance can be acquired under field conditions, most likely because of a severe fitness defect of resistant strains. Further, the additional target site resistances against anilinopyrimidines, fenhexamid, iprodion or benomyl lead to a reduced or total loss of pathogen control for these fungicides. This development is known for vineyard populations of *B. cinerea*, but recently it has also been shown for other crops (Leroch *et al.*, 2011; Leroch, unpublished). This combination of target site mutations and MDR mechanisms can

also be found in *M. graminicola* in the case of resistance against DMI or strobilurin fungicides (Roohparvar *et al.*, 2008; Leroux and Walker, 2011).

The performance of non-target fungicides can also be affected by MDR, and possibly MDR strains are selected by them. Carboxin fungicides usually are not used against *B. cinerea*, but MDR strains are also more resistant against these fungicides (Leroux *et al.*, 2010). Until now, it has not been known if these compounds lead to a selection of *B. cinerea* MDR strains. However, it shows that MDR mechanisms are also effective against other toxic compounds which are not intended to control *B. cinerea*. Thus, MDR mechanisms possibly could also interfere with the development of new fungicides, because it is impossible to predict if new compounds are substrates for the MDR efflux transporters.

For *B. cinerea*, a combination of an MDR phenotype and a resistance against the recently introduced carboxamide boscalid has been most uncommon, until now. Boscalid clearly favours the selection of a target site resistance, even if it is exported by BcmfsM2 (Fig. 20.2; Kretschmer *et al.*, 2009; Leroux *et al.*, 2010; Leroch *et al.*, 2011). An alteration or combination of boscalid with MDR-selecting fungicides like fludioxonil, cyprodinil or fenhexamid could delay the development of MDR. It is also possible to supplement fungicides with transporter inhibitors to modulate the export activity of the fungicide exporters. Although this is successful under *in vitro* conditions, it is most unlikely that such an attempt would be successful *in vivo* under

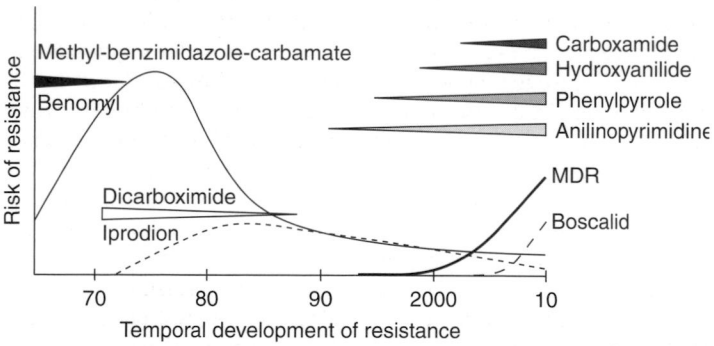

Fig. 20.2. Schematic representation of fungicide resistance development in *B. cinerea*.

field conditions because of the nature of these inhibitors.

The nature of MDR in plant pathogenic fungi is the constitutive overexpression of one or multiple exporters of the ABC or MFS transporter classes caused by different mutations; for example, in transcription factors or promoter regions (Fig. 20.3; Kretschmer *et al.*, 2009). These MDR strains are selected by the actual fungicide treatment regime and are able to migrate over far distances (Fig. 20.3). Until now, the resulting fungicide resistance levels have often not been very high (Table 20.2), but it is unknown if further mutations, including compensatory mutations to compensate for fitness disadvantages, can lead to higher expression of the affected transporters or to the overexpression of additional exporters. This could result in higher resistance factors and further broaden the resistance spectra. Thus, fungicide performance

and disease control could be strongly affected. Such a development is already obvious for *B. cinerea* MDR3 strains. First, these strains show the overexpression of at least four efflux transporters (*bcatrB, bcmfsM2, bcatrK* and *bMR3*), and further, the constitutive expression of *bcatrB* is already on average three times higher than that known from MDR1 strains. This already results in higher fungicide-resistance factors compared to MDR1 strains.

At the moment, MDR strains can still be controlled to a sufficient degree with the fungicide treatment regime used. However, the development of MDR in filamentous plant pathogenic fungi could cause a threat for fungicide performance in the future if additional mutations lead to higher overexpression of the affected transporters, or in the case that additional transporters become overexpressed.

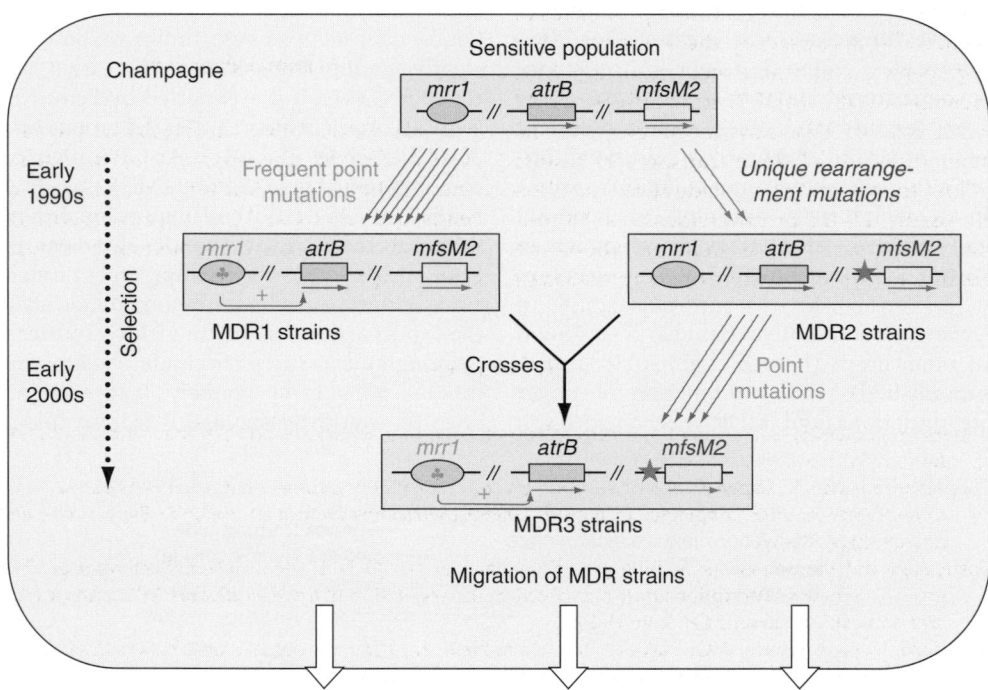

Fig. 20.3. Model of the appearance of MDR phenotypes in *B. cinerea* vineyard populations (Kretschmer *et al.*, 2009). Regular alternating treatments with modern fungicides, in particular the anilinopyrimidines pyrimethanil and cyprodinil (since 1990), the phenylpyrrole fludioxonil (since 1995) and the hydroxyanilide fenhexamid (since 2000), are assumed to be responsible for the selection of MDR phenotypes in Champagne vineyards. Repeatedly occurring point mutations in the transcription factor gene *bcmrr1* (leaf symbol) lead to overexpression of the ABC transporter gene *bcatrB* (unlinked to *bcmrr1*) and thus to the MDR1 phenotype.

20.7 Conclusions

Plant pathogenic fungi display a threat to crop production and thus have to be controlled by synthetic fungicides. Recently, a new fungicide-resistance mechanism, multidrug resistance (MDR), was found to emerge in field populations of plant pathogenic fungi. In at least three different fungi, evidence of MDR can be found. However, MDR is best characterized for the necrotrophic pathogen *B. cinerea*. MDR strains of this pathogen are found frequently in Champagne, France, and in the German wine growing areas. The actual fungicide treatment regime leads to selection of these strains. MDR strains can be subdivided into three different MDR phenotypes according to their fungicide-resistance spectra, leading to the designation MDR1, MDR2 and MDR3. MDR3 strains are natural occurring combinations of *B. cinerea* MDR1 and MDR2 strains and thus possess the combined properties of both. MDR is caused by the overexpression of plasma membrane exporters of the ABC (BcatrB = MDR1, MDR3) and MFS (BcmfsM2 = MDR2, MDR3) transporter classes. The overexpression of these transporters results in low to moderate fungicide resistance levels against all major fungicides used to control *B. cinerea*. Two different mutations are leading to the constitutive overexpression of these efflux transporters. The expression of *bcatrB* is controlled by the transcription factor Bcmrr1. In MDR1 and MDR3 strains, several amino acid sequence variations can be found in Bcmrr1 compared to sensitive strains, leading to the overexpression of *bcatrB*. In the case of MDR2 and MDR3 strains, an insertion of a new retrotransposable element in the promoter region of *bcmfsM2* leads to a constitutive overexpression of *bcmfsM2* and results in the MDR2 and MDR3 phenotypes.

Recent developments to higher constitutive transporter expression levels, the combined overexpression of several efflux transporters, reduced fungicide performance under field conditions and the combination of MDR mechanisms with single fungicide resistances (target site resistance) suggest that MDR of plant pathogenic fungi could become a potential threat for fungicide performance under field conditions.

Acknowledgements

I am thankful to M. Leroch, M. Hahn and A.S. Walker for the opportunity to discuss unpublished data. Further, I would like to thank B. Cadieux and J. Kronstad for discussions and constructive comments on the manuscript.

References

Anderson, J.B. (2005) Evolution of antifungal-drug resistance: mechanisms and pathogen fitness. *Nature Reviews Microbiology* 3, 547–556.

Chapeland, F., Fritz, R., Lanen, C., Gredt, M. and Leroux, P. (1999) Inheritance and mechanisms of resistance to anilinopyrimidine fungicides in *Botrytis cinerea* (*Botryotinia fuckeliana*). *Pesticide Biochemistry and Physiology* 64, 85–100.

Coste, A., Karababa, M., Ischer, F., Bille, J. and Sanglard, D. (2004) Tac1, transcriptional activator of CDR genes, is a new transcriptions factor involved in the regulation of *Candida albicans* ABC transporters *cdr1* and *cdr2*. *Eukaryotic Cell* 2, 1639–1652.

De Waard, M.A., Andrade, A.C., Hayashi, K., Schoonbeek, H.J., Stergiopoulos, I. and Zwiers, L.H. (2006) Impact of fungal drug transporters on fungicide sensitivity, multidrug resistance and virulence. *Pest Management Science* 62, 195–207.

Gulshan, K. and Moye-Rowley, W.S. (2007) Multidrug resistance in fungi. *Eukaryotic Cell* 6, 1933–1942.

Gupta, A. and Chattoo, B.B. (2008) Functional analysis of a novel ABC transporter ABC4 from *Magnaporthe grisea*. *FEMS Microbiology Letters* 278, 22–28.

Hamamoto, H., Nawata, O., Hasegawa, K., Nakaune, R., Lee, Y.J., Makizumi, Y., *et al.* (2001) The role of the ABC transporter gene PMR1 in demethylation inhibitor resistance in *Penicillium digitatum*. *Pesticide Biochemistry and Physiology* 70, 19–26.

Hayashi, K., Schoonbeek, H.J., Sugiura, H. and de Waard, M.A. (2002) Bcmfs1 a novel major facilitator superfamily transporter from *Botrytis cinerea* provides tolerance towards the natural toxic compounds camptothecin and cercosporin and towards fungicides. *Applied and Environmental Microbiology* 68, 4996–5004.

Hayashi, K., Schoonbeek, H.J. and de Waard, M.A. (2003) Modulators of membrane drug transporters potentiate the activity of the DMI fungicide oxpoconazole against *Botrytis cinerea*. *Pest Management Science* 59, 294–302.

Kolaczkowski, M., Kolaczkowska, A., Luczynski, J., Witek, S. and Goffeau, A. (1998) *In vivo* characterization of the drug resistance profile of the major ABC transporters and other components of the yeast pleiotropic drug resistance network. *Microbial Drug Resistance* 4, 143–158.

Kretschmer, M. and Hahn, M. (2008) Fungicide resistance and genetic diversity of *Botrytis cinerea* isolates from a vineyard in Germany. *Journal of Plant Diseases and Protection* 115, 214–219.

Kretschmer, M., Leroch, M., Mosbach, A., Walker, A.S., Fillinger, S., Mernke, D., *et al.* (2009) Fungicide-driven evolution and molecular basis of multidrug resistance in field populations of the grey mould fungus *Botrytis cinerea*. *PLoS Pathogen* 5(12), e1000696.

Lee, Y.J., Yamamoto, K., Hamamoto, H., Nakaune, R. and Hibi, T. (2005) A novel ABC transporter gene ABC2 involved in multidrug susceptibility but not pathogenicity in rice blast fungus, *Magnaporthe grisea*. *Pesticide Biochemistry and Physiology* 81, 13–23.

Leroch, M., Kretschmer, M. and Hahn, M. (2011) Fungicide resistance phenotypes of *Botrytis cinerea* isolates from commercial vineyards in South West Germany. *Journal of Phytopathology* 159(1), 63–65.

Leroux, P. and Clerjeau, M. (1985) Resistance of *Botrytis cinerea* Pers and *Plasmopara viticola* (Berk and Curt.) Berl and de Toni to fungicides in French vineyards. *Crop Protection* 4,137–160.

Leroux, P. and Walker, A.S. (2011) Multiple mechanisms account for resistance to sterol 14α-demethylation inhibitors in field isolates of *Mycosphaerella graminicola*. *Pest Management Science* 67, 44–59.

Leroux, P., Fritz, R., Debieu, D., Albertini, C., Lanen, C., Bach, J., *et al.* (2002) Mechanisms of resistance to fungicides in field strains of *Botrytis cinerea*. *Pest Management Science* 58, 876–888.

Leroux, P., Gredt, M., Leroch, M. and Walker, A.S. (2010) Exploring mechanisms of resistance to respiratory inhibitors in field strains of *Botrytis cinerea*, the causal agent of gray mold. *Applied and Environmental Microbiology* 76, 6615–6630.

Ma, Z. and Michailides, T.J. (2005) Advances in understanding molecular mechanisms of fungicide resistance and molecular detection of resistant genotypes in phytopathogenic fungi. *Crop Protection* 24, 853–863.

Morschhäuser, J., Barker, K.S., Liu, T.T., Blaß-Warmuth, J., Homayouni, R., *et al.* (2007) The transcription factor Mrr1p controls expression of the MDR1 efflux pump and mediates multidrug resistance in *Candida albicans*. *PLoS Pathogen* 3(11), e164. doi:10.1371/journal.ppat.0030164.

Nakaune, R., Adachi, K., Nawata, O., Tomiyama, M., Akutsu, K. and Hibi, T. (1998) A novel ATP-binding cassette transporter involved in multidrug resistance in the phytopathogenic fungus *Penicillium digitatum*. *Applied and Environmental Microbiology* 64, 3983–3988.

Nakaune, R., Hamamoto, H., Imada, J., Akutsu, K. and Hibi, T. (2002) A novel ABC transporter gene, PMR5, is involved in multidrug resistance in the phytopathogenic fungus *Penicillium digitatum*. *Molecular Genetics and Genomics* 267, 179–185.

Oerke, E.C. (2006) Crop losses to pests. *Journal of Agricultural Science* 144, 31–43.

Oerke, E.C. (2007) Crop losses to animal pests, plant pathogens, and weeds. In: Pimentel, D. (ed.) *Encyclopedia of Pest Management*, Volume II. CRC Press, New York, pp. 116–120.

Park, S.Y., Jung, O.J., Chung, Y.R. and Lee, C.W. (1997) Isolation and characterization of a benomyl-resistant form of beta-tubulin-encoding gene from the phytopathogenic fungus *Botryotinia fuckeliana*. *Molecules and Cells* 7, 104–109.

Roohparvar, R., De Waard, M.A., Kema, G.H. and Zwiers, L.H. (2007a) MgMfs1, a major facilitator superfamily transporter from the fungal wheat pathogen *Mycosphaerella graminicola*, is a strong protectant against natural toxic compounds and fungicides. *Fungal Genetics and Biology* 44, 378–388.

Roohparvar, R., Huser, A., Zwiers, L.H. and de Waard, M.A. (2007b) Control of *Mycosphaerella graminicola* on wheat seedlings by medical drugs known to modulate the activity of ATP-binding cassette transporters. *Applied and Environmental Microbiology* 73, 5011–5019.

Roohparvar, R., Mehrabi, R., Van Nistelrooy, J.G., Zwiers, L.H. and De Waard, M.A. (2008) The drug transporter MgMfs1 can modulate sensitivity of field strains of the fungal wheat pathogen *Mycosphaerella graminicola* to the strobilurin fungicide trifloxystrobin. *Pest Management Science* 64, 685–693.

Schoonbeek, H.J., Del Sorbo, G. and de Waard, M.A. (2001) The ABC-transporter BcatrB affects the sensitivity of *Botrytis cinerea* to the phytoalexin resveratrol and the fungicide fenpiclonil. *Molecular Plant–Microbe Interactions* 14, 562–571.

Schoonbeek, H.J., van Nistelrooy, J.G.M. and de Waard, M.A. (2003) Functional analysis of ABC transporter genes from *Botrytis cinerea* identifies BcatrB as a transporter of eugenol. *European Journal of Plant Pathology* 109, 1003–1011.

Staub, T. (1991) Fungicide resistance – practical experience with antiresistance strategies and the role of integrated use. *Annual Review of Phytopathology* 29, 421–442.

Stefanato, F.L., Abou-Mansour, E., Buchala, A., Kretschmer, M., Mosbach, A., Hahn, M., *et al.* (2009) The ABC transporter BcatrB from *Botrytis cinerea* exports camalexin and is a virulence factor on *Arabidopsis thaliana*. *The Plant Journal* 58, 499–510.

Sun, C.B., Suresh, A., Deng, Y.Z. and Naqvi, N.I. (2006) A multidrug resistance transporter in *Magnaporthe* is required for host penetration and for survival during oxidative stress. *Plant Cell* 18, 3686–3705.

Urban, M., Bhargava, T. and Hamer, J.E. (1999) An ATP-driven efflux pump is a novel pathogenicity factor in rice blast disease. *EMBO Journal* 18, 512–521.

Vermeulen, T., Schoonbeek, H.J. and de Waard, M.A. (2001) The ABC transporter BcatrB from *Botrytis cinerea* is a determinant of the activity of the phenylpyrrole fungicide fludioxonil. *Pest Management Science* 57, 393–402.

Vivier, M.A. and Pretorius, I.S. (2002) Genetically tailored grapevines for the wine industry. *Trends in Biotechnology* 20, 472–478.

Zhang, Y. and Yew, W.W. (2009) Mechanisms of drug resistance in *Mycobacterium tuberculosis*. *International Journal of Tuberculosis and Lung Disease* 13, 1320–1330.

Zwiers, L.H., Roohparvar, R. and de Waard, M.A. (2007) MgAtr7, a new type of ABC transporter from *Mycosphaerella graminicola* involved in iron homeostasis. *Fungal Genetics and Biology* 44, 853–863.

Part V

Role of FRAC

21 The Role of the Fungicide Resistance Action Committee in Managing Resistance

Andy Leadbeater

Syngenta Crop Protection AG, Basel, Switzerland

21.1 Introduction

Fungicide resistance is the naturally occurring, inheritable adjustment in the ability of individuals in a population to survive a plant protection product treatment that would normally give effective control (OEPP/EPPO, 1999). Although resistance can often be demonstrated in the laboratory, this does not necessarily mean that pest control in the field is reduced. 'Practical resistance' is the term used for loss of field control due to a shift in sensitivity. When it occurs in the field, fungicide resistance affects all those concerned with crop health; the farmers, growers, advisors and the industry that provides these with the advice and products necessary to ensure a healthy, productive crop. Without successful resistance management, the effectiveness, and eventually the number, of modern fungicides available to the farmer and grower will diminish rapidly, leading to poor yields and reduced crop quality. Such a scenario could lead quickly to overuse of affected fungicides as users strive to get products to work, leading in turn to increased and undesirable loading on the environment.

In recognition of the challenge set by the possibility of fungicide resistance development, FRAC, the Fungicide Resistance Action Committee, was formed in 1981 following an industry seminar in Brussels. It is an industry-based and -financed organization reporting to CropLife International (formerly GIFAP and GCPF). FRAC was designed to discuss resistance problems and formulate plans for cooperative efforts to avoid or manage fungicide resistance.

The purpose of FRAC is to provide fungicide resistance management guidelines to prolong the effectiveness of 'at risk' fungicides and to limit crop losses should resistance occur. In more detail, the main aims of FRAC are to:

- Identify existing and potential resistance problems.
- Collate information on fungicide resistance frequency in the field, levels and spatial patterns and distribute it to those involved in the research, development, distribution, registration and use of fungicides.
- Provide guidelines and advice on the use of fungicides to reduce the risk of selection for resistance and how to manage it should it occur.
- Recommend procedures for fungicide resistance studies.
- Stimulate open liaison and collaboration with universities, government agencies, advisors, extension workers, distributors and farmers.

FRAC is charged with providing advice to all producers, suppliers and users of fungicides on how best to use fungicides in order to avoid, delay and manage fungicide resistance in crops worldwide. They are also providers of educational material to raise awareness and train more people in the science of fungicide resistance and the art of its control. FRAC does not, however, provide advice on individual products – this is the domain of individual companies – rather, it produces information, advice and strategies on fungicide resistance in general and relating to classes of active ingredients.

FRAC members are recognized experts in the field of fungicide resistance and are frequent contributors at scientific conferences and symposia. FRAC does not work in isolation – cooperation between the members and researchers, advisors and officials in public and private bodies in order to share scientific information enables all involved to come to agreed and supported conclusions, practical advice and recommendations. The key communication route for all FRAC information is the website, www.frac.info, which contains information relevant to the advisor, grower, researcher and student.

21.2   The FRAC Steering Committee

The overall activities of FRAC are managed by the Steering Committee. FRAC is governed by its terms of reference and operates according to the standards required by CropLife International. The Steering Committee comprises senior technical people from the R&D functions of agrochemical manufacturers (the committee may appoint a member from another area of expertise in industry providing that person is suitably technically qualified). The committee members are experienced and known for their expertise in fungicide resistance management matters, both within industry and in the scientific and advisory areas. All committee members are elected to their positions. The committee consists of a chairperson, vice-chairperson, secretary, treasurer, communications officer and other members. The chairpersons of the Working Groups and Expert Fora (see below)

are members of the Steering Committee that ensures coordination and peer review of the groups. The Steering Committee is kept relatively small to facilitate discussions and decision making. The members of the Steering Committee make a commitment to undertake an active and constructive role in resistance management matters – all members of FRAC are required to contribute actively with scientific data and discussions.

The Steering Committee is responsible for the strategic direction of FRAC and each year the group develops its key objectives, an operating plan and a budget, which is reviewed by CropLife International and then implemented by the Steering Committee and Working Groups. The Steering Committee approves expenditure in line with the operating plan, which may include sponsorship of relevant symposia, conferences and research studies. In addition, the Steering Committee is tasked with resolving technical issues which cannot be resolved in the Working Groups and for technical review of any fungicides which are not covered by the Working Groups. A key responsibility in this area is the production of the FRAC Mode of Action Classification List and the related Mode of Action Wallchart. The committee reviews and addresses any resistance-related issues which may arise; most recently, this has included informing policy makers and other stakeholders of the likely impacts of the review of EU directive 91/414.

To encourage further knowledge sharing across industry, the chairperson of any other Resistance Action Committee (Herbicide Resistance Action Committee, Insecticide Resistance Action Committee, Rodenticide Resistance Action Committee) may attend FRAC meetings, and vice versa.

21.3   FRAC Working Groups

When fungicidal active substances from different manufacturers have the same mode of action and where this mode of action bears, at the same time, a significant resistance risk, a FRAC Working Group (WG) is usually established to analyse resistance risk and to develop and publish common resistance management recommendations. As already mentioned, FRAC does not provide advice on individual products: it produces information, advice and strategies on fungicide resistance in general and classes of active ingredients, i.e. mode of action. Each WG comprises technical representatives from companies with a relevant fungicide in the market or in registration – this can include manufacturers of generic products having the mode of action relevant to the WG. This is especially important where significant modifications to product class use and recommendations need to be made, to ensure a high level of consistency across the industry in product labels, use recommendations and technical advice. Members of each WG define which information and data are to be collected, pool relevant scientific information for their group, define the scope of any problem(s) and assess the risks (all the time working in full compliance with anti-trust regulations). As is the case in the Steering Committee, 'active contribution' is expected from WG members, meaning that the member company will contribute information and participate in active,

constructive discussion that will lead to the formulation of resistance management strategies for the fungicide mode of action group.

There are currently FRAC Working Groups for sterol biosynthesis inhibitors (SBIs, which include the triazoles and amines), QoI fungicides, anilinopyrimidines (APs), carboxylic acid amides (CAAs), succinate dehydrogenase inhibitors (SDHIs) and azanaphthalene fungicides (AZN) – see Fig. 21.1. The groups meet regularly and publish, at least annually, updated reports on the status of resistance and suitable resistance management recommendations.

After calling on experts for technical and legal advice, the WGs recommend technical strategies aimed at ensuring the continued effectiveness of the group of fungicides at risk and encourage their implementation. This is done, with the help of the committee, to inform appropriate personnel (users, advisory services, registration authorities and independent distributors) of the strategies and the reasons for them.

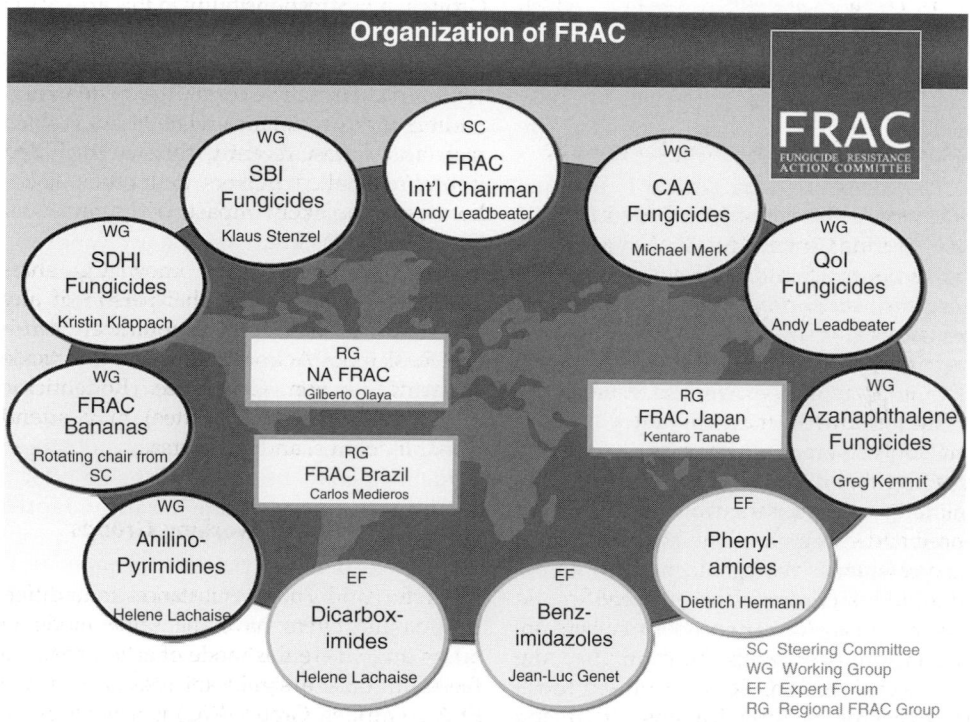

Fig. 21.1. The current organization of FRAC.

WGs establish and direct collaboration on:

- Monitoring programmes – share methodologies and definitions (in some cases, developing common methodologies), establish baseline data, agree joint interpretation of results.
- Research – define areas of basic and applied research that needs to be done, guide the direction, methodology and reporting of work already done.
- Educational programmes – encourage the publication of research results and provide relevant information to the committee.
- Verification of reports on resistance, statements, investigations, remedies.
- Formulation of use strategies for the fungicide mode of action group in terms of resistance prevention or delay and checking on the implementation of such strategies.
- Monitoring the activity of national/regional subgroups and supporting them with general guidance.

Guidelines and other publications must be agreed unanimously by all WG members before adoption. These guidelines are published on the FRAC web pages and are reviewed regularly.

The FRAC Banana Group is unusual in that it is the only crop-based WG of FRAC and deals with all modes of action used to control banana diseases. The WG is responsible for defining fungicide resistance strategies in banana cropping and aims to represent all major banana growing regions globally. Alongside members of the crop protection industry experienced in resistance matters in banana diseases, representatives from some of the banana grower associations and institutions with activities in the areas of fungicide resistance research, monitoring or strategies participate as 'invited members'. They are encouraged to contribute information for consideration by the group. The Banana Group meets once every 2 years and the chairperson has a 2-year tenure, rotating among the manufacturing companies.

Expert Fora

In time, a WG will be stopped, due to various reasons such as an overall stable situation with no new issues or the decline in importance of the chemical/mode of action group, or simply a lack of new information. In this event, consideration is given to the formation of an Expert Forum for the chemical group. The Expert Forum consists of a (usually small) number of technical people able to advise on all aspects of resistance management and its history for that group. Expert Fora do not hold regular meetings unless requested by a member. The Expert Forum is a source of information and advice, providing information published on the FRAC web page and a list of expert contacts able to provide further information if required. There are currently Expert Fora for benzimidazoles, dicarboximides and phenylamides.

21.4   Resistance Management at Regional Level

The FRAC Steering Committee and the WGs produce the core resistance management guidelines which form the basis for resistance management worldwide. The guidelines are crop and pathogen specific for all major crops, but due to the wide variety of crops treated with many fungicide classes, the WGs also produce general guidelines applicable to additional regional crops. The guidelines are communicated to all FRAC member companies and their subsidiaries and are available for consultation on the FRAC website. However, it is quite possible that some adaptation of these guidelines may be required to suit local conditions. Provided the locally adapted guidance is consistent with, and not in conflict with, the resistance management strategies agreed by the WG, this is supported. Responsibility for such adaptation is being taken increasingly by local organizations in consultation with the FRAC WGs. The two key organization types acting at the local level are the local FRAC groups, which are industry based, and the local Resistance

Action Groups (RAGs), which include members from industry, research institutes, advisory services, universities and the regulatory authorities. FRAC at the international level encourages the formation of these groups to manage resistance issues more effectively and proactively at grower level and to promote constructive interaction between industry and independent advisors.

Regional FRAC groups

The regional FRAC groups in Japan, North America and Brazil are key components of the FRAC structure, the chairperson of each being a member of the FRAC Steering Committee. These groups consist of members of companies with a technical and commercial presence in the respective countries and in this way expand the number of manufacturing companies that are members of FRAC. These regional groups act to coordinate the gathering and dissemination of resistance management information at the local level. As the member companies may not have the research facilities that are available to their international parents, the local FRAC groups may sponsor research studies with local organizations. Such studies can include the generation of local fungicide sensitivity baselines and fungicide sensitivity monitoring surveys.

FRAC Brazil is active in sponsoring studies in key crops of local importance such as soybeans and work with the Brazilian Agricultural Research Corporation (Embrapa) and Instituto Biológico to generate further knowledge in the management of crop diseases and provide resistance management recommendations. The group is involved in local training programmes concerning resistance management and has succeeded in making the FRAC Monographs (see later) available in Portuguese to ensure their effectiveness at the local level. Members are also regular contributors at the Brazilian Plant Pathology conferences.

FRAC North America (NAFRAC) is organized in a similar way to the parent FRAC, with WGs for the QoI, SBI, AP and CAA group fungicides. They have excellent working relationships with the American Phytopathological Society (APS) and FRAC members are frequent contributors to the annual APS meetings.

FRAC Japan is very successful in bringing the Japanese companies together with the local organizations of multinational companies to discuss common challenges in resistance management. Regular meetings of the full committee are held, plus some specialist meetings for individual areas of chemistry.

In addition to the above-mentioned FRAC groups, CropLife Australia have an active Fungicide Resistance Management Review Group (FRMRG), which works closely with local industry researchers to produce resistance management strategies. FRAC India has been formed to guide crop growers on the proper and judicious use of fungicides.

Regional Resistance Action Groups (RAGs)

Very active Resistance Action Groups operate in the UK (FRAG-UK), the Netherlands (FRAG-NL) and also relatively recently in Germany (ECPR-F) and the Nordic and Baltic regions (NORBARAG). The work of the RAGs varies according to the countries, but all represent a very important link between industry and independent advisors and officials. FRAG-UK and NORBARAG in particular are very active in promoting awareness on fungicide resistance issues; for example, by producing educational material and publishing advice on resistance management at the local level.

21.5 FRAC Liaison Activities

FRAC is an inter-industry organization with the possibility to pool technical information and expertise across the international research and development based companies, as well as providing agreed and consistent technical recommendations. Links to national groups, independent advisory groups, research institutes, officials and policy makers are important to create the FRAC recommendations and implement them. In addition, FRAC is in dialogue with many international organizations such as FAO and OECD. FRAC members contribute to several committees, an example being the European and Mediterranean Plant Protection Organization (EPPO), which has its own Resistance Panel on Plant Protection Products. Members of FRAC, and indeed the other Resistance Action Committees, have been involved closely in the work of this panel, which has included the production of the EPPO Standard PP1/213 Resistance Risk Analysis.

21.6   FRAC Publications

FRAC is very active in the publication of documents and scientific methodologies. The FRAC publications pages on the website contain much information relevant to all those interested in fungicide resistance and its management, including advisors, teachers and students.

21.7   The Mode of Action Code List

The FRAC Mode of Action (MOA) Code List is the most popular of all FRAC documents accessed from the FRAC web pages. It includes full details of the MOA codes, the group name and code, chemical group names for individual molecules and their common names, and comments on resistance risk. This code list is now the accepted international standard and is reviewed at regular intervals. The code list is maintained by experts on the FRAC Steering Committee, who seek additional scientific advice from researchers inside and outside the industry. Manufacturers of new active substances are always consulted, either to encourage them to propose an MOA classification based on available data, or to check with them the proposal made by FRAC. These codes are included on product labels in the USA, Australia and Canada, and gradually this trend is increasing globally

since it is viewed as vital information for those advising on the use of agricultural products, as well as for growers themselves to be aware of the resistance management classification of the products they are using. This enables them to make better choices of the use of products in an overall disease control programme.

21.8   The FRAC Mode of Action Poster

As a complementary publication to the MOA Code List, FRAC produces and updates a poster. The poster shows the structures of fungicides included in the MOA list, grouped by the respective MOA groups and codes. This provides a very visual and easy-to-understand representation of the Code List in a single view. It is updated every 1–2 years and is available as a high-resolution downloadable file from the FRAC website.

21.9   The FRAC List of Plant Pathogenic Organisms Resistant to Disease Control Agents

This list is documentation of confirmed cases of resistance by plant pathogens to modes of action covered by the FRAC Code List, together with references. Wherever possible, the case recorded is the first case published, but additional references are frequently added to cover key information on, for example, mode of action and resistance mechanism. Distinctions are made between instances of field resistance, mutation studies and resistance generated by forced laboratory selection. The list is updated regularly and information on new cases is always welcome for inclusion, but only verified cases (published in a refereed journal or substantiated in a similar way) are considered. This differentiates the list from some other lists that can be found in circulation, which may include all suspected resistance cases as 'resistance', even if they are not verified by repeat tests, molecular methods or further studies.

21.10   FRAC Methods Folder

An issue that can happen in resistance monitoring programmes is the lack of a valid test, or of a test that can allow comparison of the data with other studies. Having a scientifically correct, verified method for monitoring fungicide sensitivity is crucial for any fungicide resistance monitoring programme. The FRAC WGs have collated the methods used within their own member companies, together with methods used by recognized experts in academia and research institutes, to produce a library of recommended testing methods. The focus so far has been on methods that can be carried out easily in any standard laboratory without the need for high-technology equipment. In future, it is probable that molecular methods will be included. As well as being valuable for fungicide resistance research, the methods are also suitable for plant pathology teaching purposes. The library currently contains 32 methods, organized for individual pathogens; more methods are being added regularly.

21.11   The FRAC Monographs

The FRAC Monographs are educational publications about fungicide resistance covering all aspects from definition of resistance, how it arises and increases, how to test for resistance, how to predict the risk of resistance arising, how resistance management strategies are designed and implemented, and so on. All the monographs have been written by recognized experts and have been updated recently. As mentioned before, they have also been translated into Portuguese. The monographs are available as free downloads from the FRAC website and are also available as printed hard copies and on CD. The monographs are:

1. Monograph 1: Fungicide resistance in crop pathogens: How can it be managed? (2nd revised edition. K.J. Brent and D.W. Hollomon.)
2. Monograph 2: Fungicide resistance: The assessment of risk. (2nd revised edition. K.J. Brent and D.W. Hollomon.)

3. Monograph 3: Sensitivity baselines in fungicide resistance research and management. (P.E. Russell.)

21.12   Conclusions

Fungicide resistance is recognized as a widespread problem in global agriculture. Resistance problems in the field have been documented for more than 100 diseases (crop–pathogen combinations) and it is clear that managing resistance to the range of fungicides available today, as well as those developed in the future, is key for the sustainability of disease control in agriculture, horticulture and amenity use. The cost involved in bringing a new pesticide from research and development to the market is currently estimated as greater than US$250 million on average (Philips McDougal, 2010). For industry to continue to make investments at this level, and thus to be able to provide growers with the tools that are essential to their livelihoods, it is highly important to sustain the high levels of effectiveness of products for as long as possible. A major contributor to this is intensive research into understanding the inherent resistance risk of fungicide and the fungicide–pathogen combined risk (Kuck and Russell, 2006), carrying out monitoring surveys to establish baseline sensitivities and potential changes in populations with time and designing effective resistance management strategies.

Despite resistance to several important groups of fungicides being detected in economically important crops, there are very good examples of such fungicides remaining effective and making valuable contributions to crop disease management for many years. Examples are the phenylamides, which remain important 30 years after the first reports of resistance, and the QoI fungicides, which remain central to many disease management programmes worldwide, despite extensive resistance in many diseases. These fungicides are covered in separate chapters in this book. The work of FRAC has been vital in ensuring the continued success of these and other fungicides prone to resistance.

By ensuring that the most up-to-date scientific data and knowledge on fungicide classes is shared and interpreted, and recommendations and messages to the growers are aligned, by ensuring open communication with other scientists and advisors, and by proactive education in resistance matters, FRAC is uniquely positioned to contribute to the longevity of the effectiveness, and therefore success, of fungicides.

References

FRAC (www.frac.info).

Kuck, K.H. and Russell, P.E. (2006) FRAC: combined resistance risk assessment. In: Bryson, R.J., Burnett, F.J., Foster, V., Fraaije, B.A. and Kennedy, R. (eds) *Fungicide Resistance: Are We Winning the Battle but Losing the War? Aspects of Applied Biology 78*. Scottish Agricultural College, Edinburgh, UK, pp. 3–10.

OEPP/EPPO (1999) EPPO Standard PP 1/213, Resistance Risk Analysis. *OEPP/EPPO Bulletin* 29, 325–347.

Philips McDougal (2010) The cost of new agrochemical product discovery, development and registration in 1995, 2000 and 2005–2008 (http://www.croplife.org/files/documentspublished/1/en-us/REP/5344_REP_2010_03_04_Phillips_McDougal_Research_and_Development_study.pdf (www.croplife.org/view_document.aspx?docld=2478, accessed 1 November 2010).

Index